P9-CJT-073

RENEWALS 458-4574
DATE DUE

WITHDRAWN
UTSA LIBRARIES

LIBRARY IN A BOOK

BIOTECHNOLOGY AND GENETIC ENGINEERING

Third Edition

Lisa Yount

An imprint of Infobase Publishing

To the scientists who will create the "Biotech Century"
and the children who will have to deal with it

Biotechnology and Genetic Engineering, Third Edition

Copyright © 2008, 2004, 2000 by Lisa Yount

All rights reserved. No part of this book may be reproduced or utilized in any form or by any means, electronic or mechanical, including photocopying, recording, or by any information storage or retrieval systems, without permission in writing from the publisher. For information contact:

Facts On File, Inc.
An imprint of Infobase Publishing
132 West 31st Street
New York NY 10001

Library of Congress Cataloging-in-Publication Data

Yount, Lisa.
  Biotechnology and genetic engineering / Lisa Yount. — 3rd ed.
    p. ; cm. — (Library in a book)
  Includes bibliographical references and index.
  ISBN 978-0-8160-7217-0 (alk. paper)
  1. Biotechnology—Social aspects. 2. Genetic engineering—Social aspects.
I. Title. II. Series.
  [DNLM: 1. Genetic Engineering. 2. Biotechnology. QU 450 Y81b 2008]
  TP248.23.Y684 2008
  303.48'3—dc22   2007041313

Facts On File books are available at special discounts when purchased in bulk quantities for businesses, associations, institutions, or sales promotions. Please call our Special Sales Department in New York at (212) 967-8800 or (800) 322-8755.

You can find Facts On File on the World Wide Web at http://www.factsonfile.com

Text design by Ron Monteleone

Printed in the United States of America

Bang Hermitage 10 9 8 7 6 5 4 3 2 1

This book is printed on acid-free paper.

**Library
University of Texas
at San Antonio**

CONTENTS

PART III
APPENDICES

PART I

OVERVIEW OF THE TOPIC

CHAPTER 1

ISSUES IN BIOTECHNOLOGY AND GENETIC ENGINEERING

In April 2003, the International Human Genome Sequencing Consortium announced completion of the Human Genome Project, a huge effort—lasting more than a decade—to obtain the complete readout of the code, some 3 billion "letters" worth, that contains all the inherited information specifying the characteristics of a human being. *Genomics & Genetics Weekly* called this work "one of the most ambitious scientific undertakings of all time."[1] Appropriately, April 2003 also marked the 50th anniversary of the publication of the structure of DNA, the key discovery that allowed scientists to learn the nature of the genetic code and the way it is transmitted, and approximately the 30th anniversary of the birth of genetic engineering, which enables researchers to change the genetic makeup of living things.

As great an achievement as completion of the Human Genome Project is, the researchers involved in it know that they are only at the beginning of their work. Possessing a blueprint is not the same as having a finished building. Scientists may have the "text" for all of humanity's genes, but they do not know what most of those genes do nor how they interact with each other and the environment to produce a living, changing individual. Genetic science and the biotechnology industry it has spawned are still in their infancy, surrounded by a cloud of hopes and fears. Like many of the hopes and fears surrounding space exploration, some of these will never come to pass. Nonetheless, genetic engineering and biotechnology are already starting to produce massive changes in the way people live, their health, their environment, and, perhaps, in the very nature of humanity. Supporters and critics alike expect the 21st century to be what one critic, Jeremy Rifkin, has dubbed "the biotech century."[2]

The Human Genome Project, along with other achievements in genetics, genetic engineering, and biotechnology, raises many social, legal, and ethical issues. When scientists and businesspeople tinker with the very stuff of life, everyone is potentially affected. This book focuses on six areas in

3

which such issues are especially sensitive: agricultural biotechnology and genetically modified (GM) food; the patenting of living things, tissues, cells, and genes; DNA profiling for identification in criminal cases; genetic testing for susceptibility to disease and the possibility of genetic discrimination in employment and insurance; alteration of human genes, including gene-based treatments for disease and the possible cloning of human beings; and research on stem cells, especially embryonic stem cells. The remainder of this introduction will describe these technologies, the ethical issues they raise, and the way governments, courts, and the public have perceived and reacted to them. It will conclude by considering the future development of the technologies and of new forms of biotechnology beginning in the 21st century: industrial biotechnology, nanobiotechnology, and synthetic biology.

CRACKING THE CODE OF LIFE

In some senses, biotechnology—the use or alteration of other living things in processes that benefit humankind—is almost as old as humanity itself. Biotechnology certainly has existed as long as people have raised crops and domesticated animals, practices they began some 10,000 years ago. Although they were unaware that they were doing so, people used microorganisms in the processes of making bread, cheese, alcoholic drinks, and tanned leather. When people began sowing only the seeds from the best (most productive, most disease-resistant, and so on) of the previous year's crop plants, and breeding only the strongest or most productive cattle and other domestic animals, they also, again unknowingly, altered the genes of living things to better meet their needs. When they learned that they could breed certain plants from cuttings, they discovered cloning (the word *clone* means "twig" in Greek).

Only in the 19th century did people begin to learn what was really happening during these technological efforts. Starting in the late 1850s, French chemist Louis Pasteur showed that living yeast or other microorganisms were necessary for the fermentation that produced wine, beer, and other products. More important to the genetic side of modern biotechnology, in 1866 an obscure Austrian monk named Gregor Mendel published a paper describing the mathematical rules of inheritance of physical traits. He had worked out these rules by breeding and observing pea plants in his monastery garden. Mendel's rules showed the statistical pattern in which such characteristics as height and seed color were inherited by the hybrid offspring of two plants that differed from each other in these characteristics.

The form of a characteristic that an offspring would receive was determined by what Mendel called factors, one of which (for each characteristic) the offspring inherited from its male parent and one from its female parent. Some factors were more powerful, or dominant, than others. If a plant inherited one dominant factor and one weaker (recessive) factor for a trait, the plant would exhibit the dominant form of the trait.

Few knew of Mendel's work at the time, let alone realized that it went far toward explaining the theory of Charles Darwin, described in his *On the Origin of Species* (first published in 1859), which stated that nature acted like traditional plant and animal breeders to select and preserve the inherited characteristics that helped species survive in their environment. Darwin called his theory "evolution by natural selection."

Earlier in the 19th century, Matthias Schleiden and Theodor Schwann had proposed that microscopic bodies called cells were the basic units from which all living things are formed. Improvements in microscopes allowed scientists to begin exploring the cell's inner structures toward the end of the century. Mendel's rules of heredity and the study of cells converged in the early 1900s, when Mendel's forgotten work was rediscovered and scientists began speculating that his "factors" were somehow contained in threadlike chromosomes, or "colored bodies," which German biologist Walther Flemming had discovered in the nuclei (central bodies) of cells in 1875. (All cells except those of bacteria and other so-called prokaryotes contain nuclei and chromosomes.)

Drawing on experiments with fruit flies, Thomas Hunt Morgan of Columbia University and his coworkers proved the link between chromosomes and heredity in 1910. By then Mendel's factors had taken on the new name of *genes*, and the science of genetics had been born. In the decades that followed, Morgan's group and others began to make maps of chromosomes showing approximately where the genes that determined certain characteristics were located.

Geneticists assumed that inherited information must be coded somehow in the structure of either proteins or nucleic acids, the only types of chemical in the chromosomes. At first they thought proteins the more likely information carriers because these chemicals were known to have a complex structure. In 1944, however, Oswald Avery and his coworkers showed that a harmless strain of certain bacteria became capable of causing disease when they took up deoxyribonucleic acid (DNA) from a disease-causing strain of the same bacteria. The bacteria's descendants retained the new disease-causing ability, indicating that information for producing this trait must have been contained in the DNA. Avery's work turned the genetic spotlight from proteins to nucleic acids, especially DNA.

At that time DNA's chemical composition was known, but key aspects of its structure were not. Biologists knew that the DNA molecule had a long "backbone" composed of smaller molecules of phosphate and a sugar called deoxyribose. Attached to this backbone were many units of four kinds of small molecules called bases: adenine (A), thymine (T), guanine

(G), and cytosine (C). X-ray crystallography photographs suggested that the DNA molecule had the general shape of a helix, or coil. However, no one knew exactly how many backbone chains the molecule had, how the bases were attached to them, or how all these units were packed within the molecule.

Geneticists and biochemists in the late 1940s and early 1950s increasingly realized that the structure of DNA must hold the secret of its ability to transmit biological instructions from one generation to the next, and several groups of scientists began trying to figure out that structure. The ones who succeeded first were a young American, James Watson, and a somewhat older Englishman, Francis Crick, who worked together at England's prestigious Cambridge University. Drawing on molecular models and X-ray photographs, they concluded that each DNA molecule contained two sugar-phosphate backbones that entwined one another in a corkscrew shape that came to be known as the "double helix." Pairs of bases stretched between the two backbones like rungs on a twisted ladder. A molecule of adenine always paired with one of thymine, and cytosine paired with guanine.

Watson and Crick published their groundbreaking account of DNA's structure in the British science journal *Nature* on April 25, 1953. Five weeks later, they published a second paper describing how this structure could allow DNA to reproduce itself so that a complete copy of hereditary instructions was given to each of the "daughter" cells when a cell divided. The hydrogen bonds that joined the bases in each pair were weak, Watson and Crick explained, so the bonds might dissolve as the cell prepared to divide. Each DNA molecule would thus split down its length, like a zipper unzipping. Each base could then attract a molecule of the complementary base, bearing with it a sugar-phosphate "backbone" unit, from free-floating materials in the cell, and the bonds joining the bases would re-form. The result would be two separate DNA molecules identical to the original one. This theory was proved correct in 1958.

DNA's structure also revealed the "code" in which inherited information is stored. Molecular biologists had known since 1941 that genes contain instructions for making proteins, the complex chemicals that do most of the work of the cell. Indeed, scientists had come to define a gene as the portion of a DNA molecule that carries the instructions for making one protein. (This proved to be an oversimplification. Some genes make molecules of ribonucleic acid [RNA] or modify the activity of other genes.) The most likely candidate for the code in which these instructions were carried was the order of the bases in the DNA molecule. DNA has only four kinds of bases, however, whereas proteins consist of 20 types of simpler substances called amino acids. In 1961, Crick suggested that the "code letter" specifying a particular type of amino acid could be a sequence of three bases. The 64 ($4 \times 4 \times 4$) possible triads were more than enough to specify all of the amino acids.

During the next five years, several groups of scientists painstakingly deciphered this genetic code, determining which amino acid each of the 64 triads stood for. (Some amino acids could be represented by any of several different triads.) Meanwhile, others worked out the procedure by which information in the DNA code was translated into protein. They discovered that the DNA is copied into a form of RNA, a related chemical that differs from DNA only in having uracil rather than thymine as one of its four bases. Unlike DNA, RNA can leave the cell's nucleus and move out into the cytoplasm, the main substance of the cell body. There the RNA guides the assembly of amino acids, in the order specified by the original DNA, into a protein molecule. This, too, has proved to be a vastly oversimplified description, but it was correct enough to provide a foundation for the amazing advances that followed.

THE BIRTH OF GENETIC ENGINEERING

Building on these basic discoveries, scientists in the early 1970s developed the ability to change genes instead of merely deciphering them. Genetic engineering—the direct alteration of genes or transfer of genes from one type of organism to another—works because (except in the case of a few viruses with genes made of RNA rather than DNA) all genes have the same basic structure and work in the same way. Thus, they are in a sense interchangeable. Once a gene is placed in a cell's genome, or collection of genes, that cell becomes able to make the protein for which the gene codes, regardless of whether it ever made that substance in nature. As an example, bacteria or cattle can be made to produce human hormones by inserting the appropriate human genes into them.

Genetic engineering began in the laboratory of Paul Berg, a Stanford University biochemist, in the winter of 1972–73. Berg removed a gene from SV40 (Simian Virus 40), a monkey virus that could cause cancer in mice. Through laborious chemical manipulation, he attached to it a short piece of single-stranded DNA. He then opened up the small, circular genome of another virus, lambda, and attached a chain of single-stranded DNA with a base sequence complementary to that on the added SV40 piece to one of the lambda genome's open ends. He termed these single-stranded chains "sticky ends" because any such chain will attach itself to another strand with a complementary base sequence (a sequence in which all the bases in the second chain will pair with those in the first). For example, a chain with the sequence T-A-G-C will attach itself to another with the sequence A-T-C-G. Taking advantage of this "stickiness," Berg spliced the gene from SV40 into the lambda genome. This was the first production of recombinant DNA, or DNA into which other DNA from a different type of organism has been inserted.

Biotechnology and Genetic Engineering

Berg's technique for gene splicing would have been hard to apply on a mass scale, but the same was not true of a second technique developed shortly afterward by another Stanford scientist, Stanley Cohen, and Herbert Boyer, from the University of California–San Francisco. Appropriately, the idea for this technique—which would allow researchers to (in effect) slice and dice genes, sandwich them in any order, and pack them "to go"— was born in a delicatessen. Cohen and Boyer met at this delicatessen one evening in November 1972 after attending a scientific meeting in Honolulu, Hawaii. As they discussed their work, each discovered that the other's research held a missing piece of his own puzzle.

Unlike Berg, Cohen and Boyer were working with bacteria. A bacterium's genome is large compared to that of a virus, containing an average of 2,500 genes rather than a few hundred, although it is far smaller than those of cells with nuclei (the human genome, for instance, probably contains about 30,000 genes). Most of a bacterium's genes are carried in a single large ring of DNA, but some are contained in smaller rings called plasmids, which the bacteria can exchange in several ways. Cohen had invented a technique for removing plasmids from one bacterial cell and inserting them into another.

Boyer, for his part, was working with bacteria called *Escherichia coli* (*E. coli* for short), which commonly and usually harmlessly live in the human intestine. He was studying enzymes (proteins that catalyze chemical reactions in living cells) called restriction endonucleases, which *E. coli* and some other bacteria produce as a defense against viruses. These enzymes snip DNA into pieces. Each restriction enzyme breaks a DNA molecule apart at every spot where it encounters a particular sequence of bases, and the process conveniently produces the same kind of single-stranded "sticky ends" that Berg had so painstakingly created with his chemicals. Any two pieces of DNA cut by the same restriction enzyme, therefore, can be joined together, even if they come from very different species. Different restriction enzymes act on different base sequences. Hundreds of restriction enzymes are known today, and they have become basic tools of gene splicing. So have the ligases, a type of enzyme that both Boyer and Berg used to weld together their recombined DNA fragments.

Chatting over sandwiches, Cohen and Boyer realized that if they applied Boyer's *E. coli* restriction enzyme, EcoR1, to Cohen's plasmids, they would have a way to cut open a plasmid from one species of bacteria and, with the help of a ligase, attach it to a plasmid from another species. They proceeded to try this in spring 1973. One of their two plasmids contained a gene that conferred resistance to a certain antibiotic. Boyer and Cohen allowed a type of bacterium that was not naturally resistant to that antibiotic to take up the combined plasmids, then placed the bacteria in a culture medium containing the antibiotic. Some of the bacteria survived, which showed that the resistance gene in the engineered plasmids was still making its signature

8

protein. Later, the two scientists joined some DNA from a frog to an *E. coli* plasmid and proved that it functioned there as well.

In addition to plasmids, Boyer, Cohen, and other scientists were soon using viruses (such as lambda, which infects bacteria) as vectors, or transmission agents, for inserting foreign genes into bacteria and, later, plant and animal cells. Viruses—nature's own genetic engineers—reproduce by inserting their genes into cells, which then reproduce the viral genomes along with their own; when scientists add genes to the viruses, these are inserted and reproduced as well. As one scientist commented when he heard about the new techniques, "Now we can put together any DNA we want to."[3] The gene age had begun.

Although traditional biotechnology continues, the term has now become almost synonymous with genetic engineering. Together, biotechnology and genetic engineering represent not only a scientific field but an industry. In the decades since the "gene deli" opened, companies based on the technology of decoding and altering genes have multiplied, seemingly as fast as the bacteria in their founders' test tubes.

Biotechnology today is unquestionably big business. According to business advisory company Ernst & Young's 2006 Global Biotechnology Report, the revenues of publicly traded biotechnology firms worldwide in that year reached $63.2 billion, a growth of 18 percent above the previous year and a new record. "Biotechnology is delivering on its promises," Mike Hildreth, Ernst & Young's Americas director for biotechnology, said in connection with the equally encouraging 2005 report.[4]

The United States remains the leader in biotechnology business, but competition from Europe is increasing. Countries in Asia, especially Japan, India, and China, are also showing very rapid growth in biotechnology. Biotech investors are by no means assured of profits, however. For example, when Davis Skaggs Investment Management ranked the stock performance of 302 companies in the San Francisco Bay area, the birthplace of biotechnology, during 2006, both the top five and the bottom four performers were companies specializing in medical biotechnology.

SAFETY CONCERNS

Concerns about the safety of genetic engineering experiments and genetically altered organisms began even before genetic engineering itself. In 1971, when Robert Pollack, a geneticist at Cold Spring Harbor Laboratory on Long Island, New York, learned that Paul Berg planned to allow lambda containing SV40 genes to infect *E. coli*, he telephoned Berg to express his worries about the possible consequences of inserting genes from a cancer-causing virus into one that could infect bacteria that live in the human intestine. It might be possible, Pollack said, for bacteria containing cancer genes to escape from the laboratory and infect people. Berg decided that Pollack might be right. He transferred genes from SV40 to lambda, but he did not allow the altered lambda to infect bacteria.

Other scientists, including Boyer and Cohen, came to share Berg's and Pollack's uneasiness. As some of them pointed out in a discussion at the Gordon Conference on Nucleic Acids in mid-1973, *E. coli*, the "workhorse" bacterium that many experimenters in the new field used, not only could infect humans itself but could exchange plasmids with other bacteria, including ones that cause human disease. The result of this and similar discussions was two letters to *Science*, the prestigious journal of the American Association for the Advancement of Science, warning of the possible dangers of recombinant DNA research. The second letter, signed by top scientists in the field, including Berg, Boyer, and Cohen, and published in the July 26, 1974, issue of *Science*, was the more detailed of the two. It recommended halting several types of recombinant experiments, including those involving genes for antibiotic resistance or tumor formation, until possible hazards could be more thoroughly evaluated and guidelines for safe procedures could be established. This was the first time a group of scientists had voluntarily proposed halting a certain type of experiment because of potential dangers.

Fear of possible harm from accidental release of genetically engineered microorganisms brought 140 geneticists and molecular biologists together for a historic conference at Asilomar, a seaside retreat center in central California, in February 1975. Michael Rogers, a journalist writing in *Rolling Stone*, called it "the Pandora's Box Congress."[5] The Asilomar scientists concluded that most work on construction of recombinant DNA molecules should proceed, provided that appropriate safeguards were employed, but they agreed not to perform some potentially risky types of experiments until new laboratories with special facilities for containing potentially dangerous microbes were built. Then and later, the Asilomar scientists were praised for their restraint. Cynical observers and some of the scientists themselves, however, pointed out that their actions were at least partly an attempt to avoid public criticism and government regulation so that they could maintain control of this exciting new area of research.

The Asilomar guidelines became the blueprint for the first federal government regulations on genetic engineering. These regulations, issued in June 1976, were drafted by the Recombinant DNA Advisory Committee (RAC), a panel of molecular biologists and other scientists that the National Institutes of Health (NIH) had established in October 1974. The RAC had to approve all recombinant DNA experiments funded by the NIH, which at the time meant most of such experiments in the United States. The RAC/NIH regulations were binding on all federally funded research, and most private (usually university-based) researchers agreed to abide by them as well. About this same time, Britain published similar guidelines and established a government oversight group much like the RAC. The European Molecular Biology Organization (EMBO) recommended that researchers in other European countries follow either the U.S. or the British guidelines.

Worry about the possible escape of genetically engineered "superbugs" was not limited to scientists. Congress held its first hearings on the issue in April 1975. Sensationalist media articles, often blending fact with speculation, stirred up the public's fears. Critics, including several environmental groups, complained that the NIH guidelines were not strict enough, accused RAC scientists of conflict of interest (many of the researchers carried out the kind of experiments they were supposed to regulate), and demanded that ethicists and environmentalists have a hand in drawing up future regulations. Debate about the new science was so intense in the mid-1970s that Rockefeller University professor Norton Zinder called it the "Recombinant DNA War."[6]

Friends and foes alike were concerned that the NIH had little, if any, legal authority to regulate research other than that which it funded. Indeed, a federal interagency committee reported to U.S. Health, Education, and Welfare Secretary Joseph Califano on March 15, 1977, that, as far as it could tell, no existing federal agency had the authority to handle all problems raised by recombinant DNA research. Sixteen bills to regulate recombinant DNA research were introduced into Congress that year in attempts to remedy this situation, but none became law, partly because public concern had waned by then and partly because most scientists in the field had come to fear restrictive legislation even more than runaway bacteria. They lobbied against the regulatory bills and convinced the legislators that most recombinant DNA experiments presented no threat to public health.

As time passed and no evidence of actual harm emerged, the NIH relaxed its guidelines, eventually exempting most experiments from regulation. By the early 1980s, most scientific and regulatory bodies had ceased to regard the process of genetic engineering as inherently risky. Controversy arose again in mid-decade, however, when Steven Lindow, a plant pathologist at the University of California–Berkeley, proposed to release genetically engineered organisms—a common and harmless type of bacteria that he had engineered to make citrus and certain other plants resistant to frost damage—into the environment for the first time, as part of a small field test. Jeremy Rifkin, director of the Foundation on Economic Trends and a foe of genetic engineering since the "recombinant DNA war" days, opposed the test, calling it "ecological roulette," and several environmental groups joined him.[7] They succeeded in delaying the test for four years, but when Lindow finally carried it out on April 24, 1987, no detectable environmental effects occurred.

By this time, genetic engineering was being applied to plants and animals as well as bacteria, and the RAC was no longer the only federal body regulating it. In June 1986, the U.S. Office of Science and Technology Policy had published the Coordinated Framework for Regulation of Biotechnology, which divided regulation of genetic engineering research and products among five agencies: the NIH, the National Science Foundation (NSF), the U.S. Department of Agriculture (USDA), the Environmental Protection Agency (EPA), and the Food and Drug Administration (FDA). According to the framework, the first two groups would evaluate research supported by grants; the USDA, specifically its Animal and Plant Health Inspection Service (APHIS), would regulate genetically altered agricultural plants and animals; the FDA would cover engineered drugs, medical treatments, and foods; and the EPA would handle anything related to pesticides and other chemicals. In that same year, the Toxic Substances Control Act (TSCA) was amended to require a permit from the EPA to release most genetically engineered organisms into the environment, use them in manufacturing, or distribute them commercially for intended release. (Lindow thus had to obtain a permit from the EPA as well as the RAC before carrying out his tests.) This division of labor still applies today.

AGRICULTURAL BIOTECHNOLOGY

Most of the products that followed Steven Lindow's lead into open-field testing were plants rather than bacteria. By 1994, more than 2,000 field tests of transgenic plants (those containing genes from another species) had taken place worldwide. No obvious environmental disasters resulted from the tests, and regulation of field testing of genetically altered plants, like regulation of experiments on altered microorganisms before it, slowly relaxed as a result. In 1993, the USDA agreed to require only notification of the agency, not a permit, for testing genetically altered strains of corn, cotton, potatoes, soybeans, tomatoes, and tobacco, which together represented 80 to 90 percent of applications for plant testing.

On the whole, regulatory agencies in the United States have apparently agreed with the claims of agricultural biotechnology companies that genetically engineered plants and animals are not substantially different from those changed by traditional breeding techniques and therefore require no additional regulation. Indeed, supporters say genetic engineering is safer than traditional breeding because it allows more precise control of genetic transfer. Critics, however, point out that biotechnologists sometimes combine genes from species so different from one another that they could never be blended by conventional breeding. They believe, therefore, that transgenic organisms should receive more intense scrutiny than those produced by conventional breeding.

GENETICALLY ENGINEERED PLANTS

In 1996, when genetically modified crops were first grown commercially in the United States, they covered a mere 4 million acres. In 2006, however, according to the International Service for the Acquisition of Agri-biotech Applications (ISAAA), more than 100 million hectares (247,105,381 acres) of genetically altered crops were planted in 22 countries around the world. (The billionth acre of GM crops was said to have been planted in May 2005.) The ISAAA stated that this amount was an increase of 13 percent from the previous year and a 60-fold increase since 1996, the fastest adoption rate of any crop technology. More than half of the crops (53 percent) were in the United States, but 40 percent of the total acreage was in developing countries, chiefly Argentina, Brazil, China, and South Africa. Citigroup, an international financial conglomerate, reported in early 2006 that the global market in agricultural biotechnology had reached $5.6 billion in 2005, with a 21-percent growth rate.

Plants have been engineered to resist insects, disease, drought, salt, and herbicides. Most of these added traits benefit farmers who grow a small number of popular crops, particularly soybeans, corn, and cotton. (Soybeans make up 57 percent of the total acreage of genetically modified crops, followed by corn or maize at 25 percent, cotton at 13 percent, and canola at 5 percent.) One of the most widespread types of genetic alteration, for instance, confers resistance to herbicides such as Roundup, Monsanto Corporation's brand of the herbicide glyphosate. Agribusiness companies sell seeds for these crops and the herbicide to which they have been made resistant as a package. They often require farmers to sign contracts promising to use only that company's brand of the chemical.

Herbicide-resistant crops let farmers use herbicides more efficiently to control weeds. A farmer who has planted "Roundup Ready" soybeans, for example, needs to spray heavily with that herbicide only once, rather than using other, possibly stronger, compounds repeatedly. The crops therefore reduce the use of herbicides overall, supporters say. The crops also free farmers from the need to plow, reducing loss of topsoil to erosion. These effects help the environment as well as save farmers work and money.

Many environmentalists, however, claim that herbicide-resistant crops provide more threat than benefit. Their chief concern is gene flow, through which genes inserted into crop plants travel to wild relatives. Gene flow is especially easy with a crop like corn (maize), which releases its pollen (male sex cells) into the wind. Wild plants that acquire herbicide-resistance genes, these critics say, could become "superweeds," requiring the use of stronger herbicides that might endanger wild plants on which animals and birds depend for food. The spread of these or other engineered genes could also reduce biological diversity, which is already seriously threatened by the worldwide tendency to plant large areas in genetically identical or near-identical natural crops. This narrow genetic base can turn a plant disease epidemic into a major disaster.

Gene flow may already be occurring. Ignacio Chapela and David Quist of the University of California–Berkeley claimed in the November 29, 2001, issue of the respected science journal *Nature* that they had found engineered genes in corn from remote fields in Mexico, even though Mexico does not allow planting of genetically altered corn. This was seen as particularly threatening because Mexico is the native home of maize and a center of biodiversity for this crop. Other researchers vigorously attacked the Berkeley scientists' conclusion, even persuading *Nature* to take the unprecedented step of issuing a sort of apology for having printed their paper, but later reports by other scientists seemed to confirm it, and many experts in the field feel that gene flow is bound to occur sooner or later. This may not be much of a problem in the United States, where few crop plants have wild relatives growing nearby, but it could be a significant threat in developing countries, where crops and weeds often are related.

Another popular type of genetically altered crop may also cause unexpected environmental damage. These crops have been given a gene from the bacterium *Bacillus thuringensis* (Bt for short) that codes for a natural insecticide. They therefore can produce their own pesticides. (Indeed, the EPA classifies the altered plants as pesticides.) Like herbicide-resistant crops, crops containing Bt have become commonplace on American farms. Farmers who plant these crops can greatly reduce their use of expensive, environment-damaging chemical pesticides, supporters say, but critics have questioned whether the crops really reduce pesticide use significantly.

Organic farmers have used sprays of Bt bacteria as "natural" pesticides for decades, but sunlight destroys the bacteria a few days after spraying. Crop plants containing Bt genes, by contrast, produce the bacterial toxin all the time. This means that pest insects are exposed to it for longer periods and thus have more chance to develop resistance. Some evidence of increased Bt resistance in pest insects has already appeared. The EPA requires farmers to plant nonengineered crops of the same kind near the engineered ones so that a population of nonresistant pests will be maintained to mate with any resistant ones that may develop, thereby slowing the spread of resistance.

Furthermore, although Bt spray normally does not harm nonpest insects, the same may not be true of the toxin produced by the engineered plants. In May 1999, entomologists at Cornell University published the results of a laboratory study suggesting that pollen from Bt-engineered corn could be carried onto milkweed plants by wind and could then poison monarch butterfly caterpillars feeding on those plants. Later tests by the USDA's Agricultural Research Service and other researchers, under more natural conditions, however, showed that milkweed plants do not grow very near corn crops and therefore do not usually collect enough pollen to harm monarch caterpillars. Agricultural Research Service entomologist (insect specialist) Richard Hellmich said that chemical insecticides killed far more monarch caterpillars and other nontarget insects than Bt corn did.

Some critics say that genetically modified food crops pose threats to human health as well as to the environment. The first genetically engineered food, a type of tomato called Flavr Savr that had been altered to rot more slowly after ripening and thus be less liable to spoil during shipping and storage, was FDA-approved for sale in 1994. The Flavr Savr failed in the marketplace, however, because consumer advocates claimed that a marker gene inserted into the tomato, which conferred resistance to an antibiotic, might be passed on to humans and make them resistant to drugs prescribed by their doctors.

Ever since, groups have demanded the banning, or at least the labeling, of genetically modified foods. However, defenders of the altered foods insist that most are nutritionally no different from their natural cousins and therefore need no special labeling or regulation. The FDA agreed with this point of view, ruling in May 1992 that genetically altered foods do not need to be reviewed by the agency or labeled as long as they contain no new, toxic, or foreign substances that might cause health problems. Similarly, a report issued by the United Nations Food and Agricultural Organisation (FAO) in 2004 stated that genetically modified crops currently on the market had never been shown to cause verifiable harm to human health.

The chief health risk from genetically engineered foods probably lies in the possibility that individuals might have allergic reactions (which can be life-threatening) to novel proteins they contain. People allergic to Brazil nuts, for instance, proved also to have allergic reactions to soybeans containing a gene from the nuts. (This problem was detected before marketing, and the nut-containing soybeans were never sold.) Recognizing this potential problem, the FDA requires labels on products to which genes from known food allergens, such as peanuts, have been added.

Warnings of allergies and other health risks from GM food rose dramatically in fall 2000, when environmentalists found traces of a Bt-containing corn variety called StarLink in numerous corn products used as human foods, ranging from taco shells to corn chips. Although several kinds of Bt corn had been approved for human food, the EPA had approved StarLink only for animal feed.

Aventis CropScience, the seller of StarLink corn seed, was supposed to tell farmers to keep it separate from corn intended for food use. Nonetheless, the types of corn somehow became mixed during the storage and milling processes. Eventually more than 430 million bushels of corn were found to be contaminated with StarLink and had to be removed from the food processing chain, and more than 300 types of corn products were recalled. No one who ate StarLink-tainted corn was provably harmed by it, but the incident was an economic and public relations disaster for everyone involved. The recall alone cost more than $1 billion. Opponents of genetically engineered foods claimed that the episode confirmed their worst fears about the difficulty of segregating genetically modified from nonmodified material.

Meanwhile, plenty of people are eating genetically modified material that has been approved for human consumption. Estimates in the mid-2000s state that 60 to 70 percent of foods in the United States and Canada contained genetically altered ingredients. No health problems convincingly linked to these foods have been reported, but critics say that this may be because no one has looked for them systematically. Furthermore, since GM foods in North America are not labeled as such, determining whether people with, say, unusual allergies have eaten these foods is almost impossible.

Many environmental and health groups say that regulation of genetically modified crops and foods in the United States is far too lax and too favorable to the biotechnology industry. Numerous recommendations for tightening current regulations have been made, including clarification and streamlining of the division of responsibility among the three agencies involved, making the FDA's premarket reviews of new food crops mandatory instead of voluntary, making reviews open to public input and monitoring, and adding postmarket reviews to detect unexpected health or environmental effects. At the very least, critics say, GM foods should be labeled as such so consumers can decide whether to eat them. Biotechnology spokespeople, however, point out that natural foods can also cause allergies and other health problems, yet they do not undergo the rigorous testing that GM foods receive even under the present system. In new guidelines issued in June 2006, the FDA asked creators of genetically modified food crops containing new and possibly allergenic proteins to give the agency information about the food safety of the proteins at an early stage in the crops' development, but submission of this information remains voluntary.

Demands for increased regulation of GM foods in America are nothing compared to the outcry against such foods in Europe, where they are often called "Frankenfoods." The European Union (EU) began to require labeling of all foods that might contain genetically modified organisms in 1997, and in 1998 it placed an informal moratorium on approval of new genetically modified crops and on importation of foods that might contain unapproved GM products. The moratorium did not end until May 2004, when the European Commission approved the importation (from the United States) of a Bt-containing strain of sweet corn. The commission stated that it would approve other modified food products on a case-by-case basis.

At about the same time (April 18, 2004), stringent new rules approved by the European Parliament on July 2, 2003, went into effect. These regulations require importing countries to provide traceability of all food products "from farm to fork" and verification that foods not labeled as

GM-containing have no more than 0.9 percent accidental GM contamination. The rules cover even highly refined products such as vegetable oils, which contain no DNA or genetically modified protein even when they come from GM plants.

Biotechnology companies and U.S. farmers have protested that the EU's standards are unscientific, unnecessary, and impossible to meet. Calling the EU's attitude protectionism in disguise, the United States, Argentina, and Canada asked the World Trade Organization (WTO) in May 2003 to declare the European moratorium on importation of new genetically engineered food crops to be an illegal restraint of trade. The WTO did so in October 2006.

Many commentators say that the WTO ruling will have little effect. Even if the EU decides, or is forced, to import genetically modified foods, few European customers are likely to buy them. A 2005 Eurobarometer poll showed that 54 percent of the continent's consumers thought that genetically modified food was dangerous, and other surveys have indicated that even higher numbers will refuse to buy such foods, dangerous or not. (Their confidence was not improved by incidents such as one in September 2006, when the European Commission reported that 20 percent of European rice was contaminated with an unapproved genetically modified rice strain accidentally released in the United States and exported.) Similarly, potential consumer rejection (and farmers' fear of losing their markets) has been held to be the reason why Monsanto declared in May 2004 that it would cease development of its Roundup Ready spring wheat.

This conflict between the United States and Europe has had global effects. The European stance has encouraged other international groups to reject or strictly regulate GM foods. For example, the Cartagena Biosafety Protocol, drafted in Montreal in January 2000 as an addendum to the United Nations Convention on Biological Diversity and signed by more than 130 countries, follows Europe's "precautionary principle" approach to regulation of international shipment of GM crops, assuming the crops to be risky until they are definitely proven otherwise. It requires exporters to obtain permission from importing countries before importing live genetically altered material. This treaty went into effect as international law on September 11, 2003.

Meanwhile, numerous countries have refused to allow importation or planting of GM seeds because they fear that they will not be able to export their crops to Europe if they allow possible sources of GM contamination to enter their supplies. Even the African nations of Zambia, Zimbabwe, Mozambique, and Malawi, which were on the brink of famine, rejected donations of U.S. corn from the United Nations Food Programme in August 2002 because they feared that farmers might plant some of the corn and that, if it proved to be genetically modified, the countries would no longer be able to sell corn to Europe. Some officials also expressed concern about health risks. "We would rather starve than get something toxic," Zambian president Levy Mwanawasa announced.[8]

The African rejection particularly surprised and dismayed biotechnology supporters because they claim that developing nations have even more to gain from the new technology than developed ones. Biotech boosters both outside and within developing countries say that genetically engineered crops can improve food yields and therefore will help feed the countries' poor and hungry populations while, at the same time, preserving the environment by decreasing soil erosion, energy use, and dependence on pesticides and other chemicals.

Some crops offer nutritional and medical benefits as well, biotech supporters maintain. Their favorite example is a crop called golden rice, created in 1999, which contains added daffodil and bacterial genes that allow it to produce beta-carotene, a precursor of vitamin A. Vitamin A deficiency affects between 100 million and 140 million children in poor countries each year and makes about 500,000 of them go blind.

Opponents of altered foods said after the announcement of golden rice that the effectiveness of the rice in preventing vitamin A deficiency was considerably exaggerated. The environmental group Greenpeace, for instance, claimed that an adult would have to eat at least 12 times the normal amount of rice in order to obtain the recommended daily dose of vitamin A. In early 2005, however, researchers announced that they had created a version of golden rice that produced 23 times more beta-carotene than the original strain. Golden rice supporters also say that they never intended the rice to provide all the vitamin A in a person's diet and that even a portion of the daily requirement is enough to prevent blindness. After golden rice was developed, other scientists created rice varieties that are rich in iron, another nutrient often missing from poor Asians' diets. A Vietnamese rice research institute reported in early 2006 that it had developed a genetically modified rice resistant to insect pests and rich in vitamins A and E, iron, zinc, and another nutrient, oryzanol.

Critics also reject the overall arguments of biotechnology supporters on both sociological and scientific grounds. Widespread hunger, they say, is not caused by lack of food—some countries with large populations of hungry people, such as India, in fact have grain surpluses—but by political, social, and economic problems such as poverty and war, which prevent equitable food distribution. Increased crop yield will bring little benefit unless these problems are corrected, they maintain. Furthermore, they point out, most of the crops that have been genetically engineered so far are ones that well-to-do farmers grow for export rather than ones that poor people grow to feed their families. In spite of these problems, many poor farmers in the developing world have embraced genetically modified crops eagerly—and benefited from doing so, biotech advocates say. The International Service for the Acquisition of Agri-biotech Applications (ISAAA) claims that more than 90 percent of the people who have gained economically from planting genetically altered crops are resource-poor farmers from developing countries. The ISAAA stated in its 2006 report on worldwide agricultural biotechnology that GM crop adoption in the developing world grew by 21 percent in that year, as compared to a growth of only 9 percent in developed nations. This group maintains that, even though genetically modified seeds cost more than traditional ones and must be repurchased every year, the higher yields of the crops and the money saved by being able to use smaller amounts of herbicides and insecticides outweighs this cost, resulting in overall higher income for the farmers. A study released in fall 2005 by PG Economics stated that developing countries had gained $15 billion in accumulated economic benefits from genetically engineered crops between 1996 and 2004, as compared to only $12 billion in benefits for industrialized countries. The 2004 FAO report also concluded that genetically modified crops offer considerable benefits to poor farmers.

GENETICALLY ALTERED ANIMALS

Transgenic farm animals have existed since 1985, although the techniques for producing them are still experimental and often fail. Some have been altered to increase their value as food (to produce leaner meat, for example), but many are intended for the branch of the biotechnology industry often called "pharming." They have been given genes to induce production of human hormones, drugs, or other medically useful substances in milk (cattle and rabbits), eggs (chickens), or even urine (mice). Producing such substances in animals costs less than 10 percent of the amount needed to make them in laboratories or factories, and the substances often can be made more quickly as well. In mid-2006, a drug called ATryn, which is made from the milk of goats engineered to produce a human protein that prevents blood clotting, was given a "positive opinion" by the European Medicines Agency, a stage of approval just short of allowing the compound to be sold.

Pharming is not the only use of genetically altered animals. Mice, for instance, have been given human genes associated with cancer or other diseases, or have had genes of their own made inactive or "knocked out," to make them more useful in experiments aimed at understanding the diseases or in testing drugs. Pigs have been given human genes in the hope that the animals may become compatible donors for organ transplants. Even a primate, a rhesus macaque monkey named ANDi (short for "inserted DNA" backward), was successfully made transgenic in January 2001. ANDi merely contained a jellyfish gene used as a marker, but scientists hope that future transgenic primates, like transgenic mice, will be useful in medical research.

Biotechnology and Genetic Engineering

The technology for creating genetically altered animals received a large boost in February 1997, when Ian Wilmut and his coworkers at Scotland's Roslin Institute announced that they had cloned a lamb from an udder cell of a six-year-old Finn Dorset ewe. They named the lamb Dolly, after country-western singer/actress Dolly Parton. The amazing thing about Dolly, from a scientific point of view, was not that she was a clone—a sort of delayed twin of her "mother," with exactly the same genetic makeup—but that she had been made from a mature adult somatic (body) cell, something many scientists had thought could not be done with mammals.

Scientists cloned amphibians in 1952. In 1986 they first cloned mammals with a new technique called nuclear transfer, in which a single cell is fused by electricity with an unfertilized egg from which the nucleus has been removed. The technique was normally applied to embryo cells, but Wilmut and his team found they could use it on a mature cell if they first deprived the cell of nutrients for five days. This starvation put the cell into a resting state, in which it did not divide and most of its genes were turned off. After the cell was fused with an enucleated egg, the cytoplasm of the egg cell somehow reprogrammed the adult cell's nucleus. The combined cell eventually produced a whole animal.

Wilmut's chief aim, and that of many scientists who followed him, was to create a more efficient way to produce herds of identical, genetically altered "pharm" animals. Cloning is potentially better than breeding for producing such animals because with cloning a researcher can be sure that a particular gene or trait existing in the original animal will be preserved, whereas it may or may not be kept in breeding. Cloning may also be a useful way to preserve desirable qualities such as leaner meat or disease resistance in nontransgenic animals.

Within months, Dolly was followed by other types of cloned mammals, including the first cloned transgenic farm animals (which were also sheep). The first cloned cat was born in 2001 and the first cloned dog in 2005, leading to discussion of cloning beloved house pets. (A company that had offered to produce cloned cats on demand, however, shut down in February 2007 due to lack of business. The process would have cost owners $32,000 per animal.) Most cloned animals were created with variations of Wilmut's technique. The cloning technique, like techniques for inserting foreign genes into animal germ (reproductive) cells, is still in the early stages of development and very inefficient, however. Dolly, for instance, was the only success out of 277 tries. Ten years after her birth, Ian Wilmut said in a *Time* magazine interview that cloning's success rate was still very poor: Only 2 to 5 percent of cloned eggs resulted in live births.

Many later-cloned animals have died mysteriously before or soon after birth or have been born oversized or deformed. Some studies have suggested that the survivors age more quickly than normal animals. These problems may be due to flaws in the way the genes in the altered egg reprogram themselves. Dolly herself was euthanized in February 2003, at the age of eight years, because she had developed a virus-induced lung tumor. Roslin scientists say this kind of infection is common among older sheep and had nothing to do with Dolly's being a clone, but she may have developed other health problems, such as arthritis, at an unusually early age because of her background.

Nonetheless, some scientists have suggested that cloning could be used to preserve endangered species or even, perhaps, revive extinct ones, such as mammoths. An endangered cattlelike creature called a Javan banteng was cloned in April 2003, the first clone of an endangered animal to survive more than a few days beyond birth. Many conservationists have said, however, that reviving an extinct or nearly extinct species is of little use if the animals' habitat has been destroyed. They believe that money spent on cloning endangered species would be better used for habitat preservation.

So far, transgenic animals have produced much less controversy than transgenic plants, perhaps because these animals have been made in such small numbers. Most complaints have focused on health risks to the animals themselves. For example, animal rights groups have pointed out that a "fast-growing" pig containing growth hormone genes from cows (created by the USDA in the 1980s) was deformed, crippled by arthritis at a young age, and unable to keep itself warm because its body contained so little fat. Such groups also question the ethics of treating animals as mere factories or machines for the production of drugs.

The first major safety controversy on the animal side of agricultural biotechnology centered, not on animals that contain altered genes, but on a genetically engineered substance given to normal animals. Beginning in 1985, Monsanto developed a process for making genetically altered bacteria produce bovine growth hormone (BGH, also called bovine somatotropin, or BST). The recombinant hormone (rBGH), trade-named Posilac, is often given to dairy cattle in the United States to increase their milk production. It won approval from the FDA in 1993, thereby becoming the first genetically engineered animal hormone approved for use in the United States.

The FDA has ruled that rBGH is not an additive and therefore does not have to be listed on the label of milk taken from treated cows. A number of groups in the United States and elsewhere have questioned that decision, however. They point out that the hormone is associated with an increased risk of mastitis (inflammation of the udder) and other health problems in treated cows. To prevent or treat the mastitis, dairy farmers give cows antibiotics, some of which can remain in their milk and be drunk by humans. Constant exposure to low doses of antibiotics encourages bacteria that infect humans to become resistant to the drugs. Some critics also worry that insulinlike growth factor 1 (IGF-1), a hormone related to BGH, may be absorbed from the milk of rBGH-treated cows. This substance may increase the risk of breast and prostate cancer.

Use of rBGH continues to be controversial. Brazil, Costa Rica, Egypt, Israel, Korea, and Mexico, among others, have approved its use, but Australia, Canada, the European Union, and Japan have not. The FDA still maintains that treatment of cows with rBGH has no effect on the content or human health impact of the animals' milk, but some makers of dairy products have appealed to customers concerned about "unnatural" additives by labeling their products as rBGH-free. These companies include Ben & Jerry's Home-made Ice Cream in Vermont and Oregon's Tillamook County Creamery Association, a cooperative that makes the well-known Tillamook cheddar cheese.

An advocacy group called Food & Water Watch launched a national "Hold the Hormones" campaign in January 2007, and shortly afterward, the national coffee chain Starbucks announced that its outlets in a number of states, comprising 37 percent of the company's total production, in future would use only dairy products from cows not treated with rBGH. At about the same time, California Dairies Inc., the second-largest dairy cooperative in the United States, responded to requests from the Safeway/Von's super-market chain and other major customers by asking the cooperative's mem-bers to stop using the hormone. Products from companies that continued to use the substance would be segregated from other products, spokespeople for the cooperative said, and those companies would have to pay the coop-erative a surcharge.

In 2007, however, arguments about rBGH were dwarfed by a new debate about the safety of food products from cloned animals or their offspring. On December 28, 2006, the Food and Drug Administration announced a tentative ruling that meat and milk from cloned cattle, pigs, and goats or their offspring were safe to eat. (Because cloned animals are very expensive, only their offspring—produced in the natural way—are likely to be used for milk and meat.) During the public comment period that followed, which ended on May 3, 2007, thousands of consumers wrote the agency to express fear and disgust at the idea of eating products from clones. Most scientists who expressed opinions during the comment period, however, including an international delegation from the Federa-tion of Animal Science Societies, sided with the FDA. Opponents of cloning insisted that, at the very least, products from cloned animals should be labeled. Several bills introduced into Congress in 2006 and 2007 would require such labeling.

Despite protests, the FDA was expected to give final approval to the sale of products from cloned farm animals later in 2007. Supporters of cloning such as Scott Simplot, head of a large beef-producing company headquar-tered in Boise, Idaho, say that the process will improve the quality of meat and milk by allowing the genes of the healthiest, best-producing animals to be duplicated. The same consumers who reject rBGH, however, may decide that they do not want even the most perfect steak if it comes in such a form. Cloning animals is illegal in the European Union (though offspring of cloned animals may be imported), and sale of products from cloned live-stock is also banned there.

As for transgenic animals, the FDA's Center for Veterinary Medicine announced in October 2002 that it will regulate such animals and their products as "new animal drugs" under the Food, Drug, and Cosmetic Act and therefore will give them the strict review required for all new medicines.

As with transgenic plants, transgenic animals may present a greater threat to the environment than to health. Environmentalists are particularly worried about a transgenic salmon, developed in 2000, which has been given growth hormone genes from a different kind of fish that allow it to grow twice as fast as normal. The creators of the salmon, Aqua Bounty Farms of Massachusetts, say that all their fish will be sterile females, but critics fear that fertile fish might occasionally escape into the wild and then outgrow and outbreed natural salmon, causing an ecological disaster. The owners of Aqua Bounty were still seeking FDA approval for the salmon in 2007. In 2004, reports from the FAO and the U.S. National Research Council both expressed ongoing concern about possible environmental damage caued by transgenic animals released into the wild.

PATENTING LIFE

Herbert Boyer and Stanley Cohen did not think of patenting their revolutionary gene-splicing technique until their universities insisted that they do so. When they finally applied for the patent in 1974, they donated all proceeds from it to the universities. This altruistic approach did not remain the norm for long, however. Profit motives entered the scene in 1976, when Robert Swanson, a young venture capitalist, persuaded Boyer to join him in founding Genentech (GENetic ENgineering TECHnology), the first biotechnology company based on genetic engineering.

Like many similar businesses that sprang up in the following years, Genentech took advantage of the remarkable reproductive powers of bacteria, the first genetically engineered organisms, and the fact that techniques for growing bacteria in factories already existed. Doubling in number every 20 minutes, bacteria could produce billions of duplicates of themselves—including any genes that had been inserted into them—in just a day or so. All these clones could make the proteins for which the inserted genes coded.

Biotechnology and Genetic Engineering

Swanson and Boyer planned to use genetically altered bacteria to make human hormones or other medically useful substances that had been in short supply. They started with insulin, the sugar-controlling hormone that many diabetics must take daily. Insulin from cattle or pigs was easily available, but the animal forms of the hormone differ slightly from the human version, and about 5 percent of diabetics are allergic to them. The product of bacteria containing the human insulin gene, by contrast, was identical to the human hormone and therefore almost never caused an allergic reaction.

When Genentech offered its stock to the public for the first time in October 1980, it had produced genetically engineered human insulin experimentally (starting in 1978) but had not yet won federal permission to sell it. Nonetheless, belief in the company's future was so strong that the price of its stock jumped from $35 to $89 a share in the first few minutes of trading. By the time Genentech finally won FDA approval to sell its genetically engineered insulin in 1982, companies that had followed its lead were using modified bacteria to produce some 48 different hormones or other substances from the human body, as well as several drugs and vaccines. The business side of genetic engineering was well under way.

Where money goes, lawyers and lawsuits soon follow, and biotechnology has been no exception. Most of the suits and other legal disputes in the industry have concerned patents and other ways to protect intellectual property. Genetic engineering has broken new ground in the patent field by raising the question of when—or, indeed, whether—living things, body parts, cells, genes, or DNA sequences should be patented. Is it possible or right to "own" a piece of life?

Patents are a time-honored way of encouraging and rewarding inventiveness by giving an inventor the exclusive rights to an invention's use and sale. The doges of Venice and the rulers of England, among others, traditionally granted monopolies, or rights of exclusive sale, to inventors. The term *patent* came from the "letters patent," or open letters addressed to the general public, that kings used to proclaim these monopolies.

Article 1, section 8 of the U.S. Constitution grants Congress the right to "promote the progress of science and useful arts, by securing for limited times to authors and inventors the exclusive right to their respective writings and discoveries." Patents are issued by the Patent and Trademark Office (PTO), which is part of the Department of Commerce. In 1793, Congress defined a patentable invention as "any new and useful art, machine, manufacture or composition of matter." The Patent Act of 1952 changed the word *art* to *process* and added the criterion that, as well as being new and useful, an invention must not be obvious "to a person of ordinary skill in the art."[9]

Traditionally, products of nature, including living things, were not held to be patentable because they were not inventions, or objects created by human ingenuity. An exception of sorts was made for plant varieties, which were protected in the United States by the Plant Patent Act (1930) and the Plant Variety Protection Act (1970). These laws allowed breeders who had developed new varieties to block others from reproducing them.

Genetic engineering brought a major change to the U.S. patent system in 1980, when the Supreme Court ruled by a 5-4 vote that a patent could be issued to Ananda Chakrabarty, a scientist working for General Electric Corporation, for a genetically altered bacterium that digested petroleum and could be used to help clean up oil spills. In his majority opinion, Chief Justice Warren Burger quoted Congressman P. J. Federico as saying that Congress had intended the 1952 patent law to apply to "anything under the sun that is made by man"—which, Burger decided, could include living things, if humans had altered them.[10] Chakrabarty's patent was the first patent issued for a living organism.

The *Diamond v. Chakrabarty* ruling was soon extended to more complex organisms. Plants, seeds, and plant tissue cultures became patentable in October 1985, when Molecular Genetics, Inc., obtained a patent on a type of genetically engineered corn. A year and a half later, in April 1987, the PTO ruled that genetically engineered animals (except humans), as well as human genes, cells, and organs, could also be patented. The first genetically engineered animal to be patented was the Harvard Oncomouse, a type of mouse designed to be a test animal in cancer research, which was patented in April 1988 by Harvard University. The PTO received 1,502 patent applications for transgenic animals in the decade that followed and approved more than 90. Most were genetically altered mice intended for medical research, but the list ranged from worms to sheep. Since 1998, the PTO has approved about 7,000 biotechnology-related patents each year, and more than 400 genetically altered animals have been patented.

Not everyone has agreed that patenting living things is ethical. In 2000, for example, the Council for Responsible Genetics drafted a "Genetic Bill of Rights," which states that "all people have the right to a world in which living organisms cannot be patented."[11] The European Union's Directive 98/44, on the legal protection of biotechnological inventions, states that patents may be denied or restricted if they are considered to violate public morality, and groups that oppose genetic engineering and patents on living things, such as the European Green Party, have frequently (although not always successfully) called on this feature of the directive when protesting particular biotechnology patents.

Other criticisms of the biotechnology industry's stress on patents, including those on living things, have come from scientists. They point out that, in the United States at least, an invention cannot be patented if details of it have been published more than a year before the patent application is filed. Researchers thus often withhold data until applications can be submitted, impeding the free flow of information that has been traditional in science. This complaint is ironic, considering that patents—which require a description of an invention detailed enough to allow anyone skilled in its field to make the device—were intended to increase access to information about technology. Furthermore, as more and more scientists, including those whose main work is done at universities, become involved in biotechnology ventures, the urge to patent may affect the objectivity of their work. At the very least, such conflicts affect others' perception of the scientists' objectivity. For example, the objectivity of a scientist's study of the effectiveness of a drug might be questioned if the scientist is part owner of a patent on the drug or has stock in a company that manufactured it.

Patents also create monopolies that can reduce the supply of products, increase their price, and limit competition. This is potentially a particular problem in agricultural biotechnology, where giant companies such as Monsanto and DuPont often control both genetically engineered crop seeds and the chemicals on which they depend. Some farmers, especially in the developing world, may not be able to afford these products or to compete against those who can. Furthermore, critics of agricultural biotechnology patents say, such patents force farmers to buy new seeds from the companies every year at prices they can ill afford, rather than planting seeds from the previous year's crop as they traditionally do. Some companies have even developed genetic techniques that automatically enforce this limit by making plants that cannot reproduce. (Countries around the world have informally agreed not to use these "Terminator" seeds, however. This de facto moratorium was unanimously upheld by representatives attending the eighth meeting of the United Nations Convention on Biological Diversity in March 2006.) "Patents create a monopoly that threatens food security and the livelihood of farmers," says Anuradha Mittal, cofounder of the Institute for Food and Development Policy in California.[12] Biotechnology companies, however, point out that on occasion they have donated patented products to poorer countries at reduced or no cost.

A court case involving seed saving that particularly disturbed critics of agricultural biotechnology, especially those concerned with potential abuse of poor farmers by multinational companies, began in 1997 when a Canadian farmer named Percy Schmeiser sprayed the Monsanto herbicide Roundup on his canola crop and found to his surprise that 60 percent of the plants survived the spraying. Their hardiness was probably due to herbicide-resistance genes accidentally transferred to them from nearby farms that raised genetically altered "Roundup Ready" canola. Schmeiser kept seeds from some of the resistant plants and planted them the following year, and in August 1998, Monsanto filed a patent infringement suit against Schmeiser for using the seeds without a license from the company. Several courts upheld the suit, including the Supreme Court of Canada in May

2004, in spite of the fact that Schmeiser had taken no deliberate action to acquire the altered genes and did not sell them or make any added profit from them. Anti-biotech organizations fear that other farmers who save and reuse genetically altered seeds might face similar legal action.

The question of patents on living things becomes even more politically charged when the patents are for altered—or even, sometimes, unaltered—creatures that originally came from the poverty-stricken Tropics. Critics such as the ETC Group (Action Group on Erosion Technology and Concentration, formerly Rural Advancement Foundation International) claim that large companies perform acts of "biopiracy," raiding the developing world—the source of most of the planet's genetic diversity and home to most of its undiscovered species—for specimens or genetic samples that may prove useful. These companies, the critical groups say, provide little or nothing in return to the land that nurtured the plants or the native farmers and healers who used or developed them. They may even sell finished drugs or other products made from these plants back at high prices to the countries that originated them.

A widely discussed example of alleged biopiracy involves the neem tree, which the people of India have used to fight bacteria, fungi, and pest insects for millennia. In 1994, the USDA and W. R. Grace, a large multinational corporation, obtained a European patent for using neem oil against fungi that infest plants. Indian farmers, fearing that the patent would keep them from using their traditional remedies, raised massive protests. (Biotechnology spokespeople claim that the patent would not have affected practices in existence before it was granted.) The European Patent Office revoked the patent on May 10, 2000, and crusaders against such patents, such as Indian activist Vandana Shiva, claimed a victory. Similarly, a European patent for using neem extract as an antiseptic was repealed in 2005.

Since that time, India and some other countries have taken steps to protect their indigenous knowledge. In the mid-2000s, for example, India was compiling the Traditional Knowledge Digital Library, an online database that includes translations of medical texts in Sanskrit, Hindi, Arabic, and Persian, some of which are thousands of years old. The database itself, containing thousands of entries, has been translated into five languages so that patent officers around the world can search it before granting patents on processes or medicines that may prove to have ancient roots. Other countries are beginning or considering similar efforts.

International treaties have also attempted to safeguard the rights of the developing world against patent abuse. Agreements such as the Convention on Biological Diversity and Intellectual Property Rights, signed by 157 nations (not including the United States) at the Earth Summit in Rio de Janeiro in June 1992, have included sections on protecting countries' biological and genetic resources from biopiracy. The United Nations International Treaty on Plant Genetic Resources for Food and Agriculture, ratified in late March 2004 and put into effect 90 days later, guards against potential hiding of information regarding agricultural biotechnology by stating that any public or private research or breeding institution from a signatory country can obtain seeds of crop species covered by the treaty from a public institution in any other contracting country, at no charge.

Protests have also arisen over exceptionally broad patents, beginning with the "process" and "product" patents granted to Boyer and Cohen, which seemed to cover all of recombinant DNA technology. A later dispute of this kind related to patents granted in the early 1990s to Agracetus, a subsidiary of W. R. Grace, for all genetically engineered cotton and soybeans, regardless of the method by which they were produced. "It was as if the inventor of the assembly line had won property rights to all mass-produced goods," complained Jerry Caulder, chief executive officer (CEO) of a rival company, Mycogen.[13] Broad patents, particularly on basic processes or materials necessary for a whole field of research, can make research in that field so expensive that small companies cannot take part in it. Critics say that they drive up the prices that consumers pay for drugs and other biotechnology products as well.

A related subject of dispute is so-called reach-through rights, which give the owner of a patented technology royalties on all commercial products created by that technology, regardless of who develops them. Companies often try to obtain reach-through rights from licensees as a condition of licensing their technologies. Complaints from other companies or, sometimes, government bodies such as NIH have caused the PTO to reverse its decisions on some broad patents, such as one of those granted to Agracetus, and have forced some patent-holding companies to moderate their demands for reach-through rights.

In answer to these complaints, biotechnology industry spokespeople insist that patents on genetically altered living things and the processes that produce them are both ethical and necessary. Patents, they note, traditionally have been held to benefit society by encouraging invention and giving inventors an inducement to risk the time and money necessary to bring inventions to the marketplace. Patents also forced inventors to make public the details of their technology, thus (once the patent expired) encouraging business expansion and competition. Patents, supporters claim, are necessary to give modern biotechnology investors an incentive to support the research and testing needed to develop, say, a new drug. Such a development process can take seven to 10 years and cost up to $500 million. Biotech boosters claim that society ultimately benefits from industry patents because patented research can produce such things as increases in world food supply and life-saving medicines.

As with labeling of genetically engineered foods, differences in public feeling and law between the United States and other parts of the world have caused conflicts in regard to patenting living things. For instance, in 1998 the European Union Parliament and Council decreed in Directive 98/44 that processes for genetically modifying animals that might cause them suffering cannot be patented unless "substantial medical benefit" is likely to be derived from them, although altered plants and processes for making them can be patented. The Supreme Court of Canada decreed in December 2002 that "higher life-forms," defined as multicellular differentiated organisms including both plants and animals, cannot be patented under that country's current patent law. On the other hand, Japan, like the United States, permits patenting of higher life-forms.

The United States has repeatedly pressured other nations to strengthen their legal protection of intellectual property in biotechnology, including patented living things. In 1995, largely at U.S. instigation, the World Trade Organization (WTO) formulated a controversial agreement called the Trade Related Intellectual Property Rights Agreement (TRIPS), which requires WTO members to make their national patent laws similar those in the United States. The supposed aim of TRIPS is to reduce impediments to international trade so that all countries can obtain the benefits of biotechnology (companies will not deal with countries that do not offer them patent protection), but opponents say that the agreement's real purpose is to guarantee protection and high profits to multinational companies. Many developing countries have expressed resentment of TRIPS, saying that installing a U.S.-type patent system would be prohibitively expensive and would bring little, if any, benefit to them.

Types of human cells and natural human body chemicals have also been patented in the United States and Europe by companies that developed processes for isolating them, even if the cells or chemicals themselves were not changed. This includes human stem cell lines, at least in the United States and Britain; stem cell lines or any other uses of human embryos for commercial purposes cannot be patented in the European Union. When a line of laboratory cells can be traced to the body of a particular individual, however, disputes have arisen about who owns the rights to it.

The best known case of this kind in the United States began in 1984, when Seattle businessman John Moore sued his doctor, the University of California, and several drug companies. Moore's suit claimed that the drug companies had made and patented a profitable laboratory cell line from his spleen, which had been surgically removed as a cancer treatment in 1976. Moore claimed that, although he had consented to the operation, he had not been told about the commercial use of his spleen cells, which he said had been planned even before the surgery. He maintained that the cells were still his property even after they had been removed from his body and that they should not have been used without his permission.

A lower court supported Moore, but in 1990 the California Supreme Court ruled against him, stating that patients do not have property rights to their tissues. The court's majority opinion said that granting such rights would "destroy the economic incentive to conduct important medical research."[14] Not all states have followed California's policy, however. In July 1997, for instance, Oregon passed a law granting ownership of tissue and all information derived from it to the person from whom the tissue came.

In the 1990s the cutting edge of biotechnology patenting moved from organisms and cells to pieces of the genetic code itself. Aided by such inventions as the automatic gene sequencer and the polymerase chain reaction (PCR), which allows tiny pieces of DNA to be rapidly copied many times for analysis, scientists in the late 1980s began to work out the base sequences of the genomes, or complete collections of genes, of living things. The most ambitious genome sequencing project of all has been the Human Genome Project, launched in 1990. Even as that project was being completed, the government and private groups involved in it were arguing about who would have rights to the results. They eventually agreed that the raw sequence data would be published on the Internet, where anyone could view it.

The U.S. Patent and Trademark Office (PTO) has permitted the patenting of unaltered genes or DNA sequences, including human ones, since 1987, provided that they have been isolated and purified and that they fulfill the standard patent requirements of novelty, usefulness, and unobviousness. There has been considerable argument, however, about the degree of usefulness that an applicant for a gene or sequence patent must demonstrate. PTO guidelines issued on January 5, 2001, state that applicants must demonstrate "specific, substantial, and credible" utility. Europe also permits patenting of unaltered human genes, but the European Court of Justice ruled in 2001 that this allows "only inventions combining a natural element with a technical process making it possible to isolate that element or reproduce it with a view to an industrial application" to be patented, which means that genes and gene sequences can be patented only for particular applications.[15] Researchers at the University of Sussex in Britain reported in early 2007 that 15,600 applications for patents on human DNA sequences were filed with patent offices in Europe, Japan, and the United States between 1980 and 2003, but they said that only about a third of those requests were granted, mostly in the United States.

Critics of gene and DNA sequence patenting question whether isolation and purification should be considered sufficient to qualify naturally occurring genes or gene segments as "inventions." They also claim that gene patents tie up research information, including information obtained with public funds, and block the development of tests or cures for diseases. Some patient groups, too, fear that patents on genes such as *BRCA1* and *BRCA2*, involved in human breast cancer, will make diagnostic tests using these genes too expensive for most people to afford. Commentators concerned about excessive patenting received support from the U.S. Supreme Court in a unanimous decision in a key case, *KSR International Co. v. Teleflex Inc.*, which the court announced on April 30, 2007. The court's decision will limit patenting by making it easier for the U.S. Patent Office to reject applications on the grounds that the innovations cited are "obvious."

DNA "FINGERPRINTING"

DNA identification or "fingerprinting" has been used to identify lost children and straying fathers, reveal the genetic makeup of endangered species, and trace the origins of poached elephant tusks. It proved that bones found in Siberia belonged to the last Russian czar and disproved the claims of a woman who had insisted for decades that she was his surviving daughter. More important to most people, it has convicted—or exonerated—hundreds of people accused of rape and murder. DNA identification is the most direct application of biotechnology to the courtroom.

DNA profiling, or identification testing, was developed by Alec Jeffreys, a geneticist at the University of Leicester in Britain, in the early 1980s. In 1985 Jeffreys published a paper describing it in the journal *Nature*. He based his test on the fact that certain regions in human DNA consist of short base sequences repeated over and over. These regions are not in the protein-coding part of genes (exons) but rather are part of the so-called junk DNA (introns), the biological purpose of which is still mostly unknown. Jeffreys first used regions called minisatellites, but he soon changed his focus to other stretches of DNA called variable numbers of tandem repeats (VNTRs). Unlike most genes, which exist in only a few different forms, VNTRs vary considerably from person to person, even within the same family, in the number of repeated sequences they possess. Jeffreys concluded that the chance that two people, other than identical twins (who have exactly the same genetic makeup), would have the same numbers of repeats at, say, five different spots was vanishingly small. Comparison of DNA from a blob of semen or a bloodstain left at a crime scene with the blood of a suspect, therefore, should be able to show with high accuracy whether the crime evidence came from the suspect.

Jeffreys's DNA profiling technique was first used to verify family relationships in immigration and paternity disputes. It entered the criminal court in 1987, when, investigating the rape and murder of two teenaged girls, police in Leicestershire took the unprecedented step of asking all men between the ages 13 and 30 years in three villages near the crime scenes— about 5,000 people—to give blood samples for comparison with semen found on the bodies of the girls. Many of the samples were quickly eliminated because they showed blood types other than the one revealed in the semen. Police scientists using Jeffreys's techniques tested the rest, a laborious process that took months.

None of the samples matched the semen from the crime scene. Police were bewildered until they heard about a conversation in a local pub in which a man had bragged about having given a blood sample in a coworker's stead. They questioned the man, and he identified the reluctant coworker as a 27-year-old baker named Colin Pitchfork. Faced with evidence of his falsification, Pitchfork confessed to the crimes, and when his blood was finally tested, its DNA matched that in the semen samples. Pitchfork went to trial in September 1987, was found guilty, and was sentenced to life in prison, thus becoming the first criminal convicted by DNA evidence. DNA testing showed its other side—its ability to clear innocent people—in the same case by demonstrating that 17-year-old Rodney Buckland, who had confessed to one of the murders, could not have committed it because the DNA in his blood did not match that in the semen samples.

A mere month after Pitchfork's trial, DNA evidence was used in a rape case in the United States as well (*Florida v. Andrews*). There, too, it eventually led to a conviction. Police and prosecutors began to regard DNA testing as the greatest aid to criminal identification since the introduction of fingerprinting a century before. Judges and juries, for their part, were awed to hear experts testify that the odds of a suspect's DNA matching that in a crime scene sample by chance were, say, one in a trillion. As one juror in an early DNA case said, "You can't argue with science."[16]

Within a year or two, however, people did begin to argue with science. No one, then or since, disagreed with the basic principles behind DNA testing, but lawyers and scientists questioned both the accuracy with which the tests were carried out and evaluated and the methods used to derive the statistics that sounded so impressive in a courtroom.

From DNA profiling's first uses to the present day, the accuracy with which the tests are performed has been questioned in certain cases. Sometimes accuracy is fatally compromised by the mishandling of samples before they even reach the testing laboratories, as the defense suggested so successfully in the famous O. J. Simpson murder trial in 1995. Testing laboratories and testing kits also vary in quality and, until 1994, despite repeated requests from scientists in the field, there was no federal program for licensing or giving proficiency tests to laboratories that performed DNA analysis. In that year, Congress passed the DNA Identification Act, which gave responsibility for training, funding, and testing DNA profiling laboratories to the Federal Bureau of Investigation (FBI). The FBI has established quality assurance standards and an accreditation program for forensic DNA testing laboratories, but it does not test the laboratories directly. In 1997, furthermore, the agency's own DNA testing laboratory was shown to have, as Deputy Attorney General Jamie Gorelick admitted, "a serious set of problems" in the accuracy of its work.[17]

FBI spokespeople say that the laboratory subsequently improved, but an investigation by the Justice Department inspector general, an independent watchdog, in 2004 revealed a number of continuing problems. For example, the inspector general's office wrote, some procedures in the FBI forensics laboratory's DNA analysis unit were vulnerable to inadvertent or willful noncompliance; one technician there failed to follow proper testing procedure for two years before being detected. Widespread delays in testing of samples were also uncovered. In a report issued in November 2004, the Office of the Inspector General listed the quality of forensic laboratories as one of the top management challenges facing the Department of Justice in that year. Investigations such as one in Houston, Texas, in 2003 have shown that some state DNA testing laboratories also have major quality control problems.

Even when DNA samples are properly handled, difficulties can arise. Confusion can occur, for example, when DNA samples from several people are mixed together, as when more than one person has touched or worn the same article of clothing. If DNA is diluted, some of the base sequences used in standard profiling may be missing in the samples. Techniques used to "amplify" tiny DNA samples so that they can be tested may introduce errors. Even the computer on which test results are analyzed can occasionally make a difference: In one case, reported by Alan Dove in the June 1, 2004, issue of *Genetics and Proteomics*, a PC found a match between suspect and crime scene DNA, but a Macintosh, running the same software and analyzing the same samples, did not. Finally, as defense lawyers delight in pointing out, even an accurate DNA match proves only that a suspect was at a crime scene—not necessarily that he or she committed the crime. By the same token, absence of a suspect's DNA at a scene does not conclusively prove innocence.

In the early 1990s, some scientists and lawyers also questioned the statistics used in court to show the probability of an accidental match between suspect and crime scene samples. To begin with, as critics have pointed out, the frequently used term "DNA fingerprinting" (coined and trademarked by Jeffreys) is somewhat misleading. Each person's fingerprint is unique, and so (unless the person has an identical twin) is each person's DNA. DNA identification tests, however, do not examine a person's whole genome, but only tiny fragments of it. The chance that someone else will have the same sequence in those fragments is usually very small, but, it is not zero, especially if two suspects belong to the same family or small subset of an ethnic group. Thus, although lack of a match in a DNA test could conclusively prove innocence (assuming that the test was properly carried out), critics maintained that a match did not conclusively prove guilt.

In the mid-2000s, although mandatory national standards for DNA testing laboratories still have not been established, the overall accuracy of DNA profiling has improved enough that it is seldom questioned in court cases if law enforcement personnel and laboratories have followed proper procedures for evidence preservation and testing. Modern testing is usually automated, which reduces the risk of error. The standard DNA identification test used in the United States today checks sequences called short tandem repeats (STRs) at 13 different spots in the genome, or loci, where the sequences are known to vary considerably. The chances that two unrelated people have the same DNA patterns at all 13 of the tested loci are said to be less than one in 100 billion—perhaps as low as one in a quadrillion. For all practical purposes, DNA identification has become the unique "fingerprint" that Alec Jeffreys claimed. Furthermore, unlike the earliest DNA tests, which could take several months and required relatively large samples in good condition, today's tests can be performed in hours, can use degraded samples, and can obtain accurate identification from a bit of saliva, a single hair root, or even a few skin cells left behind by a person typing on a computer or clutching a cell phone.

Today the chief subject of debate related to DNA testing is the preservation of blood or tissue samples or identification profiles in large databases. Britain established the first national database for DNA profiles of convicted felons, the Criminal Justice DNA Database, in 1995. It was to contain profiles of everyone convicted of a crime serious enough to warrant imprisonment. Australia, Canada, China, France, Germany, South Africa, and a number of other countries now also have databases of DNA taken from convicts.

In the United States, all 50 states have authorized DNA collection from at least some groups of convicted criminals and establishment of state databases, although not all the databases are actually functional. At the federal level, the FBI opened its National DNA Index System (NDIS) in 1998. The bureau also manages the Combined DNA Index System (CODIS), software that coordinates the national, state, and local databases so that police can match samples at crime scenes with DNA profiles of convicted felons anywhere in the country. In January 2007, the FBI web site stated that NDIS contained 4,274,700 profiles, mostly from convicted offenders (it also includes profiles from some arrestees and some missing or unidentified persons). NDIS/CODIS is the second-largest DNA databank in the world, exceeded only by that of Great Britain.

Both prosecutors and defense attorneys have found DNA testing and DNA databases invaluable. The technology has allowed police to solve "cold cases," in which crimes were committed years or even decades before. In California, for instance, DNA testing led to the conviction of Larry Graham in August 2002 for raping and murdering a five-year-old girl, Angela Bugay, in 1983. DNA testing has also conclusively proven the innocence of prisoners held for decades, resulting in their release. By March 2007 the Innocence Project, established in 1991 by New York City lawyers Barry Scheck and Peter Neufeld to obtain DNA testing for convicted felons who declared their innocence, had freed 196 people, some of whom had been on death row. (The first death-row inmate to be freed through DNA testing was Kirk Noble Bloodsworth, accused of the sex murder of a nine-year-old girl in 1984 and exonerated in 1993.) Such results have made some states place a moratorium on enforcing death sentences. In cases such as *Harvey v. Horan* (2001, 2002), some convicts have even asserted a constitutional due process right to postconviction DNA testing, but as of early 2007, no higher court has agreed with them.

Because DNA databases have proven so useful, many states and countries have expanded their forensic databases considerably. All states now require DNA samples from convicted sex offenders, and most take samples from people convicted of any felony. Most people convicted of felonies in federal or military courts must give up DNA as well, and some states collect it from individuals convicted of certain classes of misdemeanors. As of early 2007, seven states (California, Kansas, Louisiana, Minnesota, New Mexico, Texas, and Virginia) even have laws authorizing collection of DNA samples from at least some people who have merely been arrested. The Justice for All Act, a federal law that took effect on October 30, 2004, made DNA profiles of people convicted of any felony eligible for being uploaded to CODIS, and the federal DNA Fingerprinting Act, signed into law on January 5, 2006, also allows authorities to collect samples from anyone arrested for federal crimes, as well as "non-United States persons who are detained under the authority of the United States."

If nothing else, such expansions are guaranteed to worsen an already severe problem: the fact that state and national databases possess hundreds of thousands of DNA samples that they lack money and facilities to test. This backlog has allowed some criminals to remain free long enough to commit additional crimes. In June 2004, for instance, Robert N. Patton, Jr., was arrested in Columbus, Ohio, for committing 37 rapes—13 of which had occurred during two years in which Patton's DNA, collected when he was first suspected of the crimes, languished untested in an overworked police laboratory. The 2004 Justice for All Act expanded federal grants for DNA analysis and other forensics programs, and some states are trying to meet their needs by outsourcing testing to commercial laboratories, but backlogs in many places are still large.

A few supporters of DNA databases, such as Akhil Reed Amar, a professor at Yale Law School, have even proposed that national databases be created to contain genetic records from all citizens, perhaps obtained at birth. As of mid-2006, no country had created such a database, but Portugal announced in 2005 its intention to do so, and Britain also seems to be moving in this direction. In March 2003, the British government changed the name of its database to the National DNA Database and announced that police would have the right to keep indefinitely DNA identity information for anyone from whom samples were taken, even if the people were never charged with a crime. A 2004 law gave police the power to take samples from anyone arrested for a "recordable offense," and a further expansion in 2005 covered even minor crimes such as speeding and littering. By that time, Britain's DNA database was already the largest in the world, including nearly 3 million samples, and was said to be adding new ones at the rate of 10,000 to 20,000 each month.

Critics oppose the expansion of forensic databases for a variety of reasons. First, some claim that the usefulness of the databases in solving "cold cases" has been exaggerated. Mark A. Rothstein and Meghan K. Talbott, writing in the summer 2006 issue of the *Journal of Law, Medicine, and Ethics*, maintain that the effectiveness of DNA databases in solving or preventing crimes has never been properly evaluated. They and several other commentators say that, instead of being judged by the numbers of matches with crime scene samples or "investigations aided," the databases should be rated on the number of cases solved. Rothstein and Talbott report that in one study in Virginia, DNA matches resulted in convictions in less than 30 percent of cases. Other critics say that expanded databases, because they would increasingly include people not convicted of major (or, indeed, any) crimes, would be a waste of law enforcement resources and might result in an increased number of wrongful convictions from laboratory errors.

Other opponents express concern, not only about the expanding size of forensic databases, but about the broadening of uses to which such databases are put. One new use that has caused comment is the so-called familial search, in which investigations are made based on close, but not perfect, matches between DNA in crime scene samples and database profiles or in which the DNA of a member of a suspect's family is tested, sometimes without that person's knowledge, as a substitute for unobtainable DNA from the uncooperative suspect. The first person to be convicted of a crime partly through a familial DNA search was a British man, Craig Harman, who killed a driver by dropping a brick from a freeway overpass in May 2003. Police traced Harman after finding that DNA on the brick almost matched the profile of Harman's brother, which was in the British forensic database. Harman was convicted of the murder on April 19, 2004.

Familial DNA searching allegedly also helped to produce a much more highly publicized conviction—that of Dennis Rader, the "BTK [bind, torture, kill] killer," in Wichita, Kansas. According to a report quoted by Frederick Bieber in the summer 2006 *Journal of Law, Medicine, and Ethics*, when police began to suspect that Rader had committed a series of brutal murders in the Wichita area, they obtained a warrant to take a biopsy sample from a physician treating Rader's daughter. (The daughter was not told that the sample had been seized.) DNA from the biopsy material was tested and found to be a close match to that found at the scenes of some of the killings. The crime scene profile matched what paternity analysis would predict for the father of the biopsy's owner—that is, Dennis Rader. This match provided some of the evidence that led to Rader's arrest on February 25, 2005. He pleaded guilty and confessed to killing at least 10 people during his trial in June.

In spite of these successes, critics fear that familial searches will result in invasion of the privacy of people like Rader's daughter, who are not even accused of any crime, let alone convicted. They could also increase the risk of mistaken arrests or convictions of people who are guilty of nothing more than being related to known criminals. Some opponents of expanded DNA databases, such as Tania Simoncelli and Barry Steinhardt, also writing in the summer 2006 *Journal of Law, Medicine, and Ethics*, are concerned about other forms of "function creep," in which forensic databases are put to uses not associated with law enforcement, such as development of medical statistics. Misuse of information from such databases, they say, could lead to genetic discrimination.

Groups such as the American Civil Liberties Union (ACLU) and the Libertarian Party see DNA databases as a potentially serious threat to the right to privacy and civil liberties, particularly if they include people who have not been convicted of crimes or consist of actual tissue samples rather than just identification profiles. They say that the taking of blood samples, at least, requires an intrusion into the body that may violate Fourth Amendment protection against unreasonable searches and seizures. In addition, although DNA profiles do not reveal personal health information because the tested loci are not actually genes, preserved blood or tissue samples would contain such information, which might lead to discrimination if it fell into the hands of employers or insurers.

So far, courts have found the taking of samples from convicted felons, at least, to be constitutional. For instance, in May 2002 the Third District Court of Appeals unanimously rejected the suit of a group of female death-row inmates in California *(Alfaro v. Terhune)* who had challenged the constitutionality of a program that required convicts to donate their DNA. In a more recent case, *United States v. Kincade*, the Ninth Circuit Court of Appeals ruled on August 18, 2004, that taking samples from parolees, as mandated by the federal DNA Act (2000), was also constititutional. The Third Circuit Court of Appeals made a similar ruling in *United States v. Sczubelek* in March 2005.

The courts are likely to feel more strongly about protecting the rights of people not convicted, or perhaps not even accused, of any crime. Useful as an all-citizen database might be in capturing criminals, tracing missing persons, and perhaps in preventing crime, many people feel that it would not be worth the price. "The inherent danger to our conception of ourselves as a free and autonomous society requires that further expansion of the preventive state, represented by the creation of a universal [DNA identification] database, be vigorously opposed," writes Rebecca Sasser Peterson of the Georgetown University Law Center in *American Criminal Law Review*.[18]

The question of what is done with material collected for forensic DNA databases is just as controversial as the collection itself. One issue is the opportunity (or lack of it) for people who are not charged with a crime or who are tried and acquitted to remove, or expunge, their records from the database. A few states automatically discard samples and profiles when arrestees are cleared of charges, but in many other states, this happens only if the acquitted defendants make a written request. A written request, accompanied by court documents, is also required for expungement from the national database, NDIS. A second issue concerns what material is being kept. Many commentators see no problem with permanent storage of DNA profiles, which contain no revealing medical information, but object strongly to the preservation of blood, cheek swabs, or other cell samples, which make a person's entire genome potentially available for readout. They say that many objections to the expansion of forensic databases would be withdrawn if only profiles, not samples, were kept.

GENETIC HEALTH TESTING AND DISCRIMINATION

Leroy Hood, a pioneer in the development of automated genome analysis machines, said in 1987 that in the next century, "When a baby is born, we'll 'read out' his genetic code, and there'll be a book of things he'll have to watch for."[19] Completion of the Human Genome Project and later gene-mapping projects are bringing the future described in Hood's prediction rapidly closer by helping scientists identify variations in DNA that are associated with increased risk of various diseases. Some people wonder, however, whether such detailed knowledge of each individual's genes will be an entirely good thing. On the one hand, it could help people avoid illnesses to which their genes predispose them by, for example, making certain lifestyle choices or undergoing tests for certain conditions unusually often. On the other hand, critics say, it could also lead to a new form of discrimination, based not on race or gender but on one's genes. "Genetic discrimination is the civil-rights issue of the 21st century," says Martha Volner, health policy director for the Alliance of Genetic Support Groups, an organization for families with inherited diseases.[20]

Potentially, as the ability to analyze individual genomes increases, genetic discrimination could affect anyone. "All of us have something or other in our genes that's going to get us in trouble," says Nancy Wexler of Columbia University, a leader in research on the form of inherited brain degeneration called Huntington's disease—and at risk for the disease herself because her mother had it. "We'll all be uninsurable."[21]

Genetic discrimination is most likely to limit access to life and health insurance, the latter of which, at least in the United States, is more or less required for quality health care. It could also bar people from jobs because most large employers provide health insurance for their workers, and employers may not want to risk having their group insurance premiums raised by hiring workers who seem likely to become sick. Genetic discrimination might even affect marriages and family relationships.

Discrimination on the basis of genetic makeup, as it was shown in characteristics thought to be inherited, existed before the word *gene* even entered the English language. This discrimination arose from the so-called science of eugenics (the word means "well born"), which was founded in the late 19th century by Francis Galton, a cousin of Charles Darwin. Galton believed that intelligence and other personality characteristics were inherited. He said that people with desirable characteristics (as he and his social group defined them) should be encouraged to have children, whereas people with undesirable characteristics (often meaning characteristics of other ethnic groups or social classes) should be discouraged or even forcibly prevented from doing so. In 1872 he wrote:

> *It may become to be avowed as a paramount duty, to anticipate the slow and stubborn process of natural selection, by endeavouring to breed out feeble constitutions, and petty and ignoble instincts, and to breed in those which are vigorous and noble and social.*[22]

Galton and his followers saw such practices as the human counterpart of scientific animal and plant breeding.

Around the start of the 20th century, Galton established the Eugenics Society in Britain to carry out his aims. Similar groups were formed in the United States and Germany. Eugenics supporters of the time included both respected scientists and such well-known nonscientific figures as Theodore Roosevelt, George Bernard Shaw, and (in his youth) Winston Churchill.

Negative eugenics—forcible blocking of reproduction for those thought not fit to reproduce—was codified into law in many places. By the 1930s, some 34 states in the United States had passed laws requiring the forcible sterilization of criminals, the mentally retarded (developmentally disabled), the insane, or others considered unfit to reproduce. According to one estimate, about 60,000 Americans were sterilized as a result of these laws.

Biotechnology and Genetic Engineering

In the landmark 1927 case of *Buck v. Bell*, even the U.S. Supreme Court upheld the forced sterilization of an allegedly developmentally disabled woman, Carrie Buck. Noting that both Buck's mother and her seven-month-old daughter also seemed to have subnormal intelligence, renowned justice Oliver Wendell Holmes wrote in his majority opinion on the case, "Three generations of imbeciles are enough."[23] (In fact, scientist-writer Stephen Jay Gould reported in a famous 1984 essay that Carrie Buck had been reexamined in 1980 and was found to be of normal intelligence, and school records suggested that her daughter had been normal, too. The only "deficiencies" of the three generations of Bucks, Gould concluded, were that they were poor, uneducated, and violated contemporary sexual mores by giving birth to children out of wedlock.)

Denmark, Finland, Norway, Sweden, and some Canadian provinces also enacted eugenics laws in the 1920s or 1930s. Germany passed a eugenic sterilization law in 1933, soon after the Nazis took power, and used it to force sterilization of some 400,000 institutionalized people whose defects were presumed to be inherited. Later the Nazis carried eugenics to its extreme by trying to wipe out the genes of Jews and whole ethnic groups through mass killing.

Partly because of the Nazi excesses, the tide of public opinion began to turn against eugenics around the time of World War II. In *Skinner v. Oklahoma* in 1942, for instance, the Supreme Court struck down a state law requiring forced sterilization of convicted criminals, saying that procreation is "one of the basic civil rights of man."[24] Belief in eugenics never totally died out, however. Many eugenics laws remained on the books until the 1970s, although they were seldom enforced. As late as 1980, 44 percent of respondents in a U.S. poll favored compulsory sterilization of habitual criminals and the incurable mentally ill. Even today, China has a law that can deny permission for marriage to or force sterilization or abortions on people with certain inherited diseases.

In the 1970s, just as traditional eugenics was fading, a new form of genetic discrimination began to appear. Tests that examine DNA directly were still decades away, but sufferers from and carriers of certain inherited diseases could be identified indirectly by tests for particular substances in their blood or other body fluids. Chief among these illnesses was sickle-cell disease (also called sickle-cell anemia), a blood disease that affects mostly people of African ancestry.

Sickle-cell disease is caused by a single recessive gene that codes for an abnormal form of hemoglobin, the protein in red blood cells that gives them their color and allows them to carry oxygen through the body. Cells containing the abnormal hemoglobin take on a crescent shape and block the body's smallest blood vessels, depriving tissues of oxygen and causing considerable pain, disability, and sometimes death. Only people who inherit defective genes from both parents suffer from the disease, however. Those who inherit a defective gene from one parent and a normal one from the other—called sickle-cell carriers or possessors of the sickle-cell trait—can pass the disease to their children, but they themselves are quite healthy. About one in 500 African Americans suffers from sickle-cell disease, and one in 10 is a carrier.

Tests on hemoglobin in blood samples can detect sickle-cell disease and sickle-cell trait, and in the early 1970s African Americans were widely screened for this condition. The screening programs were intended to identify carriers so they could be counseled about the risk of having children with the disease if they married other carriers. Because of widespread lack of understanding about the difference between sickle-cell carriers and people with the full-blown disease, however, some carriers were charged high insurance premiums or denied certain kinds of employment. African Americans supported sickle-cell screening programs at first, but they soon began to protest them as racism in a new, medical guise. A number of states therefore either dropped the programs or passed laws limiting use of information obtained from them.

The same issues raised by sickle-cell screening in the 1970s reappeared in the 1990s as a growing number of tests became available to screen DNA for mutations that cause or increase the risk of particular diseases. By the mid-2000s, genetic tests existed for more than 1,300 diseases. These included not only classic inherited diseases such as Huntington's but some forms of more common and complex illnesses, such as cancer and heart disease, which can have a genetic component.

All U.S. states have established programs to screen newborn babies for certain inherited diseases, ranging from three or four to more than 30 conditions. (These tests are not, strictly speaking, genetic tests; they reveal inherited conditions by identifying certain biochemicals in the blood.) Most of the conditions tested for are ones for which early intervention, for instance in the form of diet restrictions, is essential to prevent severe damage, and bioethicists generally feel that screening of babies should be limited to such diseases. Although the screening tests can be lifesaving to a few children, some ethicists and civil libertarians worry that private genetic information could later be obtained without consent from the blood spots on the resulting "neonatal cards," which hospitals usually keep after testing.

For adults, genetic tests need not even require a visit to a doctor's office or a laboratory: Many such tests can be ordered online. After choosing and paying for the tests, people receive a kit in the mail, rub a brush on the inside of their cheeks, return the brush to the test company in a prepaid envelope, and wait for the results. Critics, including the American College of Medical Genetics, have pointed out a number of problems with genetic tests, particularly with those that consumers order through the Internet.

First, the tests may not be accurate, because little quality-control oversight covers the laboratories that perform them. The FDA considers genetic testing to be a "service" and therefore does not regulate it. The Centers for Medicare and Medicaid Services regulate medical laboratories under the Clinical Laboratory Improvement Amendments (CLIA), but this agency does not check specifically for proficiency in genetic testing, and it cannot regulate any testing that is not performed by a clinical laboratory.

Even if a test is properly carried out, it may not reveal anything useful. Associations between particular genes and the risk of developing diseases, especially common and complex illnesses such as cancer and heart disease, are far from clear, and some tests identify genes whose connection with disease is supported by little, if any, scientific evidence. Tests that are accurate and scientifically valid may be unnecessary. Most diseases clearly associated with genetic mutations are rare, so testing for such a condition is useful only if people already know that the disease is common in their family. Establishing the risk of developing a disease that cannot be prevented or cured may be undesirable as well. For example, only about 5 percent of people at risk for Huntington's disease, an incurable brain ailment caused by a mutation in a single gene, take a test that can tell them, with close to 100 percent accuracy, whether they will develop the condition.

Most people need the help of trained genetic counselors to understand the meaning and implications of the results of genetic tests. Lack of accurate counseling can make test results confusing, misleading, or even dangerous. Some online DNA testing facilities offer consultations with counselors, but many do not, or else they provide interpretations that are unclear, exaggerate the risk of disease, or focus on selling a drug or other treatment offered by the company.

Finally, genetic tests make people focus on individual genes as a cause of disease when, in fact, most illnesses—including the most common and deadly ones, such as cancer and heart disease—result from the interaction of multiple genes with each other and with environmental factors such as lifestyle choices (smoking or eating a high-fat diet) or exposure to toxic chemicals. Thus, even if a person inherits a gene that increases the risk of, say, heart disease, that person may never develop the disease, whereas someone else with a better genetic profile but a worse assortment of environmental exposures may.

Additional ethical problems arise because the results of genetic tests affect not only individuals but their families. Suppose, for instance, that a woman learns from a test that her son will develop Duchenne muscular dystrophy, an inherited condition passed from mothers (who do not show signs of illness) to sons that causes disability and early death. The woman's sister is pregnant and has said she will terminate her pregnancy if the fetus has a life-threatening condition. The woman does not want to tell her sister about the test results, even though they mean that the sister's fetus may also be affected, because she believes that abortion is wrong. A physician or genetic counselor knows both women. Should the professional break patient confidentiality to tell the pregnant sister about the test results? "You [can] find yourself in a situation in which two individuals have competing moral interests," says Logan Karns, chair of the ethics subcommittee of the National Society of Genetic Counselors.[25] Because of these complex issues, organizations of genetic counselors recommend that individual counselors discuss the implications of testing carefully with people—and sometimes with their relatives as well—before testing takes place.

Chief among the legal issues raised by genetic tests is invasion of privacy. Some people order genetic tests directly, rather than through a physician or other health care professional, because they want to keep the results of such tests out of their medical records to prevent possible discrimination by insurers or others. They seldom have any guarantee, however, that testing companies will not give or sell their information to someone who asks for it. No federal law requires genetic testing firms to keep their results private.

The feeling of privacy invasion is even stronger when people take such tests involuntarily. This happened in the 1980s and 1990s at the Lawrence Berkeley Laboratory (LBL), a California research facility funded by the Department of Energy. In February 1998, the U.S. Court of Appeals for the Ninth Circuit ruled that LBL had been wrong to test the blood of its employees for sickle-cell trait and other conditions without their informed consent. The court held that such action violated the protection against unreasonable search and seizure guaranteed by the Fourth Amendment. Because some tests were performed only on samples from blacks, Hispanics, or women, they also violated the 1964 Civil Rights Act. This was said to be the first case in which a federal appeals court recognized a constitutional right to genetic privacy.

National DNA databases such as the ones established in 1992 by the Department of Defense and in 1998 by the FBI have also been seen as threats to genetic privacy. In April 1996, John C. Mayfield, III and Joseph Vlakovsky, both U.S. Marines with exemplary military records, were convicted in a court-martial of disobeying a direct order because they refused to give blood for storage in the Department of Defense database. The soldiers said that letting the government archive their genetic information violated their right to privacy, and they feared that the information might eventually be used to discriminate against them in some way. This was the first case to challenge the right of an employer to mandate sample donation for genetic testing. The case was later dismissed as moot because both men left military service before it was settled, but the military did agree to destroy blood samples at donors' request after the donors left the service. They still maintained their right to collect samples from active personnel, however.

Similar disputes have surfaced in Iceland, where a company called de-CODE Genetics (started by a native Icelander) attempted to mine genetic data from the medical records of the country's entire population in a search for genes related to particular diseases. The Icelandic government and much of the island's people cooperated at first, but some became concerned about privacy and other issues and filed suit to remove relatives' records from deCODE's database. In 2003, Iceland's supreme court ruled in their favor, saying that the company's system did not sufficiently protect individuals' identity. Similar concerns have been expressed about blood samples collected for other large genetic database projects.

People are especially worried about genetic privacy because of fears that knowledge of genetic susceptibilities to disease might lead to discrimination in insurance and employment. While many people see nothing wrong with insurers charging higher rates to those who choose unhealthy activities such as smoking, they feel it is unfair to penalize individuals for their genes, which they cannot choose or change. Individuals have often told interviewers that they are afraid to be tested for defective genes they suspect they carry because they fear that insurers or employers will gain access to the results and then deny them or members of their families insurance and jobs. Some think that such discrimination has already occurred. For example, in a poll (cited in a March 2007 *Scientific American* article) that interviewed 587 people who had, or were closely related to people who had, either genetic diseases or nongenetic chronic illnesses, twice as many of those with genetic diseases as with other chronic conditions said they had been denied health insurance or charged prohibitive rates for it. Furthermore, 23.5 percent of people with genetic disorders were allowed only limited coverage of their conditions, as opposed to only 14.2 percent of people with other chronic diseases.

Of course, fears and beliefs are not the same as facts. Writing in the September 1, 2005, *New England Journal of Medicine*, Henry T. Greely, a professor of law and genetics at Stanford University, said that few verified examples of genetic discrimination in insurance had appeared, quite possibly at least partly because genetic heritage is only one factor in causing most common diseases, and present-day genetic tests cannot accurately forecast who will develop these conditions. Greely thought discrimination was more likely to occur in employment than in insurance. Similarly, Dean Rosen, senior vice president of policy for the Health Insurance Association of America, insisted in early 1999 that "the fears [of genetic discrimination] out there are just not reality."[26] But another insurance executive, who preferred to remain anonymous, advised people to "apply for insurance today, get [genetically] tested tomorrow" rather than the other way around.[27]

By 2005, all but three states had passed laws that limited genetic discrimination in health insurance to some degree, and more than 30 states had banned or limited genetic discrimination in employment. Laws of this kind, however, are often weakened by lack of a clear definition of genetic testing and, therefore, the kind of information that is covered. The laws also often do not cover genetic information obtained indirectly, for instance through family histories or through tests for gene products such as abnormal hemoglobin in the blood of people with sickle-cell trait. Finally, state insurance laws may not affect employers' self-funded plans, which in 1995 provided insurance for over a third of the nonelderly insured population, or insurance purchased directly by individuals.

The federal government has been less active than the states in attempting to ban genetic discrimination. Antidiscrimination legislation has been introduced into Congress several times, most recently in early January 2007, and has been passed in the Senate twice (in 2003 and 2005), as well as obtaining the support of President George W. Bush. So far, however, the House of Representatives has not approved it. President Bill Clinton signed Executive Order 13145, which prohibits genetic discrimination in federal employment, in 2000, but it does not apply to insurers or private employers.

Nonetheless, three existing federal laws do, or at least may, have an impact on genetic health testing and discrimination in employment. The Health Insurance Portability and Accountability Act (HIPAA), passed in 1996, is aimed mainly at preventing people from losing their health insurance when they change jobs, but it includes a provision that forbids health insurers issuing group plans from denying insurance on the basis of preexisting genetic conditions. It does not cover the 5 to 10 percent of Americans who have individual insurance plans, however, and it may not cover companies that insure their own workers, as opposed to those who buy group plans from insurance companies. It also does not necessarily prevent insurers from raising premiums on the basis of genetic information or protect the privacy of such information. The Employment Retirement Income Security Act (ERISA) also prohibits employers from discriminating against employees on the basis of existing or predicted health care expenses.

Finally, the Americans with Disabilities Act (ADA), passed in 1990, may apply to healthy people who have inherited "bad" genes, if they can show that the results of genetic tests have made their employers perceive them as disabled. The federal Equal Employment Opportunity Commission (EEOC) filed a suit against the Burlington Northern Santa Fe Railroad in February 2001, charging that the company had violated the ADA when it performed genetic tests on certain employees without their knowledge. The railroad settled out of court, however (it admitted no wrongdoing but agreed to stop all genetic testing, keep existing test results confidential, and pay the tested employees a total of $2.2 million in damages), so the EEOC's strategy received no judicial test.

In *Bragdon v. Abbott*, a case brought before the Supreme Court in 1998, the court ruled that the plaintiff, who had tested positive for HIV but did not yet show any symptoms of AIDS, was protected by the ADA. A person carrying, say, *BRCA1* or the gene for Huntington's disease would seem to be in a similar situation. In a June 1999 Supreme Court ruling on another ADA case (*Sutton v. United Air Lines*), however, Justice Sandra Day O'Connor wrote in her majority opinion, "We think the language [of the ADA] is properly read as requiring that a person be presently—not potentially or hypothetically—substantially limited [in a major life activity] in order to demonstrate a disability."[28] This would seem to dim hopes that the ADA will cover healthy people with genetic predispositions to illness.

Several decisions by lower courts also suggest that the ADA's protection against genetic testing and employer use of genetic information is limited at best. For one thing, commentators have pointed out, employees invoking the ADA would have to show, not only that an employer believed (correctly or otherwise) that they had a genetic defect, but that the employer regarded this as a disability (as defined by the ADA) and discriminated against the employees because of it. The ADA also does not prohibit the obtaining of genetic information, only certain uses of it. The law states that an employer may ask for any kind of medical information, genetic or otherwise, between the time of a job offer and the time a new employee begins work.

Discrimination based on genetic health testing has been an important issue in Britain, where the national health care system is increasingly supplemented by private insurance, as well as in the United States. In a 2002 poll, for example, 20 percent of women who tested positive for the breast cancer-associated genes *BRCA1* and *BRCA2* reported problems with insurance discrimination within a year of taking the test. To ease the public's fears, beginning in 1997, the British government and the Association of British Insurers, a regulatory body that covers 95 percent of the country's insurance companies, agreed to moratoria on using genetic testing data in setting rates for all but the most expensive mortgage and health insurance policies. The most recent moratorium went into effect on March 14, 2005, and will extend until November 1, 2011. Insurers reported no hardship under the previous moratoria, which suggests that genetic testing is not a very large factor in their determinations.

By early 2006, more than 40 countries on three continents (North America, Europe, and Asia) had adopted either laws against genetic discrimination or moratoria on the use of genetic information in insurance like the one in effect in Britain. Article 11 of the Convention for Human Rights and Biomedicine, adopted by the Council of Europe in November 1997, prohibits discrimination on grounds of genetic heritage, and the UNESCO General Conference's Universal Declaration on the Human Genome and Human Rights, also adopted in 1997, has a similar provision. As with laws opposing genetic discrimination in the United States, however, the meanings of terms such as *genetic testing* are not clearly defined in these rulings. The methods by which the declarations are to be enforced are also not clear.

Some commentators who are concerned about genetic discrimination in the United States say that new federal laws should be passed to protect genetic privacy and forbid employers and insurers from acquiring or using genetic information. Others say that, ultimately, genetic information is no different from any other medical information. Rather, they see the debate about insurers' use of genetic information as simply one aspect of the basic conflict between insurers (and employers who pay for employees' insurance) and the insured regarding access to and use of all medical information.

As a number of writers have pointed out, this conflict is built into the nature of life and health insurance. Insurers can stay in business and make a profit only if they and the people they insure have equal knowledge of those individuals' risks of illness or death. If insured people alone know about a risk, the insurers say, they may take out large amounts of health or life insurance because they expect to need it. This "adverse selection" throws off the statistical methods by which insurance premiums are determined and drives up premiums for everyone. Harvie Raymond, director of managed care and insurance operations at the Health Insurance Association of America, insisted in 1995 that "he who assumes a risk should have the opportunity to evaluate that risk" and that laws barring insurance companies from obtaining genetic information "could create a great deal of havoc in the industry, causing costs to go up and fewer people to ultimately be able to afford coverage."[29] Conversely, if insurers know of a person's increased risk, they have a powerful financial incentive to deny insurance or charge that individual very high premiums. This could make insurance unavailable to the very people who need it most.

Some commentators believe that, as genetic risks become easier to predict, this conflict will become soluble only by separating health care from the insurance system, probably by turning to a national health care system like the ones in Britain and Canada. Lori B. Andrews of the Chicago–Kent College of Law has predicted that "increasingly sophisticated genetic diagnostic tests may force a total rethinking of the concept of health insurance" and that the perceived injustice of genetic discrimination "will provide the impetus for the development of a national health system."[30] Another possibility, suggested by genetics expert Thomas Caskey, is to place people at high risk of genetic disease in a special insurance pool, as is sometimes done with otherwise uninsurable drivers. Alternatively, says Patrick Brockett, director of the Risk Management and Insurance Program at the University of Texas, insurers could be allowed to set rates on the basis of genetic tests, but people who do poorly on the tests could be given vouchers or other subsidies to help them pay for their insurance.

Another form of discrimination could stem from genetic determinism, the widely held belief that genetics—perhaps even single genes—are primarily or wholly responsible for such behaviors as violence, risk taking, and alcoholism. Although these behaviors may have a genetic association in certain cases, they are most likely determined by complex interactions between genes and environment. Indeed, scientists increasingly agree that such interactions determine almost all physical characteristics, including most susceptibility to disease. Nonetheless, if genetic determinism continues to be popular and genes associated with, say, violence are found, civil libertarians say, children with such genes might be put in special schools or otherwise segregated or treated differently from other children. Attributing violent behavior to genetics could also lead to a new variant of racism if "violence genes" prove to be more common in some racial or ethnic groups than in others. Perhaps the most likely and also most important danger of focusing on exclusively genetic causes of behavior such as violence, critics claim, is the diversion of attention and funding from correction of social causes of such behavior. "We know what causes violence in our society," says genetic discrimination expert Paul Billings. "[It is] poverty, discrimination, [and] the failure of our educational system."[31]

HUMAN GENE ALTERATION

Genomics, the study of human and other genomes (including the way groups of genes interact with each other and the environment), and other sciences connected to biotechnology are already beginning to help physicians diagnose and treat disease. More than 300 approved medical biotechnology products existed in mid-2004, and even more were under development.

Several gene-mapping projects have identified new disease-related genes. For example, the HapMap project, which examined the DNA of 269 people from four ethnic groups (Japanese, Yoruba [Nigeria], Han Chinese, and northern and western European [Utah]) to detect minute differences in their genomes, found a gene associated with age-related macular degeneration, the leading cause of blindness among older people. DeCODE Genetics's database of genetic variations within Icelanders, most of whom are descended from a small group of ancestors, has helped scientists pinpoint a gene related to type 2 (adult onset) diabetes.

Researchers are also zeroing in on variations in genes that affect the ways people's bodies react to and break down drugs. Gene variations can determine which drugs and which doses are most effective in treating disease in particular individuals and which will have the fewest undesirable side effects. Someone with a gene variation that makes his or her liver break down a certain drug unusually slowly might need a smaller dose of that drug than another person with the same illness whose liver destroys the drug more quickly, for instance. Similarly, gene variations in cancer cells can determine which anticancer drugs will work against particular tumors. One new anti-cancer drug, Herceptin, even comes with its own gene test kit.

Drugs and vaccines based on researchers' new understanding of genes and genetic interactions are also appearing. An article in the June 13, 2005, *Business Week* stated that 230 drugs and other products created from genetic and biotechnology techniques are already on the market, including a number of new anticancer drugs such as the previously mentioned Herceptin and Gleevec. Hundreds more are undergoing clinical trials and seem likely to win FDA approval before the end of the new century's first decade. Most amazing of all, though, scientists have designed treatments that attempt to alter or replace disease-causing genes themselves.

GENE THERAPY

On September 14, 1990, a four-year-old girl watching *Sesame Street* from her hospital bed made history. The child, Ashanthi deSilva, had been born with a mutant gene that made her body unable to produce an enzyme called adenosine deaminase (ADA). Lacking this protein, some of the cells in her immune system could not thrive. As a result, like other people with poorly functioning immune systems, she was easy prey for every microbe to which she was exposed. During all of her short life she had suffered from one infection after another.

Life changed for Ashanthi—and, potentially, for the world—on that September day. W. French Anderson, Michael Blaese, and their coworkers at the National Institutes of Health (NIH) had devised a technique for inserting a normal ADA gene into white cells (part of the immune system) taken from blood. When blood cells extracted from her earlier and treated with this method were reinjected into Ashanthi as she watched her hospital television, she became the first person to receive gene therapy, or injection of altered genes for the purpose of treating disease.

She was by no means the last. Although Ashanthi's therapy, which was repeated several times in the next two years, did not cure her disease, it was relatively successful. She has had to continue taking an injected form of ADA, but the combined treatments have allowed her to lead an almost normal life. In 2004, at age 18, Ashanthi was still a healthy young woman. Emboldened by this achievement, researchers went on to develop experimental gene treatments, not only for other inherited disorders such as hemophilia (which causes uncontrollable bleeding) and cystic fibrosis (which makes breathing difficult and increases susceptibility to lung infections) but for more common illnesses such as cancer, heart disease, and AIDS. They created a variety of techniques for altering the human genome, including correction of abnormal genes as well as insertion of normal ones.

Unfortunately, with the apparent exception of a handful of French children with another form of severe inherited immune deficiency, gene therapy has not completely cured any of the people who tried it. Furthermore, it has been blamed for at least one death and three cases of cancer—the latter in the same French children who had been freed of their immune problem. Because of the safety issues raised by these alarming events, regulations governing gene therapy in the United States have been tightened considerably, and a number of experimental trials have been stopped either temporarily or permanently.

The person whose death was blamed on gene therapy was an 18-year-old Arizona man named Jesse Gelsinger, who had an inherited liver disease. He died on September 17, 1999, just three days after scientists at the University of Pennsylvania had injected a high dose of gene-carrying viruses into the blood vessel leading to his liver. His death shook the fledgling field of gene therapy and the agencies that regulated it: the FDA, the RAC, and NIH. In early 2000, after investigating the Gelsinger experiment and others, all three agencies sharply criticized gene therapy researchers for failing to use consent forms that fully warned prospective patients of possible dangers, failing to report bad reactions to treatments, and having close ties with commercial companies that stood to profit from the treatments if they succeeded (thereby creating a possible conflict of interest). A German patient in a gene therapy trial for an inherited immune disease (a different type from the ones Ashanthi deSilva and the French children had) also died on April 10, 2006, but that death is thought to have been caused by a massive infection, not by the gene therapy.

Probably the greatest risk in gene therapy comes from the viruses that experimenters usually employ to carry new genes into cells. Some of the viruses that are the most effective at inserting potentially helpful genes into a cell's genome are also among the most feared, such as HIV, the virus that causes AIDS. Before using these viruses in a treatment, scientists remove the genes that allow the viruses to reproduce and cause disease, but concern that the viruses might somehow revert to their natural, dangerous state is bound to be high.

Even when the viruses themselves remain harmless, they can still cause problems indirectly. In the case of the French children, whose illnesses were reported in September 2002, January 2003, and early 2005, the viruses placed in cells from the children's bone marrow apparently unloaded their genetic cargo on top of a gene that controls cell growth, causing that gene to remain active longer than it should and overproduce certain blood cells. The result was a blood cancer much like leukemia. One of the children died of the cancer in October 2004, but the other two were successfully treated with chemotherapy.

Viruses caused a different difficulty for Jesse Gelsinger: His immune system apparently attacked them so vigorously that it killed him in the process. In some other experimental treatments, patients can safely be given a virus injection once, but researchers dare not repeat the process because severe reactions are likely to occur if the people's immune systems encounter the virus a second time. In still other cases, patients' immune systems have simply done what nature designed them to do—inactivated the viruses, including the added genes that the viruses carried. Experimenters are trying to develop safer and more effective ways to deliver genes, including ways that do not involve viruses.

Sometimes the genes themselves cause trouble. They may reach too few cells to be effective, may fail to integrate into cells' genomes, or may fail to make their proper proteins after they are in place. Alternatively, they may have unwanted effects along with desirable ones. For example, researchers in the late 1990s explored what seemed to be a very promising treatment for heart and blood vessel disease that used injections of a gene for a blood vessel growth factor to make new vessels grow to replace blocked ones. Cancerous tumors also benefit from the growth of blood vessels, however, and in at least one case the treatment apparently spurred the growth of a previously undetected cancer by providing the tumor with blood vessels.

Some researchers are trying to get around these problems by using virus vectors that do not insert genes permanently into cells' genomes, but rather just let the genes float freely in cells and make their proteins for a little while. Even this temporary exposure can have major effects on some diseases, just as taking a drug for a limited time can. One scientist, Stephen Rothman of the University of California–San Francisco, thinks such temporary gene treatments might someday be given as pills. When the treatment is complete, or if dangerous side effects appear, a patient could simply stop taking the pills, and the gene effects would end.

Gene therapy researchers say that designing treatments is hard when so little is known about what most genes do and how they interact with each other and the environment. They counsel patience, even when experiments have occasional distressing outcomes. They point out that almost all medical treatments have had failures, even fatal ones, while they were being developed. Some critics question whether gene therapy will ever deliver the medical miracles its boosters have promised, but supporters such as W. French Anderson still believe that it will one day do so. "No other area of medicine holds as much promise for providing cures for the many devastating diseases that now ravage mankind," Anderson wrote in 2000.[32] They point out that gene therapy encompasses hundreds of different strategies and say that even if some approaches fail, others may still succeed.

In spite of the field's setbacks, scientists are still vigorously exploring gene therapy. In 2004, more than 300 companies in the United States were said to be involved in about 500 clinical trials of various gene treatments, including treatments for cancer, heart disease, and Alzheimer's disease.

Researchers are also investigating gene-based "vaccines" that stimulate immune responses to tumors.

One gene therapy drug has even been approved for sale, though not in the United States. In October 2003, China's State Food and Drug Administration approved an injected drug called Gendicine, which uses a harmless virus to put copies of a tumor suppressor gene called p53 into cancer cells. The cells of many cancers lack functional p53 genes, which help to control cell growth. When the genes are inserted, they make a protein that forces the abnormal cells to destroy themselves. Gendicine, which also stimulates the patients' immune systems to attack the cancer cells, is currently used only to treat certain cancers of the head and neck, but some evidence indicates that it may be effective against other types of cancer as well.

GENE ENHANCEMENT

The potential social effects of human gene therapy may prove at least as important as its medical effects. Both supporters such as W. French Anderson and critics such as Jeremy Rifkin agree that Ashanthi deSilva's treatment represented not only a scientific advance but, as Anderson said at the time, "a cultural breakthrough, . . . an event that changes the way we as a society think about ourselves."[33] They and many others, however, are still arguing about whether that change is for good or ill.

Few people question the morality of altering genes of human body cells to prevent or cure a life-threatening illness like Ashanthi's (although Germany, perhaps sensitive about the eugenics aspect of its Nazi past, forbids any alteration of human genes). After that, however, the ethical ground becomes shakier. Rifkin and other critics fear that, if the technical problems that presently limit human gene alteration are solved, the definition of "disease" or "defect" may be stretched to include relatively minor problems (such as nearsightedness or obesity) or even mere differences (such as shyness or shorter-than-average height). If all these are engineered away, the result could be, at best, an undesirable loss of genetic diversity or, at worst, a new form of eugenics. Even Anderson has said he feels strongly that gene therapy should be used only to treat serious disease.

Although governments might insist on, or at least strongly encourage, alteration of "defective" genes as a way to control health care costs, many critics and supporters of human gene alteration suspect that the demand for gene changes is more likely to arise from market forces than from government fiat. People might buy gene treatments for themselves in attempts to, say, grow larger muscles and make themselves into superathletes. Well-to-do parents might try to create "designer babies" by inserting genes likely to produce intelligence or physical beauty into their unborn children, just as they now purchase orthodontic treatments or special schooling.

Researchers warn that gene alterations could backfire, failing to produce desired effects or producing undesired effects or both. For example, an experimental gene alteration in mice has apparently made the mice more intelligent, as measured by increased speed in navigating mazes, but it has also left the animals in chronic pain. Even if alterations could be made predictable and safe, say critics such as University of Chicago bioethicist Leon R. Kass, the ability to specify genetic makeup could make parents think of their children more as goods ordered out of a catalog than as independent individuals.

Routine human gene alteration probably lies far in the future, but its potential for raising "slippery slope" ethical issues is well illustrated by a technique called preimplantation genetic diagnosis (PGD), which fertility clinics already employ. PGD does not alter genes, but it produces a similar effect by allowing prospective parents to select future children that either possess or lack genes for certain inherited traits. In this technique, a couple donate eggs and sperm to create fertilized eggs that will grow into embryos, just as they would for in vitro ("test tube") fertilization. When about a dozen embryos have developed to the size of a few cells, clinic workers remove a cell from each (at this stage, doing so does not harm the embryo) and examine the cell's genes. Based on the examination, they and the parents decide which embryos to implant in the mother's womb for further growth (several embryos are often implanted to be sure of obtaining a single full-term baby). PGD is most commonly used when both parents carry a gene that can cause a serious illness, such as sickle-cell anemia or early-onset Alzheimer's disease. Only embryos that do not contain the defective gene are implanted, guaranteeing that the couple's children will be free of the disease.

Few people argue with this use of PGD, but other applications are ethically murkier. In August 2000, for example, a couple in Denver, Colorado, gave birth to a son whom they had selected by PGD, not only to be free of Fanconi's anemia, an inherited disease that ran in both their families, but to be of a tissue type that was compatible with that of his older sister, who already had the disease. This compatibility meant that cells from his umbilical cord and bone marrow could be transplanted into his sister, potentially giving her a healthy blood system. (The transplant took place and was apparently a success.) A British couple obtained permission from their country's Fertilisation and Embryology Authority in 2002 to do something similar. In July 2004, this same British agency relaxed its rules on PGD, for the first time allowing the procedure to be done for reasons other than preventing the birth of children with serious inherited diseases.

A still more controversial use of (indirect) genetic selection also occurred in 2002, when a deaf lesbian couple deliberately chose a sperm donor with a strong strain of inherited deafness in his family in an attempt to ensure that they would produce a deaf child. They did not use PGD, but other deaf couples have said they would like to use that technique for the same purpose. Most hearing people find the idea of deliberately creating a child with what they see as a disability (an evaluation with which some deaf people strongly disagree) to be strange or even indefensible, but supporters such as Julian Savulescu of the Murdoch Childrens Research Institute in Britain say that if parents are allowed to select against inherited deafness when choosing an embryo to implant, they should have an equal right to select for deafness.

Present-day genetic selection has not been confined to medical conditions, just as many predict that future gene alteration will not be. Couples in Australia and the United States have used PGD and other gene-based diagnostic techniques for gender selection, for example. The American Medical Association (AMA) and the American Society for Reproductive Medicine have issued statements opposing selection of embryos for non-medical reasons, but some parents apparently disagree.

The ethical questions raised by human gene alteration will become especially great if germ-line genes—those in the sex cells (the cells that become sperm and eggs), whose genetic information is passed on to offspring—are ever altered. Germ-line gene alteration has not yet been deliberately performed in humans, though it has been done in animals; all present gene therapy is intended to affect only somatic (body) cells.

Supporters of human germ-line gene modification say that it could eliminate inherited diseases completely, avoiding the necessity to choose or treat offspring in each generation. Critics, however, claim that PGD makes such risky action unnecessary. They also point out that even removal of a gene known to be associated with a serious disease might not be as clear-cut a good as it might seem. For example, people who inherit a single copy of the "defective" gene that causes sickle cell (in people who inherit this form of the gene from both parents) not only are healthy but seem to be more resistant to malaria (a serious blood disease, caused by a parasite, that is very common in the areas where the gene originated) than people who have only the normal form of the gene. By permanently deleting such a gene, therapists might unknowingly remove some characteristic that the human species will need at a future time. Altering genes associated with behavior would be even more likely to produce unpredictable effects. "Like Midas, bioengineered man will be cursed to acquire precisely what he wished for, only to discover—painfully and too late—that what he wished for is not exactly what he wanted," warns Leon Kass.[34]

Because of concerns like these, most people today, including most geneticists, feel that the human germ line should never be changed. In February 2000, for instance, nearly 250 concerned environmentalists and other leaders signed an open letter warning that human germ-line genetic engineering "is an unneeded technology that poses horrific risks" and urging that it be banned.[35] As of mid-2004, at least 17 countries had passed laws forbidding germ-line gene alteration.

Nonetheless, germ-line alteration may already be occurring, at least inadvertently. In early 2001, a fertility clinic reported that, to avoid diseases carried in the mothers' mitochondrial DNA, some 30 children worldwide had been created by a technique in which an egg nucleus from the mother was transferred into an enucleated egg from a donor (with cytoplasm containing healthy mitochondria) and then fertilized with sperm from the father. All the cells of such babies appeared to contain DNA from all three people. This DNA is expected to be inheritable, so, in effect, the germ line was changed by addition of the donor's mitochondrial DNA, producing what the report called "the first case of human germ-line modification resulting in normal healthy children."[36] Some proposed applications of gene therapy to fetuses also present a small chance of altering the fetuses' germ cells, although that is not the treatments' purpose.

HUMAN CLONING AND STEM CELL RESEARCH

The idea of human cloning has generated even more heated debate than other proposed forms of human gene alteration. It has haunted movies and novels as well as ethical discussions since genetic engineering began, and it took on new relevance after the appearance of Dolly, the cloned sheep, in 1997. Two types of human cloning have been discussed: reproductive cloning, or cloning with the aim of producing a fully developed child, and so-called research or therapeutic cloning. In this type of cloning, human embryos are created by somatic cell nuclear transfer, the same technique that produced Dolly, and allowed to multiply only up to the blastocyst stage, a ball of about 200 cells. They are then destroyed so that certain cells within the ball, called embryonic stem cells, can be harvested.

Scientists have known about stem cells in the adult body for some 40 years. Unlike most body cells, stem cells retain the ability to reproduce and do not completely specialize into particular types of tissue. They are the body's factories for making cells to replace those that wear out and die. Stem cells in the bone marrow generate both the oxygen-carrying (red) and the immune system (white) cells in the blood, for instance. Other stem cells replace skin cells and, to a limited extent, nerve cells. These so-called adult stem cells are also found in fetuses, newborns, and the blood of the placenta and umbilical cord. Adult stem cells (including umbilical cord blood cells) have already been used in a few cancer therapies, and they are the reason that bone marrow transplants are an effective, though drastic, treatment for some diseases.

Stem cells in human embryos, however, were discovered only in 1998. James Thomson of the University of Wisconsin–Madison, and scientists at Geron, a California biotechnology company, purified the cells and developed a way to maintain them in the laboratory. They, like all other researchers on embryonic stem cells so far, obtained the cells from excess embryos created in fertility clinics and donated to research by the couples who had had them made. About 400,000 such embryos are said to exist in the United States. Those not used in research are usually discarded.

Most researchers in the field feel that embryonic stem cells have a far greater capacity to mature into multiple types of tissue than adult stem cells do. Boosters of embryonic stem cell research say that these cells could revolutionize medicine by providing sources of specialized cells to replace those damaged by disease and age, such as neurons to replace those destroyed by Parkinson's disease or Alzheimer's disease, or pancreatic islet cells to replace those destroyed by diabetes. Research cloning could allow larger numbers of such cells to be obtained than would be possible with leftover embryos. Furthermore, if replacement cells come from cloned embryos made from the body cells of a particular patient (as Dolly was made from a body cell of an adult ewe), they theoretically should not be rejected by the patient's immune system, making risky immunosuppressive drugs unnecessary. Research on cloned embryos and embryonic stem cells could also reveal much about how normal embryos develop and how diseases express themselves in individual cells, supporters say.

Other people, however, oppose the cloning of embryos for stem cell or other research. Some see all research on embryos as wrong because they believe that independent life and personhood begin at conception. (They also object to the creation of the "surplus" fertility clinic embryos that stem cell researchers want to use.) Others, such as the American Association for the Advancement of Science, do not object to research on excess embryos from fertility clinics but dislike the idea of creating new embryos (cloned or otherwise) specifically for the purpose of experimenting on and destroying them. They feel that doing so comes too close to treating human life—even if it is not fully developed life—as a commodity, a mere "thing." Some feminists also fear that, if embryo cloning became widespread and women were paid to donate eggs, a great demand would arise for the eggs, leading to the exploitation of poor women, who, desperate for money, would consent to undergo the risky hormone treatments and surgical procedures necessary for egg donation.

Some critics also attack embryonic stem cell research on scientific grounds, saying that adult stem cells are likely to work just as well or that the promised use of cloned embryos to produce tissue transplants for individual patients probably would remain too expensive to be practical, even if it could be made to work from a technical standpoint. Some scientists also doubt whether the use of stem cells from embryos cloned from a patient's cells would necessarily overcome transplant problems, because the embryo cells would contain different mitochondrial DNA, from the donor egg. Furthermore, studies of stem cells from cloned animal embryos have uncovered numerous genetic defects, suggesting that human embryonic stem cells might be dangerous to use as a medical treatment, perhaps inducing cancer, even if they could be produced. (Some researchers even think that renegade adult stem cells lie behind ordinary cancers.)

A major breakthrough in embryonic stem cell research seemed to occur in February 2004, when South Korean scientist Woo Suk Hwang announced that his laboratory had cloned human embryos, extracted embryonic stem cells from them, and kept the cells alive in a laboratory to form a new cell line—the first time these things had been done with cloned embryos. Hwang proclaimed a second triumph in May 2005: 11 stem cell lines descended from embryos cloned from different patients. By that time, however, doubts had begun to arise about his achievements. In December, Seoul National University, where Hwang worked, published a statement saying that most of Hwang's work on human cells had been faked. (His August 2005 claim to have cloned a dog, however, apparently stood up to examination.) Further controversy was raised by the fact that Hwang had accepted eggs for his research from female laboratory employees and had paid the women for their eggs—both considered questionable ethical practices.

Competing researchers were at first discouraged that Hwang appeared to have beaten them and then shocked by the scandal of his deception, but the news of his failure simply made them redouble their own efforts to derive and use stem cells from human embryos, cloned or otherwise. In mid-2006, both Advanced Cell Technology, in Alameda, California, and Geron Corporation, based in nearby Menlo Park, were said to be close to asking the FDA for permission to begin the first clinical trials of therapies based on embryonic stem cells. Geron's treatment might ultimately aid people with spinal cord injuries and multiple sclerosis, a nerve disease; the therapies from Advanced Cell Technology are aimed at age-related macular degeneration, heart disease, and skin damage.

Reaching a consensus on the ethics of therapeutic cloning and embryonic stem cell research has so far proved impossible, and attempts to create a legal compromise have been frustrating for both sides. On August 9, 2001, President George W. Bush announced that federal funding could be provided for research on human embryonic cell lines that already existed, but funding for creating or experimenting on new cell lines was forbidden. Bush claimed that more than 60 cell lines were available, but scientists in the mid-2000s say that the figure is really closer to 20. These older cells now contain abnormal chromosomes and do not grow well. They also need to be cultured with outdated methods that involve mouse cells, which have introduced material into the stem cells that may make human immune systems attack the cells or could infect patients with mouse viruses. Those who believe that all experimentation on embryos is wrong, on the other hand, feel that Bush's measure did not go far enough, since it does not prevent research funded by private companies or other nonfederal sources.

Private funding certainly has not been lacking—and a few states have stepped into the breach as well. The most widely publicized contribution has come from California. In November 2004, 59 percent of the state's voters approved Proposition 71 (the California Stem Cell Research and Cures Initiative), an amendment to the state constitution that earmarked $3 billion for stem cell research, including research on embryonic stem cells, over a 10-year period. The measure also established a new facility, the California Institute for Regenerative Medicine, to manage the work.

A lawsuit from pro-life groups, claiming that the agency set up to distribute the money suffered from a conflict of interest and lack of state supervision, tied up most of the ambitious California program's funding for several years. The First District Court of Appeals in San Francisco voted 3-0 to reject the suit on February 26, 2007, however, and the state's Supreme Court voted not to review the case on May 16 of that year, allowing the lower court's decision to stand and permanently ending the suit. The agency distributing funds from the measure issued more than $50 million in grants in June and stated that it expected to distribute an additional $222 million early in 2008. Meanwhile, Massachusetts, New Jersey, Wisconsin, and several other states have proposed similar initiatives, and some have begun providing funding to embryonic stem cell researchers.

Congress has attempted to overcome the Bush administration's opposition and restore federal funding for embryonic stem cell research as well, but so far its efforts have not been successful. The Stem Cell Research Enhancement Act, which would have allowed funding for stem cell lines created from excess fertility clinic embryos (but not cloned embryos created for the purpose of research), was passed by the House of Representatives in 2005 and the Senate in 2006. Even such pro-life figures as Utah Republican senator Orrin Hatch and then-House majority leader Bill Frist of Tennessee supported the bill, but on July 19, 2006, President Bush issued his first presidential veto to prevent the measure from becoming law. Congress passed a similar bill on June 7, 2007, and Bush again vetoed it on June 20, 2007. In both cases, supporters of the bill lacked the two-thirds majority necessary to override the presidential veto.

Some commentators see the government's lack of support as a blessing in disguise because it has encouraged private industry, states, and other countries to fund stem cell research. Others, however, say that it is hampering the United States in the worldwide race to commercialize stem cell technologies (some U.S. scientists have even moved to Britain or other countries with a more encouraging attitude) and keeps the research from having the sort of oversight that the National Institutes of Health's Recombinant DNA Advisory Committee provided for genetic engineering and gene therapy experiments. The National Research Council drew up guidelines for embryonic stem cell research in April 2005, including the recommendation of an oversight committee to review all research plans, but these guidelines have no legal force.

Other countries have developed different approaches to therapeutic cloning and embryonic stem cell research. On March 29, 2004, the Canadian Parliament passed the Assisted Human Reproduction Act, which prohibits human cloning (either reproductive or therapeutic), creation of embryos for research purposes (although embryos from fertility clinics may be used), or paying egg or embryo donors. Britain, by contrast, not only permits but actively supports embryonic stem cell research and therapeutic cloning. The country legalized such cloning in 2001 (reproductive cloning is not legal, however) and granted the first two licenses to carry it out in August 2004 and February 2005. The first license went to Alison Murdoch and other researchers at the University of Newcastle upon Tyne, and the second was awarded to a group coheaded by Ian Wilmut, creator of Dolly the cloned sheep. Both teams want to clone cells from people with particular diseases (diabetes for Murdoch's group; amyotrophic lateral sclerosis, or motor neuron disease, for Wilmut's) in order to study the conditions and, they hope, eventually derive treatments for them.

Several Asian countries, including South Korea (whose very active stem cell program was by no means halted by the Woo Suk Hwang scandal), Singapore, Japan, India, and China, also encourage therapeutic cloning and embryo research. Europe, Latin America, and Africa have been lukewarm to the procedure at best, however. The European Union adopted a measure in July 2006 that would fund embryonic stem cell research between 2007 and 2013, but the research must use only surplus embryos from fertility clinics; cloning or otherwise creating them for the purpose of destruction is not permitted. Despite this limitation, a number of member states and nongovernmental organizations, including some Green (environmentalist) and socialist as well as pro-life groups, objected to the legislation. Sweden and Belgium have joined Britain in permitting embryonic stem cell research, but Australia, France, Germany, Italy, Norway, South Africa, and Switzerland all have banned such procedures, as have Argentina, Brazil, Costa Rica, Mexico, and numerous other South and Central American countries.

Overall, opponents of therapeutic cloning show some signs of prevailing over supporters. On March 23, 2005, acceding to proposals from Costa Rica and the United States, the United Nations General Assembly (by a vote of 84 to 32) adopted the Declaration on Human Cloning, which calls on member states to ban all forms of human cloning. The declaration is not a binding law, but its passage has greatly encouraged foes of cloning.

A number of scientists are trying to work around the ethical objections to embryonic stem cell research by devising ways to collect stem cells without killing embryos—or by redefining *embryo*. Some, including Woo Suk Hwang (in research that does not appear to have been faked), have derived stem cells (in mice) from parthenotes, embryolike structures that sometimes develop from unfertilized eggs but do not go on to form fetuses. In August 2006, Robert Lanza and other researchers at Advanced Cell Technology claimed to have shown that a researcher could take a single cell from an eight-cell embryo, much as is done for preimplantation genetic diagnosis, and derive stem cells from it, leaving the original embryo unharmed by the procedure. Still other scientists reported in January 2007 that they found cells in amniotic fluid, the liquid that surrounds the fetus in the uterus, that have most of the properties of embryonic stem cells, and on June 13 of that year a different group announced that they had reprogrammed skin cells from an adult mouse in a way that made the cells act like embryonic stem cells. Some commentators who oppose research on embryos have hailed such efforts, but others say that they have more to do with semantics than with science.

Meanwhile, other researchers are trying to develop therapies that use the much less controversial adult stem cells. At least eight treatments involving adult stem cells were already being sold in 2006, and a number of others were in clinical trials. Scientists in Thailand and at Tufts University in Boston, for instance, have injected stem cells from bone marrow into hearts damaged by disease and report that the cells apparently grow into new blood vessels, improving blood flow and bringing relief to the sufferers.

Some scientists—and parents—are enthusiastic enough about the possibility of treatments with adult stem cells taken from umbilical cord blood that they recommend "banking" babies' cord blood for possible future use by their own families or those of others (cord blood cells from unmatched donors produce relatively little immunological reaction). A bill to authorize and fund the establishment of a national cord blood bank system was signed into law in December 2005, and in late 2006, the federal Health Resources and Services Administration awarded $12 million to six nonprofit cord blood facilities to begin setting it up. The administration hopes that the national bank will eventually contain 150,000 units of blood. Cord blood banking is expensive, however—preserving a unit of blood was estimated in late 2006 to cost about $1,500, plus about $100 a year to maintain the blood—and some commentators say that it is not likely enough to be useful to justify the expense, either for individual families or for the government.

The debate about therapeutic cloning has been so bitter perhaps partly because opinions on the ethics of the practice are fairly evenly divided. Repro-

ductive cloning—cloning with the intent of producing a child—in a way has caused less argument, because most people agree about it: Almost everyone opposes it. During the early 2000s several individuals and groups, ranging from a flying saucer cult called the Raelians to, most recently (in mid-January 2004), Panos Zavos, a retired professor of reproductive physiology from the University of Kentucky, stirred great concern by claiming to have produced, or to be about to produce, a cloned child. No such child has ever appeared, but many people feel that, sooner or later, a cloned baby will be born.

Objections to reproductive cloning begin with the practical one that, at least at present, such a procedure would be very risky to the prospective fetus. Dolly's was the only live birth in 277 tries, and only about 1 to 5 percent of subsequently created animal clones survived to birth. Many of these survivors proved to have defects, ranging from subtle to grotesque.

Ethical concerns about human cloning reach far beyond safety issues, however. Some critics say that cloning humans would be "playing God." Others fear that the parents of cloned children, or the children themselves, would regard the children as mere products rather than independent human beings. Producers of clones might expect to control the cloned children's lives completely or to see exact duplicates of themselves or whatever famous person they chose as the source of the clone. Some opponents of human cloning have pictured nightmare scenarios featuring, at one extreme, cadres of identical Hitlers (or, more benignly, Madonnas or Michael Jordans), or, at the other, armies of mindless slaves or even warehouses of headless bodies kept for possible organ donation.

Defenders of human cloning say that these fears are groundless. They point out that human cloning already occurs naturally in the form of identical twins, and they claim that an artificially produced human clone would simply be an age-delayed twin. They say that such a clone would be just as much a separate individual, with his or her own personality, as a twin is; indeed, the personality differences between a clone and his or her original would be greater than those between twins because they would be raised in different eras and therefore would be bound to have very different life experiences. Clones would be entitled to the same rights that other citizens have, including protection from slavery and forced organ donation. Cloning, these supporters say, could provide help to infertile couples who can reproduce in no other way.

Supporters, however, have been few. Most members of the public, legislators, and even scientists in North America and Europe today apparently feel that reproductive cloning should be made illegal, at least until safety questions are resolved, and perhaps permanently. The Council of Europe banned reproductive cloning on January 12, 1998, and at least two dozen individual nations, including Australia, Belgium, Britain, Denmark, Germany, India, Israel, Japan, the Netherlands, and Spain, have also banned or placed a moratorium on it. In the United States, California became the first state to outlaw reproductive cloning in January 1998, and some other states have followed.

As of mid-2007, no federal law prohibits reproductive cloning—yet. The House of Representatives rejected a bill to do so on June 6 of that year, chiefly because Republican members objected to the fact that the bill did not also prohibit therapeutic cloning. The FDA has claimed the right to regulate reproductive cloning and says that the agency will not give its approval to any such projects.

Some scientists say that the whole issue of human cloning—for any purpose—may be moot because such cloning may never be practical or even possible. Researchers attempting to clone nonhuman primates such as monkeys have so far failed, apparently because certain proteins necessary for cell reproduction are damaged in the nuclear transfer process in primate eggs in a way that does not happen with the eggs of other mammals.

FUTURE TRENDS

No one knows what advances the coming "biotech century" will bring, but they are sure to be amazing, and their effects on society will be challenging, if not wrenching. There were good reasons why James Watson, the first head of the Human Genome Project, earmarked 3 percent of the project's budget for investigation of its ethical, legal, and social implications (ELSI).

THE FUTURE OF EXISTING BIOTECHNOLOGIES

Despite the world's mixed feelings about its products, the agricultural biotechnology industry predicts continued growth. For instance, the International Service for the Acquisition of Agri-biotech Applications predicted in early 2007 that 20 million farmers will plant 200 million hectares of genetically modified (GM) crops by 2015. A report in early 2005 estimated a fivefold increase in the value of engineered crops worldwide during the following 10 years, rising to $210 billion by the end of that time. However, continued resistance to GM foods in Europe and elsewhere, perhaps triggered by incidents of food contamination or reports of environmental damage caused by GM crops, could cut into this rosy picture substantially. Environmentalists and others who oppose such crops are sure to be watching for signs that the technology is causing the kinds of harm that they have warned against.

The next generation of GM crops is likely to carry even greater benefits and risks than the present one, and the benefits are likely to be aimed at consumers rather than farmers. In early 2007, Monsanto was reported to have developed a canola product with modified oils that reduce trans fats, a dietary element considered harmful to human health. Monsanto and others are also working on food crops that contain added health-promoting nutrients, such as omega-3 fatty acid, oleic acid, and increased protein, or else have been freed of proteins that tend to cause allergies. A larger variety of crops probably will be bioengineered in the future, including fruits and vegetables as well as grain crops.

Scientists are also experimentally engineering plants to produce vaccines, drugs, and other substances useful in medicine. For instance, a scientist at the University of Tokyo announced in June 2007 that he had engineered a strain of rice to produce a protein that could vaccinate people against cholera, a severe digestive disease caused by bacteria. Such plant "pharming" is predicted to be a $200 billion industry by 2013. If successful, it could considerably lower the prices of these medicines and also make them more easily available in developing countries. (Unlike conventional vaccines, for instance, plant-based vaccines would not require refrigeration, and they could be distributed as seeds and grown locally.) Because they would not require much acreage to show a profit, these new "pharm" crops would be easier for small farms to grow than present genetically engineered crops. Crops being investigated for possible drug production include tobacco, alfalfa, rice, corn, and mosses.

On the other hand, biotechnology critics have expressed even greater concern about the possibility of pharmaceutical crops unintentionally entering the food supply than about present-day crops—a fear that some producers and users of food crops share. For example, Anheuser-Busch, which uses Missouri-grown rice in its beer, threatened in April 2005 to stop buying rice from the state if a small company called Ventria Biosciences completed a proposed plan to plant rice containing human genes (for proteins found in breast milk, tears, and saliva) in Missouri. The beer company backed down only after Ventria promised to plant its altered rice just in the northwestern portion of the state, 120 miles away from the nearest food-crop rice.

A substance that cures illness in one person could cause it in another, critics of food crops altered to produce drugs point out. The possibility that such a disaster could happen was underlined in November 2002, when an ounce of corn engineered to produce a pig vaccine was detected in soybeans destined to become human food. The contamination did not reach the food supply in this case, but biotech opponents say that consumers might not be so lucky next time. The FDA and USDA jointly issued guidelines for growing crops genetically altered to produce drugs in September 2002. Suggestions for keeping drug-producing plants separate from food crops include growing the plants underground and using plants such as the tiny aquatic plant Lemna, which can be raised in contained production facilities.

Meanwhile, some scientists, particularly those working for national and international groups that are independent of large companies, are trying to tailor biotechnology to the needs of the developing world by engineering pest and disease resistance into crops that people there grow for food. For example, Kenyan scientist and biotechnology supporter Florence Wambugu has helped to engineer a yam to resist a virus that is a major pest of this crop, widely eaten in sub-Saharan Africa; the first trial crops of the new potato were harvested in late 2001. Numerous public institutions and public-private partnerships working in or for developing countries are preparing similar crops. If more such crops are produced, particularly (says Wambugu) in projects that include the input of local farmers, claims that biotechnology brings little benefit to farmers in developing countries may be muted. Indeed, the International Service for Acquisition of Agri-biotech Applications predicts that the most significant future growth of agricultural biotechnology will occur in exactly those countries.

Development of transgenic animals is predicted to continue as well. Researchers are experimenting with transgenic insects, engineered, for example, to be unable to reproduce or to pass on diseases, as a form of pest control. Some environmental groups, however, fear that such insects might transfer their genes to other insect species or might become pests themselves. Meat and other products from cloned or genetically modified animals probably will be on supermarket shelves by the end of the 21st century's first decade, though whether consumers will choose to buy them remains to be seen. Some medicines are likely to come from transgenic animals as well. Producing drugs in engineered animals' milk could cost two or three times less than producing them in cell cultures, a report in early 2005 stated.

Heart valves from pigs and cattle are already transplanted into humans, and in the future, genetic engineering may make xenotransplantation of whole organs from pigs or other animals possible, potentially ending the present-day organ shortage. Pigs engineered to lack a molecule that triggers a destructive immune reaction when transplanted into other species have already been developed. Baboons have survived for months with organs from such pigs, whereas transplanted organs from ordinary pigs are destroyed within hours. Other trigger molecules still must be eliminated before xenotransplantation becomes practical, however, and so must the threat of accidentally transferring pig viruses into humans receiving the transplants.

Legal and trade conflicts are likely to continue over patents and other protection of intellectual property in biotechnology, particularly in regard to biologically useful material discovered in developing countries and prepared for market by large companies in industrialized nations. Biotechnology companies, for their part, feel that they need and deserve patent protection in exchange for the considerable time, effort, and money they must spend in bringing new products to market, and they fear having their work exploited without compensation in countries where intellectual property laws are weak. They are sure to continue to demand the strengthening of such laws in international treaties and trade agreements. Nonwestern countries, on the other hand, will no doubt go on protesting these companies' patenting of materials and knowledge that originally came from those countries or of products that they expect people in those countries to buy.

Differing opinions about the morality of modifying and of patenting or "owning" types of living things, body parts, cells, and genes also will surely continue to clash. Some religious groups feel that claiming ownership or creation of living things usurps the role of God. Even among people who do not have strong religious beliefs, large numbers feel a deep moral unease about these subjects.

Moral concern is greatest when the materials being patented or changed are human. This unease can only increase as the flood of information released by the Human Genome Project and related research grows and as scientists continue to develop techniques and products involving alteration of human cells, such as embryonic stem cells. Rules of the U.S. Patent and Trademark Office forbid patenting a human being, and Article 5 of the European Community Biotechnology Directive states that the human body is unpatentable. Determining whether, say, an artificially grown human organ, a chimeric animal containing human genes, or a particular type of altered human embryo is patentable may force governments to define exactly what they consider to be a human being.

As the complete sequences of more and more genomes are determined, the focus of biotechnology is shifting to the management and sale of genetic information. This focus has spawned new scientific disciplines called genomics and bioinformatics, which marry genetics and computer technology. These new sciences increase the efficiency and scope of genetic engineering, for instance speeding the development of drugs and allowing the entire metabolisms of plants, animals, and even perhaps humans to be radically altered. They also raise new ethical issues, especially if some scientists or businesspeople begin to see (or to be accused of seeing) living things as mere bundles of information, as open to modification as computer programs.

DNA identification will surely continue to be a standard feature of certain types of court cases. Indeed, DNA evidence in trials may become more common as technology for acquiring and testing samples improves. In the future, DNA chips may allow forensic detectives to do preliminary analysis at the crime scene, and improved sequencing machines may make laboratory DNA testing more detailed and precise.

Technological improvements might also increase the use of familial DNA searches and even more controversial procedures, such as a new test developed by DNAPrint Genomics Inc. of Sarasota, Florida, which uses analysis of genetic variations to identify the ethnic ancestry of a tested subject. The company is working on other tests that could predict a perpetrator's physical characteristics, such as eye and skin color, and even perhaps behavioral characteristics, such as sexual orientation or likelihood of drug addiction. Critics are bound to see such tests as a high-tech form of racial profiling or other prejudicial treatment.

Conflict about the extension of DNA databases beyond convicted felons is likely to grow, particularly if improvements in technology and funding remove the present backlog of untested samples. Another major terrorist attack in the United States could increase calls for a national database of all arrestees or even all citizens. On the other hand, continued conflicts about health insurance, the rise in identity theft, and concerns about government and corporate intrusions into privacy could spur demands that DNA samples from anyone except convicts be destroyed immediately after testing.

Unless major changes are made in the way health care is delivered and paid for in the United States, arguments about employers' and insurers' use, or misuse, of information from genetic tests are bound to grow as such tests become more widespread, more accurate, and more detailed. National laws forbidding genetic discrimination probably will be passed, but enforcing them may be difficult.

Discrimination is just one of the problems that will become more acute when precise and detailed genome readouts are routinely performed on babies at birth, which most likely will happen by midcentury at the latest. Large mapping projects aimed at isolating genes associated with particular diseases, like the Cancer Genome Atlas Project, begun by NIH's National Cancer Institute in December 2005, and technological improvements that greatly speed up gene sequencing are rapidly bringing that day closer. Much sooner—by 2010, according to an article published in *Newsweek* in December 2006—scientifically valid tests for genes associated with increased risk of developing at least some common, serious diseases such as Alzheimer's disease, heart disease, depression, and type 2 diabetes will become available. Devices called DNA chips, already in existence, will be improved so that they can quickly test for hundreds of genes at once, perhaps allowing genetic tests to be done even in doctors' offices.

Detailed information about the variations in each person's genome could allow medicine to be individualized as never before, leading to major improvements in both treatment and prevention of disease—if people are willing to abide by the health warnings they are given. On the other hand, it may also increase people's feeling that the entire path of their lives, at least in terms of health, was laid out unalterably in their genes before they were born. Both supporters and critics of biotechnology advances will have to work hard to avoid the pitfall of genetic determinism, but scientists' increasing understanding of and respect for epigenetics (the influence of the environment on genes) should help to counter this trend.

The applied science of pharmacogenomics will improve the effectiveness and safety of drugs by allowing physicians to vary treatments according to individuals' genetic makeup. It is estimated that today, a particular drug is likely to be effective in only 25 to 60 percent of patients with the medical condition that the drug is designed to treat. Learning, for instance, which variations in genes that control the speed with which a drug is broken down in the body or which mutated cancer genes a person possesses, could help a doctor determine which treatment would be safest and most effective for that particular individual. A form of DNA chip called the AmpliChip already can analyze a drop of blood and detect about 30 variations in two genes that determine how fast the liver metabolizes common types of drugs.

DNA chips, which can easily fit in a pocket, will also allow many diseases to be identified more quickly than can be done at present, an advance that could be especially valuable during an infectious disease epidemic or a bioterrorist attack. Such chips could help health care workers in developing countries, who often lack access to testing laboratories, as well.

By helping researchers understand how drugs work and how they are affected by genetic variations, genomics and related sciences can let pharmaceutical companies find new uses for existing medications and "rescue" drugs that have not been effective in the population as a whole but may work well in people with particular forms of certain genes. Advances in biotechnology and genetics will continue to spawn new drugs, vaccines, and other medical treatments as well. Some drugs, for example, may come from the phenomenon of RNA interference (RNAi), first reported in 1998 (its discoverers won the Nobel Prize for physiology or medicine in 2006). In RNAi, small pieces of double-stranded RNA are used to turn off particular genes, such as those that make a cell cancerous. Drugs using RNAi to attack age-related macular degeneration were already in clinical trials in 2006.

To be sure, as Tom Abate, a California reporter who specializes in biotechnology, points out, new—and no doubt extremely expensive—drugs may not be health care's greatest need. Abate wrote in 2003, "We have an industry that has not yet turned a profit, trying to invent costly new medicines for a clientele that's having trouble paying for the things that are already in the medicine chest. This does not seem to be an encouraging trend."[37] On the other hand, Michael J. Mandel wrote in *Business Week* in 2004 that treatments that are expensive at first often become much cheaper as time passes. He thinks that improvements in biotechnology, including more efficient mass production of medicines and more accurate prescribing made possible by genetic tests, will reduce future health care costs.

Biotechnology critics fear that drug companies' desire to make a profit may make them concentrate on products that enhance (or claim to enhance) the health and performance of basically well people rather than cure sick people (especially sick people in countries that cannot afford to pay high prices for medications). Such products might boost intelligence, endurance, or muscle strength, for example. They could prove to have unexpected side effects or be abused in various ways, such as providing an unfair advantage in athletic competitions.

The same may prove to be true with gene therapy and parents' genetic selection or alteration of their unborn children. Solving the technical problems besetting gene therapy may take several decades, but eventually alteration of genes in human body cells, probably increasingly done before birth, is likely to make inherited disease a thing of the past. It is also likely to be a part of treatments for more common conditions such as cancer and heart disease. "Health" and "disease" are parts of a continuum, however, and no one knows what degree of gene alteration society will accept. At least some wealthy parents, clandestinely if necessary, will probably attempt to enhance the genetic endowment of normal children. They may discover that producing another Albert Einstein or Marilyn Monroe is easier said than done, however, because environment shapes a child as much as genetics.

Genetic alteration or other biotechnological advances that greatly extend lifespans in developed countries or increase infant and youth survival in developing countries, desirable as these might be on an individual and humanitarian level, could exacerbate the world's already severe population problems. Furthermore, National Medal of Science winner Edward O. Wilson warns that a culture run by 150-year-olds is "likely to be a very conservative culture, one in which those who have survived and enjoyed longevity extension . . . won't be revolutionaries. They won't be bold entrepreneurs or explorers who risk their lives."[38]

There is no technical reason why germ-line genes should be any more difficult to alter than the genes of body cells, but arguments about the ethics of making such alteration in humans are guaranteed to be more acute than those surrounding gene therapy or other alteration of body cell genes. Germ-line alteration, after all, goes beyond the individual to potentially affect the evolution of the entire species. Many people are sure to doubt the wisdom of taking on such an awesome responsibility.

What is banned today, however, may be permitted tomorrow. If alteration of human germ-line genes ever does become widespread, some commentators think it will ultimately change the very nature of the species. For instance, Princeton University geneticist Lee J. Silver predicts that, because only the wealthy are likely to be able to afford gene alteration for themselves or their children (at least at first), differences between those who can buy such treatments and those who cannot will eventually be so great that the two groups may become separate species. Supporters of germ-line alteration, such as DNA pioneer James Watson and researcher Gregory Stock of the University of California–Los Angeles, say that such alteration could create "posthumans" with powers undreamed of today, such as supersenses, striking athletic ability, or perhaps even flight. "If we could make better human beings by knowing how to add genes, why shouldn't we do it?" Watson said in 2003.[39]

Reproductive human cloning will probably be banned in the United States and most other countries within the first decade of the 21st century. Nonetheless, unless the technical problems involved prove insurmountable, someone, somewhere, is almost sure to produce a cloned human child within a decade or so. Fears may die down when people realize that a cloned child is not much different from any other baby, as happened with in vitro fertilization once it became widespread. Some of the dream, or nightmare, scenarios will fade when people realize that clones are not instant adults (as in some movies), automatons, or exact personality duplicates of their "parents" or anyone else. On the other hand, if the first cloned children prove to suffer from major deformities or health problems, further attempts at reproductive cloning may cease—voluntarily or otherwise.

Even if society eventually comes to accept reproductive cloning, such activity seems unlikely to become common. As a way to produce armies of either dictators or slaves, or even as a way to "resurrect" a lost loved one, most commentators agree, cloning simply will not work. For instance, the first cloned cat, "CC" (for Carbon Copy), was NOT a "carbon copy" of her mother; the mother was a calico, but the daughter, although genetically identical, was a gray tabby. Cloning will probably be used only by a small number of infertile (including same-sex) couples and, perhaps, a few eccentrics who do not understand the interplay between genetics and environment. Laws defining family relationships will probably have to be modified if cloned human children come into existence, however, just as they have been altered in the past to accommodate surrogate motherhood, anonymous sperm donation, and other new reproductive technologies.

The fate of research cloning is harder to predict. It will probably depend on the overall progress of stem cell research. If further animal experiments and human tests fail to bear out the promise of embryonic stem cells, or if adult stem cells prove to be (or can be made) adequate for most proposed medical uses, the demand for using, or at least for creating, embryos for stem cell research will fall. On the other hand, if scientists develop a very effective new treatment for, say, cancer or Alzheimer's disease that requires embryonic stem cells, pressure from patients and their families may lead legislators to remove bans on research cloning and experiments on embryos. Demand for research cloning might also increase if leftover embryos from fertility clinics, already known to be weaker on average than those chosen for implantation, are shown to be generally too defective to be useful for stem cell treatments.

Biotechnology and Genetic Engineering

NEW KINDS OF BIOTECHNOLOGY

Many forms of biotechnology have been under development for decades, but several types are in their infancy. These include industrial biotechnology, nanobiotechnology, and synthetic biology. Boosters and opponents alike predict that these new technologies could alter humanity's future in ways that are just beginning to be imagined.

Industrial biotechnology has been called the "third wave" of biotechnology. In Europe it is known as "white" biotechnology, as contrasted to "green" (agricultural) and "red" (medical) biotechnology. It uses living things, most commonly microorganisms or plants, or components of living things, such as enzymes, to affect manufacturing. It is being applied to cleaning up the environment, developing alternative sources of energy, creating new types of plastics and other materials, and inventing manufacturing processes that are less expensive, less polluting, and more efficient than chemical-based ones. Its supporters say that it can increase use of renewable resources and reduce use of energy and water, dependence on oil, and the carbon emissions that allegedly spur global warming.

Some industrial biotech products are already on the market. Certain enzymes, substances from living cells that speed up chemical reactions, were introduced into laundry detergents in the 1970s to remove stains from clothing, substituting for polluting phosphates. More recently, other enzymes have been applied to reduce harmful trans fats in food. Corn is used to make a clear plastic called PLA (polylactide), which can be shaped into biodegradable packaging or (when broken down by genetically engineered *E. coli* bacteria) a polymer, Sorona, that is used in clothing, carpets, and packaging. Still other enzymes are beginning to be used to reduce the amount of chlorine, a dangerous pollutant, required for bleaching in pulp and paper manufacturing; they also cut the energy used in the process by 40 percent.

Agricultural biotechnology companies are working on other industrial biotech products for the near future, some of which involve genetic modification. Syngenta, for example, is developing a form of genetically engineered corn that will contain higher than normal amounts of an enzyme called amylase. The company says that this corn could reduce the cost of manufacturing ethanol, an alternative fuel that is currently the largest-volume chemical made by bioprocessing, by 10 percent. Other firms are creating genetically engineered enzymes that can turn cellulose in plant waste directly into ethanol; Iogen, a firm in Ottawa, Canada, has already built a pilot plant for this process. Similarly, in May 2007, Arcadia Biosciences, a biotechnology company in Davis, California, announced that it had made an agreement with the government of China's main rice-growing region to adapt a new strain of genetically engineered rice to that region. The rice absorbs nitrogen more efficiently than natural rice and therefore needs less nitrogen-based fertilizer. Arcadia said that the rice could help to reduce global warming because soil bacteria convert nitrogen-containing fertilizer to nitrous oxide, a greenhouse gas that has 300 times more power to induce warming than carbon dioxide.

While some scientists focus on crop plants as raw materials for industrial biotechnology, others concentrate on microorganisms such as yeasts, bacteria, and fungi. Microbes make most of the enzymes already being applied in industrial biotech. Some types of microorganisms can clean up toxic waste such as discharges from the papermaking industry, a process called bioremediation. Gene sequencing pioneer Craig Venter, who headed the private company that competed with the federal Human Genome Project in a highly publicized "race" to decode the human genome, thinks that microorganisms might also be used to reduce the cost of making ethanol or, in the more distant future, produce inexpensive hydrogen fuel. As genomic analysis reveals more about the biochemical abilities of different types of microbes, a wide range of new uses for them is bound to be discovered. In 2000, for example, the U.S. Department of Energy began the Genomics: GTL program to analyze the genomes of microorganisms in the hope of finding ways to use them in developing alternative energy sources.

Biotechnology products made up 5 percent of the chemical industry's worldwide sales in 2004, and a report in that year from McKinsey & Co. predicted that this figure would double by 2010. Another 2004 analysis, by Burrill & Co., stated that the company expected total shipments of industrial biotechnology products to reach 60 billion pounds per year by 2025, four times the amount produced today. In order to fulfill this optimistic outlook, however, industrial biotechnology companies will need to develop new markets for their products and gain more public and private funding for research.

Environmentalist and other opponent groups have expressed some of the same concerns about industrial biotechnology as they have about agricultural biotechnology, though many do not feel as strongly about "white" biotech because the microorganisms used in most of its processes are confined to bioreactors inside factories. (Indeed, biodegradable PLA packaging is said to have become quite popular in GM-phobic Europe.) Some commentators fear that altered microorganisms might escape to the environment in factory waste, but industrial biotech specialists maintain that their modified microbes could not survive in the wild.

More safety issues arise from an even newer type of biotechnology, nanobiotechnology. Nanobiotech is one form of nanotechnology, defined as technology concerned with particles 1 to 100 nanometers (billionths of a meter) in size. As with industrial biotechnology, a few products containing nanomaterials are already being sold: Nanoparticles have been used to make clear sunscreen, wrinkle- and stain-resistant clothing, and tennis balls that keep their bounce longer, for instance.

Numerous biomedical applications of nanotechnology are currently under development, and many more are likely to follow. Some focus on diagnosis. For example, researchers at the University of North Carolina–Chapel Hill are commercializing X-ray machines that use carbon nanotubes; these machines are smaller, cheaper, more energy efficient, and able to produce better images than conventional ones. Nanotech descendants of existing "smart pills" could be swallowed and used to diagnose cancer and other ailments of the digestive tract. Nanoparticles that glow when they undergo certain changes in cells are also being used experimentally in an imaging technique to detect cancer. Sensors involving carbon nanotubes could help diabetics monitor their blood sugar levels without having to draw blood.

Other likely nanobiotech advances will improve drug delivery. Gold nanoparticles are absorbed 600 times more effectively by cancer cells than by normal cells. Combined with antibodies that are attracted to cancer cells, these nanoparticles could take drugs or radiation directly to malignant cells while leaving other cells alone. Nanotubes or other nanoparticles can be engineered to deliver drugs, radiation, or heat only when they are activated by laser light or magnetic fields, another way of making them target specific kinds of cells. Drugs enclosed within nanoparticle shells often can be released steadily over longer periods than drugs delivered in other ways, making them more effective. They can also reach areas that drugs alone cannot easily penetrate, such as the brain. Nanovic, an Australian company, concluded in March 2004 that nanotechnology was involved in up to 1 percent of drug delivery strategies currently under development and said that the figure might rise to 14 percent by 2015.

More radical forms of nanotechnology, involving manufacturing at the nano level, are slated for the far future. "Nanorobots" or "nanomachines" may enter cells to detect abnormalities, deliver drugs, or even repair the cells directly, like tiny surgeons. Other nanomachines may monitor and stimulate neurons in the brain to diagnose or repair damage from illnesses such as Parkinson's disease.

Biotechnology foes are already worried about the safety of nanotechnology products because particles at the nano level behave differently from larger particles (or, for that matter, atoms and molecules) of the same substances. Some of the most commonly used nanoparticle shapes, including carbon nanotubes (which can be as thin as 1.2 nanometers, almost three times smaller than the width of a DNA molecule) and fullerenes, or "buckyballs" (hollow spheres made up of 60 carbon atoms), have been shown to be toxic to animal cells, for instance, promoting the formation of clots in blood vessels. Some nanoparticles, too, are made from poisonous elements such as selenium and lead. Ultrafine particles in pollution have already been found to cause more breathing problems than larger particles, and critics fear that nanoparticles, which are even smaller, may be still worse.

Much greater concern has been expressed about the far future, when nanomachines that can replicate themselves may be produced. Some commentators have claimed that such machines, running out of control, could reduce the entire world to "gray goo." Nanotech supporters say that this is simply a nightmarish science fiction scenario, highly unlikely to happen.

Nonetheless, even groups that support nanotechnology, such as the Foresight Nanotech Institute in Palo Alto, California, say that research into health and environmental safety aspects of this new technology is essential and should be funded by the federal government. Particular worry has been expressed about free nanoparticles (those not sealed into products), risks to workers manufacturing products containing nanoparticles, and environmental risks that may arise when such products are discarded and end up in landfills. Protesting groups such as the Natural Resources Defense Council stress the importance of early regulation, saying, for instance, that all nanomaterials should have to meet the Toxic Substances Control Act's requirements for potentially harmful new chemicals. The government has said that 4 percent of the $1 billion budget awarded to the National Nanotechnology Initiative will be devoted to research on the societal and environmental effects of nanotechnology.

Even newer (the first conference on it took place in 2003) and more science fictional-sounding than nanotechnology is still another discipline, synthetic biology. Synthetic biologists aim to turn what they see as the relatively crude alterations involved in past "genetic engineering" into true engineering: assembling biological circuits and perhaps even whole organisms to exact specifications from standardized parts, just as traditional engineers combine electronic switches and circuits to make television sets or computers. The components that synthetic biologists plan to use are stretches of DNA comprising sets of genes (often from different species) that make cells perform particular reactions. Drew Endy at the Massachusetts Institute of Technology, one of the centers of synthetic biology research, is already gathering such components, which fellow synthetic biologist Tom Knight calls BioBricks.

Craig Venter is also exploring synthetic biology by trying to create a "minimal genome," an artificial bacterial genome that will contain the smallest set of genes that permits life. Components similar to Drew Endy's BioBricks could be added to this bare-bones genome to customize it for different purposes, such as producing hydrogen fuel. As Venter's work shows, synthetic biology can overlap industrial biotechnology; some researchers are also combining it with nanotechnology.

Jay Keasling of the University of California–Berkeley, whom *Discover* magazine chose as its scientist of the year for 2006, is trying to develop a practical application of synthetic biology, using genes from several different species to re-engineer *E. coli* bacteria so that they can inexpensively produce artemisinin, a valuable antimalaria drug that currently can be made only from certain plants. Other synthetic biologists, like researchers in industrial biotechnology, are working on ways to make ethanol from cellulose.

In some ways, synthetic biology brings biotechnology and genetic engineering back in a full circle. The fears voiced about this new discipline, including concerns that artificially created organisms will escape into the environment and create untold damage or that bioterrorists might release such organisms deliberately to start epidemics, sound very much like those expressed when recombinant DNA was first invented. Like the early genetic engineers, synthetic biologists hope to deflect criticism by policing themselves, for instance by screening orders to make sure that no one is asking for parts that could be combined to make a disease-causing microbe.

In the global political climate of the mid-2000s, the issue of possible terrorist use of biotechnology looms much larger than it did when genetic engineering was born in the early 1970s. The letters containing anthrax spores mailed to U.S. media and politicians in October and November 2001 changed the view of bioterrorist attacks in the United States from a somewhat remote possibility to a reality, and from the beginning, both government and the public have feared that terrorists might use biotechnology to make deadly germs such as those that cause smallpox, bubonic plague, or ebola even more dangerous. The Soviet Union and pre-apartheid South Africa are known to have had bioweapons programs that included genetic engineering, and other countries may have done so as well. The "superbugs" created by these programs may still exist, waiting for clever or well-to-do terrorists to snap them up. Worse still, at least one FBI official has claimed that graduate or even undergraduate college microbiology students in a well-equipped laboratory could perform the same kinds of gene alteration that the programs carried out.

Two legitimate scientific experiments have shown how easy it would be for terrorists with laboratory experience to modify disease-causing microbes. In January 2001, researchers in Australia announced that about a year earlier, trying to make a contraceptive that would control rodent pests, they had inserted a gene for a protein found in the mouse (and human) immune system into viruses that cause mousepox, a weaker cousin of smallpox, and had found that the viruses became much more deadly than before. The altered viruses killed all the animals injected with them, even those that had been vaccinated against mousepox, by destroying their immune systems. "In the wrong hands, [other viruses altered in a similar way] could become biological warfare weapons of terrible proportions," said Bob Seamark, former head of the laboratory where the research was done.[40] Similarly, U.S. scientists announced in July 2002 that they had made a polio virus "from scratch," using gene sequences from a mail-order supplier and instructions downloaded from the Internet. They showed that their manufactured virus could cause disease in mice. Experts such as Gerald Epstein of the Department of Defense say that future scientists may need to work with governments to decide which experiments are too dangerous to publish or even, perhaps, to perform.

On the other hand, some legislators and officials, such as Connecticut Democratic senator Joe Lieberman, are encouraging government support of private biotechnology research in the hope that the industry can play a vital role in the country's defense against bioterrorism. Biotech companies have already invented detectors that accurately identify "signature" sequences of DNA belonging to several kinds of deadly bacteria in less than an hour, rather than in the days to weeks that traditional culturing methods require. Other firms are working on engineered vaccines that, unlike most existing ones, use only a small fragment of a protein found on the surface of a target virus or bacterium rather than whole weakened or killed microorganisms. A vaccine of this type for pneumonic plague, a form of plague that is highly contagious and spreads through the air, was tested on mice and found to be very effective, according to a report released in June 2003.

In general, genetic engineering researchers in the future will need to be more open with the media and the public if they want to regain the trust they have lost in recent years. Biotechnology companies, especially large ones, will also need to think harder about the actual and perceived impacts of their actions in order to avoid the kind of public disapproval and government bans that have blocked genetically modified foods in Europe. Both the industry and government regulators must show themselves very quick to detect and halt any environmental or health damage caused by biotechnology products. Biotechnology must also find more ways to benefit ordinary consumers, and particularly the world's disadvantaged people, if the industry wants to retain international respect and support.

In turn, the public—not to mention the public's media sources and political representatives—will need great improvements in education to deal with the ethical challenges of the biotech century. People must learn to grasp such things as the statistics of probability and risk or the complex interaction between genes and environment. Only education can help people sort through the hype and the nightmares presented by supporters and opponents of various techniques and reach reasoned conclusions about how humanity should use its wonderful and terrible new power to alter the essence of life. As Charles Weiner, emeritus professor of history of science and technology at the Massachusetts Institute of Technology, writes:

> *[Scientific] self-regulation is not adequate for today's urgent social and political choices about the directions, priorities, and limits to human and agricultural applications of genetic engineering and biotechnology. . . . They must*

be decided in the public arena and take account of concerns for social justice and moral values, as well as effects on health and environmental safety."[41]

Leon Kass—along with many other thinkers—devoutly hopes that "human beings through their political institutions can exercise at least some control over where biotechnology is taking us."[42]

[1] "International Consortium Completes Human Genome Project," *Genomics & Genetics Weekly*, May 9, 2003, p. 32.

[2] Jeremy Rifkin, *The Biotech Century* (New York: Jeremy P. Tarcher/Putnam, 1998), p. xv.

[3] Unknown scientist, quoted in Edward Shorter, *The Health Century* (New York: Doubleday, 1987), p. 238.

[4] Mike Hildreth, quoted in "Ernst & Young's 2005 Global Biotechnology Report." Almeida Capital Altassets (Alternative Assets Network). Available online. URL: http://www.altassets.com/cgi_local/MasterPFP.cgi?doc=http://www.altassets.com/casefor/sectors/2005/nz7931.php. Posted on December 20, 2005.

[5] Michael Rogers, quoted in James D. Watson and John Tooze, *The DNA Story* (San Francisco: W. H. Freeman, 1981), p. 28.

[6] Norton Zinder, quoted in Burke Zimmerman, *Biofuture* (New York: Plenum Press, 1984), p. 141.

[7] Jeremy Rifkin, quoted in Paul Ciotti, "Saving Mankind from the Great Potato Menace," *California Magazine*, October 1984, p. 97.

[8] Levy Mwanawasa, quoted in Gavin du Venage, "African Nations Ban Biofood Aid Despite Famine," *San Francisco Chronicle*, August 23, 2002, p. A15.

[9] U.S.C. 35, Section 103(a).

[10] P. J. Federico, quoted in *Diamond v. Chakrabarty*, 447 U.S. 303.

[11] Council for Responsible Genetics, quoted in David B. Resnik, "DNA Patenting and Human Dignity," *Journal of Law, Medicine, and Ethics*, Summer 2001, p. 152.

[12] Anuradha Mittal, quoted in Peg Brickley, "Payday for U.S. Plant Scientists," *The Scientist*, January 21, 2002, p. 22.

[13] Jerry Caulder, quoted in Richard Stone, "Sweeping Patents Put Biotech Companies on Warpath," *Science*, May 5, 1995, p. 656.

[14] *John Moore v. Regents of California*, 51 Cal. 3d 120.

[15] European Court of Justice, quoted in "Europe Rules on Biotech Patents," *Chemistry and Industry*, November 5, 2001, p. S684.

[16] Unknown juror, quoted in Peter J. Neufeld and Neville Coleman, "When Science Takes the Witness Stand," *Scientific American*, May 1990, p. 46.

[17] Jamie Gorelick, quoted in "Fugitive Justice," *Nation*, March 3, 1997, p. 4.

[18] Rebecca Sasser Peterson, "DNA Databases: When Fear Goes Too Far," *American Criminal Law Review*, Summer 2000, p. 1219.

[19] Leroy Hood, quoted in Joel Davis, "Leroy Hood: Automated Genetic Profiles," *Omni*, November 1987, p. 118.

[20] Martha Volner, quoted in Geoffrey Cowley, "Flunk the Gene Test and Lose Your Insurance," *Newsweek*, December 23, 1996, p. 48.

21 Nancy Wexler, quoted in Lauren Picker, "All in the Family," *American Health*, March 1994, p. 24.

22 Francis Galton, quoted in William Cookson, *The Gene Hunters* (London: Aurum Press, 1994), p. 24.

23 *Buck v. Bell*, 274 U.S. 200.

24 *Skinner v. Oklahoma*, 316 U.S. 535.

25 Logan Karns, quoted in "Genetic Testing Results: Who Has a 'Right' to Know?" *Medical Ethics Advisor*, May 2002, p. 52.

26 Dean Rosen, quoted in Christopher Hallowell, "Playing the Odds," *Time*, January 11, 1999, p. 60.

27 Anonymous insurance executive, quoted in Hallowell, p. 60.

28 *Sutton v. United Air Lines*, 527 U.S. 471.

29 Harvie Raymond, quoted in Seth Shulman, "Preventing Genetic Discrimination," *Technology Review*, July 1995, p. 17.

30 Lori B. Andrews, quoted in Rick Weiss, "Predisposition and Prejudice," *Science News*, January 21, 1989, p. 41.

31 Paul Billings, quoted in Bettyann H. Kevles and Daniel Kevles, "Scapegoat Biology," *Discover*, October 1997, p. 62.

32 W. French Anderson, "The Best of Times, the Worst of Times," *Science*, April 28, 2000, p. 629.

33 W. French Anderson, quoted in Joseph Levine and David Suzuki, *The Secret of Life* (Boston: WGBH Educational Foundation, 1993), p. 207.

34 Leon Kass, "The Moral Meaning of Genetic Technology," *Commentary*, September 1999, p. 32.

35 Bill McKibben et al., quoted in Richard Hayes, "The Quiet Campaign for Genetically Engineered Humans," *Earth Island Journal*, Spring 2001, p. 28.

36 J. A. Barritt et al., quoted in Mark S. Frankel, "Inheritable Genetic Modification in a Brave New World," *Hastings Center Report*, March–April 2003, p. 31.

37 Tom Abate, "Celebrating 50 Years of DNA," *San Francisco Chronicle*, 24 February 2003, p. E4.

38 Edward O. Wilson, "A World of Immortal Men," *Esquire*, May 1999, p. 84.

39 James Watson, quoted in "Design-a-Kid: Does Humanity Need an Upgrade?" *Christian Century*, May 17, 2003, p. 22.

40 Bob Seamark, quoted in Curtis Rist, "Genetic Tinkering Makes Bioterror Worse," *Discover*, January 2002, p. 76.

41 Charles Weiner, "Drawing the Line in Genetic Engineering," *Perspectives in Biology and Medicine*, Spring 2001, p. 208.

42 Leon R. Kass, "A Reply," *Public Interest*, Winter 2003, p. 60.

CHAPTER 2

THE LAW AND BIOTECHNOLOGY

LAWS AND REGULATIONS

Hundreds of pieces of legislation, regulations, and policy statements relating to biotechnology and genetic engineering have been issued by the U.S. Congress, the legislatures of the states, or other government bodies such as the National Institutes of Health (NIH). This section details some of the best known and most important ones. They are grouped according to four major topics discussed in Chapter 1: agricultural biotechnology and safety, patenting life, DNA "fingerprinting" and databases, and genetic health testing and discrimination. Within each topic, they are arranged by date.

Agricultural Biotechnology and Safety

RECOMBINANT DNA ADVISORY COMMITTEE CHARTER (1974)

In accordance with Section 402 (b) (6) of the Public Health Service Act (42 U.S.C. 282), the NIH in October 1974 established the Recombinant DNA Advisory Committee (RAC), consisting of 15 members. At least eight of the committee members were to be experts in recombinant DNA research, molecular biology, or similar fields, and at least four were to be experts in applicable law, standards of professional conduct and practice, public attitudes, the environment, public health, occupational health, or related fields. According to an NIH description of the RAC posted on the Internet in 2000, a third of the committee's members "represent public interests and attitudes."[1] The committee's job is to advise the NIH director concerning the current state of knowledge and technology regarding recombinant DNA and to recommend guidelines to be followed by investigators in the field. The director is required to consult it before making major changes in existing NIH guidelines.

In the early years of its existence, the RAC had to approve most new types of genetic engineering experiments. Today, however, its approval is seldom required except in experiments involving humans (gene therapy); all clinical trials involving the transfer of recombinant DNA to humans that are funded by NIH must be registered and reviewed by the RAC. The RAC's charter was most recently renewed in 1997, and NIH wrote in 2000 that the committee still "serves a critical role in the oversight of Federally funded research involving recombinant DNA."[2] The committee was still active in 2007.

COORDINATED FRAMEWORK FOR REGULATION OF BIOTECHNOLOGY (1986)

In June 1986, the federal Office of Science and Technology Policy published the Coordinated Framework for Regulation of Biotechnology, which divided regulation of genetic engineering research and technology among five agencies: the NIH, the National Science Foundation, the U.S. Department of Agriculture (USDA), the Environmental Protection Agency (EPA), and the Food and Drug Administration (FDA). The first two groups were to evaluate research supported by government grants. The latter three agencies were, and are, the chief regulators of the environmental testing and sale of biotechnology products, under the authority of several laws that were amended to include genetically altered organisms. The agencies consider whether such products are safe to grow, safe for the environment, and (if intended as human or animal food) safe to eat. State laws, such as seed certification laws, may also affect bioengineered products.

FEDERAL INSECTICIDE, FUNGICIDE, AND RODENTICIDE ACT (FIFRA) (1947, 1970)

The Federal Insecticide, Fungicide, and Rodenticide Act (7 U.S.C. 136) was passed in 1947 to regulate the distribution, sale, use, and testing of chemical and biological pesticides. After the EPA was established in 1970, it took over regulation of pesticides under FIFRA. FIFRA has been amended to include plants and microorganisms producing pesticidal substances, such as agricultural crops genetically modified to produce *Bacillus thuringensis* (Bt) toxin.

TOXIC SUBSTANCES CONTROL ACT (TSCA) (1976, 1997)

The Toxic Substances Control Act (15 U.S.C. 53), passed in 1976, gives the EPA the authority to, among other things, review new chemicals before they are introduced into commerce. Section 5 of TSCA was amended in

1986 to classify microorganisms intended for commercial use that contain or express new combinations of traits, including "intergeneric microorganisms," which contain combinations of genetic material from different genera, as "new chemicals" subject to EPA regulation under the act. The EPA believes that organisms containing genes from such widely separated groups are sufficiently likely to express new traits or new combinations of traits to justify being termed "new" and reviewed accordingly. The EPA handles this review under its Biotechnology Program. Altered microorganisms containing genetic material from two species in the same genus are not subject to regulation under TSCA.

EPA regulations under TSCA were amended on April 11, 1997, to tailor the general screening program for microbial products of biotechnology to meet the special requirements of microorganisms used commercially for such purposes as production of industrial enzymes and other specialty chemicals, creation of agricultural aids such as biofertilizers, and breakdown of chemical pollutants in the environment (bioremediation). According to the EPA, this change provides regulatory relief to those wishing to use these products of microbial biotechnology while still ensuring that the agency can identify and regulate risks associated with such products.

PLANT PEST ACT (1987, 1993, 1997, 2000)

The Animal and Plant Health Inspection Service (APHIS) is the agency within the USDA that is responsible for protecting U.S. agriculture from pests and diseases. Under the Plant Pest Act (7 U.S.C. 7B), originally passed in 1987, APHIS regulations provide procedures for obtaining a permit or for providing notification to the agency before introducing into the United States, either by import or by release from a laboratory, any "organisms and products altered or produced through genetic engineering which are plant pests or which there is reason to believe are plant pests." The law was amended in 1993, 1997, and 2000 to simplify requirements and procedures. The 1997 amendment made notification, rather than obtaining of a permit, sufficient for the release of most new types of genetically engineered plants into the environment. In August 2002, the USDA created a new unit within APHIS, the Biotechnology Regulatory Services (BRS), to focus on regulating and facilitating agricultural biotechnology.

FEDERAL FOOD, DRUG, AND COSMETIC ACT (FFDCA) (1938, 1958, 2004)

First passed in 1938, the Federal Food, Drug, and Cosmetic Act (21 U. S.C. 9) has been amended to give both the EPA and the FDA control over certain biotechnology products. The FFDCA gives the EPA the right to set tolerance limits for substances used as pesticides on and in food and feed. This includes tolerances for residues of herbicides used on food crops genetically altered to be herbicide tolerant and for pesticides in food crops that produce such substances. The FFDCA gives the FDA the power to regulate foods and feed derived from new plant varieties, including those that are genetically engineered. It requires that genetically engineered foods meet the same safety standards required of all other foods.

The FFDCA was most recently amended in 2004. The FDA's current biotechnology policy under FFDCA treats substances intentionally added to food through genetic engineering as food additives if they are significantly different in structure, function, or amount from substances currently found in food. The agency has concluded, however, that many genetically altered food crops do not contain substances significantly different from those already in the diet and thus do not require FDA approval before marketing. This ruling has been criticized by those who believe that genetically modified food crops may threaten human health.

Patenting Life

PATENT LAW IN THE CONSTITUTION (1787)

Article I, Section 8, of the U.S. Constitution gives Congress the power to enact laws relating to patents—that is, to "promote the progress of science and useful arts, by securing for limited times to authors and inventors the exclusive right to their respective writings and discoveries."

1952 PATENT LAW REVISION

Current patent law stems from Title 35 of the U.S. Code, which was revised on July 19, 1952. The law specifies the requirements for patentability and the procedure for obtaining patents. It also gives the Patent and Trademark Office the job of granting patents and administering patent regulations.

The parts of Title 35 of greatest concern to biotechnology are Sections 100 to 103, which describe the criteria that determine which inventions are patentable. Section 101 states that a patentable "process, machine, manufacture, or composition of matter, or any . . . improvement thereof" must be "new and useful." Section 102 further defines the requirement of novelty. Section 103 adds the qualification that a patent may not be obtained if "the subject matter [of the item to be patented] as a whole would have been obvious at the time the invention was made to a person having ordinary skill in the art to which said subject matter pertains." The third paragraph of Section 103, added later, specifically adds biotechnological processes, including gene alteration and production of cell lines, to the list of patentable items.

The 1952 patent law was cited in the landmark Supreme Court case *Diamond v. Chakrabarty*, which in 1980 allowed living things other than plant varieties to be patented for the first time. In his majority opinion on that case, Chief Justice Warren Burger referred to testimony accompanying the 1952 law in which Congressman P. J. Federico, a principal drafter of the legislation, stated that Congress intended it to "include anything under the sun that is made by man."

PLANT PATENT ACT (1930)

Traditionally, living things were held to be "products of nature" and thus not patentable. In 1930, however, Congress passed the Plant Patent Act (35 U.S.C. 15), which provides that "whoever invents or discovers and asexually reproduces any distinct and new variety of plant" may obtain a patent on it. The patent grants "the right to exclude others from asexually reproducing the plant or selling or using the plant so reproduced." Only asexual reproduction was mentioned in this act because, at the time, hybrids could not be made to "breed true" (reproduce sexually).

PLANT VARIETY PROTECTION ACT (1970)

By 1970, it had become possible to reproduce hybrid plant varieties sexually, that is, by seed. Congress therefore passed the Plant Variety Protection Act (7 U.S.C. 57), which extends the patent or patentlike protection of the Plant Protection Act to plants that could be reproduced in this way. The variety protected has to be new, distinct, uniform, and stable. Except for farmers, who are allowed to save and reuse seeds under certain circumstances, the act forbids unauthorized sexual reproduction of protected plant varieties. The act specifically excludes fungi and bacteria from its coverage.

DNA "Fingerprinting" and Databases

DNA IDENTIFICATION ACT (1994)

On September 13, 1994, Congress passed the DNA Identification Act (42 U.S.C. Sec. 14131). This act attempted to answer criticisms of the quality of forensic DNA testing by ordering the Federal Bureau of Investigation (FBI) to establish standards for quality assurance and proficiency testing of laboratories and analysts carrying out such testing. The FBI was supposed to set up a system of blind external proficiency testing (that is, testing done by an outside agency) for forensic DNA laboratories within two years, unless the agency concluded that such a system was not feasible.

The FBI did establish quality assurance standards and an accreditation program for forensic DNA testing laboratories, and it trains scientists to audit laboratories, but it does not directly oversee testing or require compliance except for those obtaining Department of Justice grants. (Individual DNA analysts who participate in the bureau's National DNA Index System [NDIS] are required to pass a proficiency test twice a year, but the bureau does not handle the testing.) Critics have questioned the wisdom of giving the FBI control over quality assurance for DNA testing because the bureau's own testing laboratory has been shown to have major quality control problems in at least two Justice Department investigations, one in 1997 and a second in 2004.

The DNA Identification Act also authorized the FBI to establish a national database of DNA profiles from people convicted of crimes, samples recovered at crime scenes or from unidentified human remains, and relatives of missing persons who voluntarily donate samples. This database, the National DNA Index System (NDIS), opened in October 1998. NDIS is part of the Combined DNA Index System (CODIS), which also includes software that helps state and local laboratories coordinate their databases with the national one and each other so that law enforcement personnel can match their own felon and crime scene samples against others obtained anywhere in the country. By early 2007, all 50 states and the U.S. Army, as well as the FBI, were participating in the CODIS program.

THE DNA ACT (2000)

The DNA Analysis Backlog Elimination Act, commonly known as the DNA Act, was passed on January 24, 2000, and subsequently became P.L. 106-546 (42 U.S.C. 14135). Its provisions included authorization of $170 million over fiscal years 2001 through 2004 for grants to states to help their laboratories reduce the backlog of untested forensic DNA samples, subject to satisfaction of quality assurance standards by the laboratories involved. More controversially, it made collection of DNA samples (for profiles to be included in the national forensic database) a condition of supervised release or parole for people convicted of "qualifying federal crimes" such as murder, sexual abuse, kidnapping, robbery, burglary, or arson. A convict who refused to provide such a sample would be guilty of a class A misdemeanor, and probation offices could use "such means as are reasonably necessary to detain, restrain, and collect a DNA sample" from anyone who would not donate the sample voluntarily. The constitutionality of this law was later challenged, unsuccessfully, in several court cases.

THE JUSTICE FOR ALL ACT (2004)

The Justice for All Act (P.L. 108-405), which became law on October 30, 2004, seemed to have something for everyone. The first part provided a list of rights for crime victims, and the second part (expanding on the program established by the DNA Act) authorized $755 million over five years to re-duce testing backlogs for forensic DNA samples. It also granted more than $500 million to reduce other forensic science backlogs, train criminal justice and medical personnel to collect DNA evidence, and promote the use of DNA technology to identify missing persons. The third part, termed the In-nocence Protection Act, provided access to post–conviction DNA testing in federal cases, required the preservation of DNA evidence in most federal criminal cases as long as the convicted person remained in prison, and in-creased compensation for wrongful convictions. It also established a program, named after Kirk Bloodsworth, the first death-row inmate to have a convic-tion overturned through DNA testing, that would provide $25 million over five years to help states cover the costs of post–conviction DNA testing.

Nonetheless, the act was criticized by some groups because it also expanded CODIS, the national DNA database program, to include samples from people convicted of any federal felony offense. Indeed, CODIS was authorized to ac-cept almost any DNA information that states chose to collect.

DNA FINGERPRINTING ACT (2006)

The DNA Fingerprinting Act was a small part (Title X) of the Violence Against Women and Department of Justice Reauthorization Act (P.L. 109-162), signed into law on January 5, 2006. This law further expanded CODIS, already extended by the DNA Act (which the new law amended) and the Justice for All Act, to include samples from anyone arrested for qualifying federal offenses—whether or not they were later convicted. (However, people could ask for their records to be expunged if they could provide final court orders saying that charges had not been filed or had been dropped, or that the people had been tried and acquitted.) Samples could also be required from "non-United States persons who are detained under the authority of the United States."

Genetic Health Testing and Discrimination

AMERICANS WITH DISABILITIES ACT (ADA) (1990)

Passed in 1990, the Americans with Disabilities Act (42 U.S.C. 12101–12111, 12161, 12181) was intended to increase disabled people's access to public spaces, communication, transportation, and jobs and to prevent discrimina-tion against them in employment and other areas. Section 12102 of the act defines disability as meeting one of three criteria: (1) a physical or mental impairment that substantially limits one or more of the major life activities of an individual; (2) a record of such impairment; or (3) being regarded as having such impairment. People suffering from inherited diseases would surely be considered disabled, but it is not yet clear whether people who are presently healthy but are likely to develop an inherited illness later in life (as in late-onset diseases such as Huntington's disease) or have inherited a gene associated with increased risk of an illness such as cancer can be considered disabled. The Equal Employment Opportunity Commission ruled in 1995 that using genetic test results to deny employment to people in this category was discrimination under the ADA, but its ruling does not have the force of law.

HEALTH INSURANCE PORTABILITY AND ACCOUNTABILITY ACT (1996)

The Health Insurance Portability and Accountability Act, passed on August 21, 1996 (P.L. 104–191), was intended primarily to help people keep their health insurance when they change jobs. A paragraph in Section 701, "Increased Portability Through Limitation on Preexisting Condition Exclusions," forbids considering genetic information as a preexisting condition for insurance purposes unless a person is actually suffering from an inherited disease. Thus, for instance, a woman who is shown by a test to have a mutated form of the gene *BRCA1*, which is associated with an increased risk of breast and ovarian cancer, but who does not actually have cancer could not be denied insurance.

COURT CASES

A great deal of litigation has arisen over issues related to biotechnology and genetic engineering. In biotechnology, many cases have been patent disputes. In human genetics and genetic engineering, most have related to either DNA fingerprinting and databases or genetic discrimination. This section describes some of the important court cases relating to patenting of living things (including products of the human body), forensic DNA testing, and genetics-based discrimination in insurance and employment (including eugenics laws).

Patenting Life

DIAMOND V. CHAKRABARTY, 447 U.S. 303 (1980)

Background

Ananda Chakrabarty, a scientist working for General Electric Corporation, modified a bacterium of the genus *Pseudomonas* so that it could digest crude petroleum, something no natural bacterium could do. He did so by forcing the bacterium to take up four types of plasmids from other bacteria that could digest different components of crude oil, a technique that could be classified as genetic engineering but did not involve recombinant DNA (the individual plasmids were unaltered).

In June 1972, believing that his new bacterium would be useful in cleaning up oil spills, Chakrabarty applied for patents (in the name of General Electric) on the process of making the bacteria, a mixture in which they could be spread on water, and the bacteria themselves. The Patent and Trademark Office (PTO) granted the first two patent requests but rejected the third on the grounds that, as living things, the bacteria were "products of nature" rather than "manufactures" and thus were not patentable under U.S. law. The Patent Office Board of Appeals affirmed the PTO's decision. The U.S. Court of Customs and Patent Appeal (CCPA), however, reversed it.

The government appealed the decision to the U.S. Supreme Court in 1979. At first the court refused to hear the case, telling the CCPA to reconsider its decision in light of the comment that the court had made in a 1978 case, *Parker v. Flook*, in which the judges had said they believed they should "proceed cautiously when we are asked to extend patent rights into areas wholly unforeseen by Congress." The CCPA held its ground, however, and the Court accepted the case later in 1979. It was argued on March 17, 1980, under the full name *Sidney A. Diamond, Commissioner of Patents and Trademarks, v. Ananda M. Chakrabarty.*

Legal Issues

The question before the Supreme Court was technically a narrow one: interpreting the language of 35 U.S.C. 101 (part of the revision of patent law that Congress had passed in 1952) to determine "whether a live, human-made micro-organism is patentable subject matter"—that is, whether it was a "manufacture" or "composition of matter" as Congress intended the 1952 law to be construed. The root of the question, however, was deeply significant: Could a living thing—even one whose genes had been deliberately altered—be considered to be a human invention?

Traditionally, laws of nature, physical phenomena ("products of nature"), and abstract ideas have been held not to be patentable. In testimony accompanying the 1952 law, however, Congressman P. J. Federico, a principal drafter of the legislation, stated that Congress intended it to "include anything under the sun that is made by man."

Congress had already allowed the patenting, or at least a patentlike protection, of plant varieties in the Plant Patent Act of 1930 (which permitted plant breeders to keep exclusive rights to asexual reproduction of new varieties they developed) and the Plant Variety Protection Act of 1970 (which extended the protection to sexual reproduction of new plant varieties). House and Senate reports accompanying the 1930 act said that its purpose was to "remove the existing discrimination between plant developers and industrial inventors" because the acts of developing new plant varieties and new compositions of nonliving matter were conceptually equivalent.

The CCPA, in defending its decision, claimed that Chakrabarty's altered bacterium met the patent criteria of novelty, usefulness, and nonobviousness specified in the 1952 law (35 U.S.C. 101–103).

> *We look at the facts and see things that do not exist in nature and that are man-made, clearly fitting into the plain terms "manufacture" and "compositions of matter." We look at the statute and it appears to include them. We look at legislative history and we are confirmed in that belief. We consider what the patent statutes are intended to accomplish and the constitutional authorization, and it appears to us that protecting these inventions, in the form claimed, by patents will promote progress in very useful arts.*

The CCPA further maintained that the fact that the things that had been "manufactured"—that is, deliberately altered by humans—were alive made no difference for patenting purposes: "[There is] no *legally* significant difference between active chemicals which are classified as 'dead' or organisms used for their *chemical* reactions which take place because they are 'alive'" [emphasis in original]. Counsel for the PTO argued, on the other hand, that Congress did not want bacteria to be patentable because the Plant Variety Protection Act specifically excluded them. In general, the PTO said, Congress had not resolved the question of whether living things should be considered patentable.

Amicus curiae ("friend of the court") briefs filed by various groups brought up several other issues. Biotechnology critic Jeremy Rifkin's organization, the People's Business Commission (now the Foundation on Economic Trends), for instance, filed a brief urging that Chakrabarty's patent be refused because of the possible threats genetically altered organisms posed to human health and the environment. The brief claimed that granting the Chakrabarty patent would provide an incentive for commercial exploitation of the new gene-altering technology that was not in the public interest. The biotechnology company Genentech, conversely, filed a brief calling for acceptance of the patent, pointing to the 1978 relaxation of the NIH guidelines as evidence that most scientists no longer feared recombinant DNA technology.

Other briefs questioned the ethics of "owning" organisms or claiming to have made them, saying that only God could make living things. Representatives of the rapidly growing biotechnology industry, on the other hand, stressed that patent protection was vital to the industry's growth. They said that if patents on engineered organisms were not allowed, much of the knowledge being generated in the new field would remain hidden in the form of trade secrets rather than being revealed so that others could use it once the patents expired.

Biotechnology and Genetic Engineering

Decision

On June 16, 1980, the Supreme Court decided by a 5-4 vote that "a live, human-made micro-organism is patentable subject matter under [U.S.C.] 101. Respondent's micro-organism constitutes a 'manufacture' or 'composition of matter' within that statute." Chief Justice Warren Burger, writing the court's majority opinion, stated,

> *The patentee has produced a new bacterium with markedly different characteristics from any found in nature and one having the potential for significant utility. His discovery is not nature's handiwork, but his own; accordingly it is patentable subject matter.*

The court concluded that Congress had intended the patent laws to have wide scope. Burger rejected the patent office's argument that the fact that Congress had passed the two plant patent protection acts meant that it had not intended the original 1952 statute to cover living things. He cited congressional commentary on the 1930 act stating that the work of the plant breeder "in aid of nature" was patentable invention. The distinction, he said, was not between living and nonliving things but between unaltered products of nature and things that had been invented by humans. The exclusion of bacteria from the 1970 act, Burger said, meant only that bacteria were not considered to be plants, not that they were not considered to be patentable.

The patent office's second argument against Chakrabarty's patent was that microorganisms could not be patented until Congress expressly authorized such protection because Congress had not foreseen genetic engineering technology at the time it passed the 1952 law. Somewhat reversing the court's cautious position in *Parker v. Flook*, Burger wrote that he "perceive[d] no ambiguity" in Congress's language that would exclude genetically altered organisms.

Justice William Brennan wrote the minority opinion. He claimed that in the plant patent acts, Congress chose "carefully limited language granting protection to some kinds of discoveries, but specifically excluding others," including bacteria. "If newly developed living organisms not naturally occurring had been patentable under 101, the plants included in the scope of the 1930 and 1970 Acts could have been patented without new legislation," he stated. In conclusion, recalling the recommendation for caution that the court had expressed in *Flook*, Brennan commented,

> *I should think the necessity for caution is that much greater when we are asked to extend patent rights into areas Congress has foreseen and considered but not resolved. . . . [The majority's decision] extends the patent system to*

cover living material even though Congress plainly has legislated in the belief that [present law] does not encompass living organisms. It is the role of Congress, not this Court, to broaden or narrow the reach of the patent laws.

Finally, the question of the dangers of genetically engineered organisms—the description of which Chief Justice Burger called "a gruesome parade of horribles"—was an issue beyond the court's competence, Burger concluded. "Arguments against patentability . . . , based on potential hazards that may be generated by genetic research, should be addressed to the Congress and the Executive, not to the Judiciary," he wrote. He also noted:

The grant or denial of patents on micro-organisms is not likely to put an end to genetic research or to its attendant risks. The large amount of research that has already occurred when no researcher had sure knowledge that patent protection would be available suggests that legislative or judicial fiat as to patentability will not deter the scientific mind from probing into the unknown any more than Canute could command the tides.

Impact

Commentators such as patent lawyer Mitchel Zoler have claimed that from a strict legal standpoint, the *Chakrabarty* decision was "trivial law."[3] It broke no new legal ground, but rather provided only a minor clarification of existing patent laws. Furthermore, Donald Dunner, another patent lawyer, noted in the year following the decision that the ruling was "important but . . . not life or death of the [biotechnology] industry, and even had it gone the other way, it probably would not have been."[4] Even if altered microorganisms themselves had not been considered patentable, the processes of making them would have been, or their nature could have been kept hidden as trade secrets. Furthermore, the ease of altering bacteria suggested that patents, even once obtained, could be fairly easily circumvented by further alteration.

The psychological impact of this Supreme Court decision on both supporters and critics of biotechnology, however, was enormous. By the time the court made its ruling, dozens of patent applications for recombinant and other genetically engineered organisms, mostly bacteria, had been submitted to the PTO but had not been ruled upon. The PTO now felt free to begin granting these patents. The court's decision, which was widely publicized, gave a considerable boost to the biotechnology industry and encouraged those who were thinking of investing in it. Peter Farley of Cetus Corporation, a leading biotechnology business that had been among those filing amicus briefs in the case, said afterward that "the positive impact [of the decision] was in the Court's bringing genetic engineering as a commercial enterprise to the attention of the entire country."[5] The decision also ignited public discussion about whether patenting living things was ethical and to what degree a patent implied "ownership" of a life-form.

The PTO expanded the *Chakrabarty* ruling in 1987, extending patent protection to animals, cells and cell lines, body parts, plasmids, and genes, including human ones. After that, the only type of living thing that could not be patented was a whole human being.

JOHN MOORE V. REGENTS OF CALIFORNIA, 51 CAL. 3D 120 (1990)

Background

John Moore, a Seattle businessman, learned in 1976 that he suffered from a rare form of leukemia. He went to the Medical Center of the University of California–Los Angeles (UCLA), for treatment, where his case was assigned to David W. Golde, M.D. Informing Moore "that he had reason to fear for his life," Moore later alleged, Golde recommended removing Moore's enlarged spleen (an abdominal organ that makes blood cells) as a treatment for the disease. Moore consented, and his spleen was removed on October 20.

Moore alleged later that, unknown to him, Golde had noticed even before the operation that Moore's blood cells had an extraordinary ability to make immune system chemicals called lymphokines, which have potential commercial use as medical treatments because they stimulate production of immune cells that fight bacterial infections and cancer. Golde therefore "formed the intent and made arrangements to obtain portions of [Moore's] spleen following its removal." Furthermore, Golde called Moore back from Seattle for several additional treatments between 1976 and 1983, during which he took samples of Moore's blood, bone marrow, sperm, and skin.

During this same period, Golde worked with Shirley Quan, another UCLA researcher, to develop Moore's cells into an immortal cell line that was eventually named "Mo," after Moore. The two researchers and their employers, the Regents of the University of California, applied for a patent on their cell line in 1981, and the patent was granted in 1984. They then made lucrative contracts with Genetics Institute and the drug company Sandoz for use of the cell line.

At no time did Golde or anyone else involved in the research tell Moore about the cell line or the patent; indeed, Moore alleged, Golde repeatedly denied any commercial plans when Moore directly asked him about such a possibility. Moore nonetheless somehow found out about the extremely lucrative use to which his cells were being put (one estimate placed the potential value of products from the patented line at $3 billion). He filed suit in 1984, naming Golde, Quan, the Regents, and the drug companies as defendants. He claimed that he still "owned" his removed cells, at least in the sense that he had a right to say what use was made of them, and that he had a proprietary interest in any products that these or other researchers created with the cell line made from the cells.

A California superior court denied Moore's right to sue in 1986, but he appealed the decision, and an appellate court reversed it two years later, stating that "the essence of a property interest—the ultimate right of control— . . . exists with regard to one's human body." Golde, the university, and the other defendants appealed the case to the California Supreme Court, which heard it in July 1990.

Legal Issues

The California Supreme Court agreed to rule on whether John Moore had grounds to sue the defendants for using his cells in potentially lucrative research without his permission. The basic question was whether a person has an ownership right to body cells and tissues after they have been removed and, if so, whether this right entitles the person to compensation if others develop those cells or tissues into a commercial product.

"In effect," wrote Justice Edward A. Panelli in the court's majority opinion, "what Moore is asking us to do is to impose a tort duty on scientists to investigate the consensual pedigree of each human cell sample used in research"—something that no court had ruled on before. This was an important issue. An Office of Technology Assessment report to Congress had noted in 1987,

> *Uncertainty about how courts will resolve disputes between specimen sources and specimen users could be detrimental to both academic researchers and the infant biotechnology industry, particularly when the rights are asserted long after the specimen was obtained. . . . The uncertainty could affect product developments as well as research. Since inventions containing human tissues and cells may be patented and licensed for commercial use, companies are unlikely to invest heavily in developing, manufacturing, or marketing a product when uncertainty about clear title exists. . . . Resolving the current uncertainty may be more important to the future of biotechnology than resolving it in any particular way.[6]*

The case also raised questions about what information a doctor has a "fiduciary responsibility" to give a patient in order to obtain truly informed consent before doing a medical procedure. To a lesser extent, it considered what financial or other responsibility secondary parties that have an interest in research done on a person's tissues but have no direct dealings with the person, such as a university or a drug company, bear to the person from whom the tissues came. Finally, as Justice Armand Arabian wrote in a separate opinion concurring with the majority, the Moore case raised "the moral issue" of whether there is "a right to sell one's own body tissue for profit."

Decision

The California Supreme Court decided by a 5-2 vote that John Moore did not have a right of ownership over his tissues or cells once they had been removed from his body. "Moore's novel claim to own the biological materials at issue in this case is problematic, at best," Justice Panelli wrote in his majority opinion. Moore therefore could not sue under tort law for a conversion—essentially, a theft—of those parts. The court defined conversion as "a tort that protects against interference with possessory and ownership interests in personal property" and concluded that "the use of excised human cells in medical research does not amount to a conversion." Moore also had no direct right to income from the cell line patent because it represented invention on the part of the researchers, not any creative effort by himself, the court ruled.

In his majority opinion, Justice Panelli also expressed concern that allowing Moore to sue on the grounds of conversion

> . . . would affect medical research of importance to all of society. . . . The extension of conversion law into this area [use of cells after they have legitimately been removed from the body] will hinder research by restricting access to the necessary raw materials. . . . Th[e] exchange of scientific materials [cell lines] . . . will surely be compromised if each cell sample becomes the potential subject matter of a lawsuit.

Furthermore, Panelli wrote, "The theory of liability that Moore urges us to endorse threatens to destroy the economic incentive to conduct important medical research" because of the danger of lawsuits from disgruntled patients whose cells had been transformed into cell lines.

Although it rejected Moore's right to sue for conversion, the court found that David Golde had violated his "fiduciary duty" to Moore by not telling him about the proposed for-profit use of his cells. Golde's commercial plans represented a potential conflict of interest with his role as Moore's physician, and his failure to inform Moore of those plans denied Moore some of the facts he needed in order to give informed consent to the spleen operation. Justice Panelli wrote:

> We hold that a physician who is seeking a patient's consent for a medical procedure must, in order to satisfy his fiduciary duty and to obtain the patient's informed consent, disclose personal interests unrelated to the patient's health, whether research or economic, that may affect his medical judgment.

The court therefore ruled that Moore could sue Golde, at least, on the grounds that Golde had violated his fiduciary duty and that he had failed to obtain Moore's properly informed consent. Grounds for suing the other defendants were more dubious, though the court did not rule out the possibility of such suits.

All the justices agreed that Golde had violated his duties to Moore, but they had differing opinions about the larger question of Moore's ownership of his body tissues. In a separate concurring opinion, Justice Arabian wrote about "the moral issue" involved in the case. He claimed that to allow Moore to sue for conversion—in other words, to support Moore's claim that he owned his tissues after their removal and should have been paid for them—would be to "recognize and enforce a right to sell one's own body tissue for profit," an idea of which Arabian clearly did not approve. The justice wrote:

> [Moore] entreats us to regard the human vessel—the single most venerated and protected subject in any civilized society—as equal with the basest commercial commodity. He urges us to commingle the sacred with the profane. He asks much.

Arabian expressed fears that supporting Moore's claim would result in "a marketplace in human body parts" and stated that the state legislature should settle the question of whether such a situation was permissible.

Justices Allen Broussard and Stanley Mosk expressed other viewpoints in separate dissenting opinions. Broussard supported Moore's right to sue for conversion because of the allegation that Golde had been planning his research before suggesting that Moore have his spleen removed. By not telling Moore about his plans, therefore, Golde had interfered with Moore's ownership rights to his cells *before* the cells had been removed. Broussard pointed out that the state's Uniform Anatomical Gift Act allowed people to specify donation of their organs for transplantation after their death and claimed that this fact supported the idea that people could say how they wanted donated parts of their body to be used. Broussard wrote:

> The act clearly recognizes that it is the donor of the body part, rather than the hospital or physician who receives the part, who has the authority to designate, within the parameters of the statutorily authorized uses, the particular use to which the part may be put.

Unlike Panelli, Broussard did not feel that allowing occasional suits like Moore's (which had the unusual feature that the commercial usefulness of his cells had been discovered, and actively concealed from Moore, before the cells had been removed from his body) would put a damper on medical research with cell lines or that, even if it did, this was sufficient reason to deny Moore's right to sue. Because most of the value of the cell line patent lay in the researchers' work, Broussard suspected that the damages Moore would receive would be relatively small even if he won his suit.

Broussard's view of the effect the court's decision would have on the possible sale of cells or body parts was exactly the opposite of Arabian's. Broussard wrote:

> *Far from elevating these biological materials above the marketplace, the majority's holding simply bars plaintiff, the source of the cells, from obtaining the benefit of the cells' value, but permits defendants, who allegedly obtained the cells from plaintiff by improper means, to retain and exploit the full economic value of their ill-gotten gains free of their ordinary common law liability for conversion.*

Justice Mosk also dissented from some of the majority's decisions and reasonings. If past judicial rulings did not cover ownership of body parts, Mosk saw no reason not to extend them:

> *If the cause of action for conversion is otherwise an appropriate remedy on these facts, we should not refrain from fashioning it simply because another court has not yet so held or because the Legislature has not yet addressed the question.*

Mosk also felt that, although Moore had contributed no creative effort toward development of the patented cell line made from his spleen, he was nonetheless a kind of "joint inventor."

> *What . . . patients [like Moore] . . . do, knowingly or unknowingly, is collaborate with the researchers by donating their body tissue. . . . By providing the researchers with unique raw materials, without which the resulting product could not exist, the donors become necessary contributors to the product.*

Because of that contribution, Mosk said, Moore should be entitled to some compensation.

Mosk agreed with Broussard that a threat to medical research on cell lines was not a sufficient reason to deny Moore's claim, though he gave different reasons for his view. Secrecy and competition in the biotechnology industry had already severely inhibited the exchange of information and research materials, Mosk wrote. Furthermore, he claimed, researchers would know where their cells came from and whether proper consent for their use had been obtained if they engaged in "appropriate record-keeping."

Above all, Mosk, like Broussard, felt that denying Moore's right to own his body parts would result in the human body being treated as a salable product and thus was morally reprehensible. Mosk quoted an earlier judicial decision that stated:

> *The dignity and sanctity with which we regard the human whole, body as well as mind and soul, are absent when we allow researchers to further their own interests without the patient's participation by using a patient's cells as the basis for a marketable product.*

Impact

John Moore eventually filed suit on the grounds that the court had left open. His suit was settled out of court.

The results of the Moore trial pleased biotechnology researchers, who had feared possible liability from working with cell lines or having to share revenue from them if Moore's right to sue was upheld. It disappointed those who disapproved of the patenting of living things or of tissues, cells, or genes taken from human beings. At the same time, it discouraged the establishment of a marketplace or "body shop" where human organs, tissues, or cells would be bought and sold; at least, they would not be sold by their original owners or while still residing in those owners' bodies.

Although the Moore case is often cited as establishing the principle that people do not own rights to their body tissues, the ruling does not apply outside California (or even necessarily to all cases within California). Some other states have different laws. In Oregon, for instance, a 1995 law (amended in 1997) specifically grants ownership rights over tissues and the genetic information derived from them to the people from whose bodies the tissues came. There have been no national rulings on this issue.

DNA "Fingerprinting" and Databases

FLORIDA V. ANDREWS, 533 SO. 2D 841 (FLORIDA FIFTH DISTRICT COURT OF APPEALS, 1988)

Background

In 1986, a number of women were raped, beaten, and cut in Orlando, Florida. The intruder entered their homes late at night, when they were alone, and evidence suggested that he had stalked them before the attacks to learn their habits. He covered each woman's head with a blanket or sheet, and only one of his victims, Nancy Hodge, saw his face. Police were able to obtain semen samples from Hodge and one other woman, a young mother.

Responding to a report of a prowler in early 1987, police captured Tommie Lee Andrews, a 24-year-old warehouse worker. Nancy Hodge picked out Andrews's picture from a photo lineup, and he was charged with her rape and that of the other woman who had provided a semen sample.

Tim Berry, Andrews's prosecutor, was reluctant to base his case entirely on Hodge's identification. A blood typing test suggested that Andrews could have committed the rapes—but so could 30 percent of the men in the United States. That summer, however, another attorney told Berry about the DNA "fingerprinting" technique for identification of individuals that British geneticist Alec Jeffreys had invented a few years before. The technique had been used in numerous immigration and paternity cases and had just made its first appearance in a British criminal court. It allowed DNA from small samples of blood, semen, or other body fluids found at a crime scene to be compared with that from a suspect's blood at certain locations in the DNA molecule that differed considerably from person to person.

Berry sent the semen samples and a little of both Andrews's and Hodge's blood to Lifecodes in Valhalla, New York, one of the few laboratories in the United States then able to perform the test. The DNA in the semen samples matched that of Andrews but not Hodge. The Lifecodes analyst said the odds of the match occurring by chance (that is, of Andrews being innocent, yet still having DNA that matched that in the semen sample) were one in 10 billion—over 4 billion more than the population of the world.

Legal Issues

At the time of Andrews's trial, DNA profiling evidence had been used in only one other criminal case in the United States, and that case had not been a close parallel to that of Andrews. The British case, however, was similar. Jeffreys's test had shown a match between the DNA in semen found on two teenage girls who had been raped and murdered in Leicestershire and that in the blood of Colin Pitchfork, a 27-year-old baker. During the hunt for the killer, the police had taken the unusual step of asking all men between ages 13 and 30 in several villages—some 5,000 people—to voluntarily give samples of their blood for testing. Fearing detection, Pitchfork had persuaded a coworker to give blood in his stead, but the man bragged about it and was overheard. When questioned, he led the police to Pitchfork. Pitchfork confessed and was convicted of the crimes. DNA testing had also exonerated another suspect in the case, Rodney Buckland. Buckland had confessed to one of the killings, but his DNA did not match the semen sample, so he could not have been guilty.

The spectacular success of DNA "fingerprinting" in the Pitchfork case had been widely publicized. The test was still extremely new to forensics, however, and there was considerable question about whether it would meet the "*Frye* rule" (based on a 1923 case, *Frye v. United States*), by which many judges decided whether evidence from a new scientific technique would be admitted in a trial. In the *Frye* case, the court had ruled that a technique had to be "sufficiently established to have gained general acceptance in the particular field in which it belongs" before evidence based on it could be used.

Decision

Andrews's trial for the rape of Nancy Hodge took place in October 1987. In a pretrial hearing on October 19, Berry brought in an expert witness who testified that the technique on which Jeffreys's test was based, although new in the courtroom, was widely accepted in genetics and molecular biology laboratories. The judge agreed on this basis that DNA profiling met the *Frye* requirement, and he allowed the DNA evidence to be presented in Andrews's trial. When a Lifecodes expert brought up the one-in-10-billion statistic, however, the defense lawyers objected. Unprepared for the challenge, the expert withdrew the statistic. Without it, even the DNA evidence combined with Hodge's identification of Andrews apparently was not enough. The jury was unable to reach a verdict, and the judge declared a mistrial.

Andrews went on trial for the young mother's rape a few weeks later, and this time the prosecutors were able to provide legal backing for the statistics that supported the DNA test results. In this case, furthermore, Andrews had left literal as well as genetic fingerprints behind. He was found guilty on November 6, becoming the first person in the United States to be convicted of a crime partly on the basis of DNA evidence. In addition, he was retried for Hodge's rape in February 1988, and this time, despite questions raised by the defense lawyers, he was convicted. His total sentences for the two convictions amounted to 100 years in prison.

Impact

Coming soon after the widely publicized success of DNA profiling in the Pitchfork case, the technique's usefulness in convicting Andrews caused prosecutors to turn to it eagerly in similar cases. It was hailed as "a prosecutor's dream," the greatest aid to identifying criminals since the development of fingerprinting a century before. Judges and juries began to accept it as well. Nonetheless, its validity usually had to be established in a separate *Frye* hearing for each case.

HARVEY V. HORAN, 278 F. 3D 370 (2001, 2002)

Background

James Harvey was convicted of rape and forcible sodomy in Virginia in April 1990. His trial included DNA evidence that had been tested in 1989, soon after forensic DNA testing was introduced to the legal arena. Testing methods of the time were crude, and the results of the test were inconclusive.

Harvey consistently claimed that he did not commit the crime. In 1993 he sought the assistance of the New York–based Innocence Project, which works to obtain new DNA testing for convicted felons who say that they are innocent and might be cleared by such testing. Lawyers for the project asked Robert F. Horan, Jr., the prosecutor in Fairfax County, for access to the rape kit used in Harvey's trial, claiming that new techniques of DNA testing might show that Harvey's semen did not match the DNA in the kit samples. Horan refused. Harvey's lawyers then sued Horan in a federal district court in Alexandria, Virginia, in July 2000, citing 42 U.S.C. 1983, a statute that permits citizens to sue state or local officials who allegedly violate their constitutional rights.

Legal Issues

From a technical standpoint, the chief issue was whether it was legally appropriate for Harvey (and, by extension, other felons) to use the Section 1983 statute to command access to DNA evidence. Since the statute refers to violation of constitutional rights, however, the more important underlying question was whether demands for access to and retesting of DNA evidence after conviction could be tied to any constitutional right. In addition to denying this possibility, Horan claimed that permitting convicts to demand DNA retesting would cost the state a crippling amount of money and time and would hamper the processing of new cases. Peter Neufeld, cofounder of the Innocence Project, denied this charge.

Decision

U.S. District Judge Albert V. Bryan, Jr., ruled on the case in Alexandria, Virginia, on April 16, 2001. He ordered Horan to provide the rape kit to Harvey's lawyers and, more important, stated clearly that, in his opinion, Horan had violated Harvey's civil rights—specifically, his right to due process under the Fourteenth and Fifteenth Amendments—and that convicted felons in general had a right to DNA testing that might prove their innocence. Bryan based his decision on a 1963 Supreme Court case, *Brady v. Maryland,* which held that prosecutors violated defendants' constitutional right to due process when they suppressed evidence. "Denying the plaintiff access to potentially powerful exculpatory evidence would result in . . . a miscarriage of justice," Bryan wrote. Contemporary news stories stated that Bryan's was the first such ruling by a federal judge in the country. Other judges in similar cases had not accepted the constitutional argument.

Horan appealed Bryan's decision, and the Fourth U.S. Circuit Court of Appeals reviewed it in January 2002. A three-judge panel voted to reverse the district court ruling. Writing the majority opinion, Chief Judge J. Harvie Wilkinson III stated that, in his opinion, Section 1983 was not the appropriate legal vehicle for Harvey to use in demanding the DNA evidence. He held that Harvey was trying to invalidate a final state conviction following a trial that, by Harvey's own admission, was fair in terms of the evidence available at the time.

Addressing the larger issue, Wilkinson wrote that he could see no grounds for claiming a constitutional right to access DNA evidence after a conviction. Like Horan, he feared that accepting Harvey's arguments would open the door to an endless round of convict lawsuits and expensive demands for re-hearing as scientific techniques advanced. "Establishing a constitutional due process right to re-test evidence with each forward step in forensic science would leave perfectly valid judgments in a perpetually unsettled state," Wilkinson wrote. He also pointed out that other evidence, including testimony from a second man convicted of the rape, implicated Harvey.

Harvey appealed the circuit court panel's decision, asking that the case be reheard by all the circuit court judges (en banc). The court denied his request in March 2002. Judge Wilkinson again stressed that he felt that Section 1983 was inappropriate in this case. He stated that Harvey should have kept his appeals in the state court system rather than attempting to sue in federal court. Federal courts, he said, should not try to decide who, if anyone, has a right to demand DNA testing when Congress and numerous state legislatures were still wrestling with that issue. In a dissenting opinion, however, Judge J. Michael Luttig maintained that, because avoidance of wrongful convictions was so important, convicts under at least some circumstances might well have a constitutional due process right to DNA testing.

Impact

Supporters of DNA testing for convicted felons hailed Bryan's decision as a landmark, even though it was not binding on other courts. The appeals court's rulings, however, dimmed their hopes for establishment of a constitutional right to DNA testing. At the same time, a number of states, including Virginia, were considering or passing laws providing some degree of testing access for convicts. In February 2002, in fact, just a month after the appeals court reversed the district court's decision, circuit judge David T. Stitt ordered Horan to turn over the evidence to Harvey's lawyers on the basis of a Virginia statute signed into law in May 2001, a month after Bryan's ruling. The new Virginia law gave people convicted of violent crimes, including death row inmates, the right to seek court orders for testing that might provide proof of their innocence. Ironically, when the new DNA test was finally performed, it confirmed Harvey's guilt.

Harvey was able to obtain his DNA evidence under the state law; his civil rights case presumably has become moot. However, the broader issue of a convict's constitutional right to demand DNA testing remains unsettled. Because this question has implications, not only for the use of forensic DNA databases, but for application of the death penalty and possibly for criminal law in general, some case similar to Harvey's probably will be taken to the U.S. Supreme Court.

UNITED STATES V. KINCADE, 379 F. 3D 813 (9TH CIR., 2004)

Background

Thomas Cameron Kincade, a decorated U.S. Navy seaman, robbed a bank using a firearm on July 20, 1993. He was arrested, pleaded guilty, and was sentenced to 97 months' imprisonment, followed by three years' supervised release. On March 25, 2002, while Kincade was still on supervised release, his probation officer asked him to submit a blood sample pursuant to the 2000 DNA Act, which required such samples of all people convicted of certain federal offenses, including robbery, whether they were presently in prison or on parole, probation, or supervised release. Kincade refused. After failing to persuade Kincade to change his mind, the probation officer informed the district court of the refusal and recommended that Kincade be sent back to prison.

In a briefing to the district court prior to his revocation hearing, Kincade challenged the constitutionality of the DNA Act on grounds that it violated the Ex Post Facto Clause, the Fourth Amendment, and the separation of powers principles embodied in Article III and the Due Process Clause. U.S. District Judge Dickran Tevrizian rejected these constitutional challenges during Kincade's revocation hearing, which took place on July 15, 2002. He sentenced Kincade to four months' imprisonment and two years' supervised release.

Back in prison, Kincade finally was forced to submit to DNA profiling. Nonetheless, he continued to challenge the constitutionality of the DNA Act in court on Fourth Amendment grounds. On October 2, 2003, a three-judge panel of the federal Ninth Circuit Court of Appeals upheld his challenge by a vote of 2-1 and reversed his conviction (345 F.3d 1095). The Justice Department, however, requested a rehearing by the full 11-member appeals court (en banc), and the request was granted. The court heard arguments in the case on March 23, 2004, and rendered its decision on August 18 of that year.

The Law and Biotechnology

Legal Issues

The Fourth Amendment to the U.S. Constitution affirms that

> [t]he right of the people to be secure in their persons, houses, papers, and effects, against unreasonable searches and seizures, shall not be violated, and no Warrants shall issue, but upon probable cause, supported by Oath or affirmation, and particularly describing the place to be searched, and the persons or things to be seized.

The question before the court was whether compulsory DNA profiling of certain conditionally released federal offenders, in the absence of grounds for suspecting that they had committed additional crimes, violated the Fourth Amendment. A number of court decisions had confirmed that collecting DNA samples from prisoners was constitutional, but the constitutionality of collection from convicts on supervised release had not been settled.

Decision

The appeals court voted 6-5 to vacate the three-judge panel's decision and uphold Kincade's resentencing—in other words, to find the DNA Act's requirement for DNA samples from convicts on supervised release constitutional. Judge Diarmuid F. O'Scannlain wrote the plurality's opinion. O'Scannlain stated that some courts had allowed warrantless searches if these searches fulfilled "special needs" unrelated to law enforcement, such as the need to supervise children in school. Several courts, he pointed out, had found the DNA Act constitutional because it met the special needs requirement. O'Scannlain, however, preferred an alternative approach taken by other courts, in which the reasonableness of the law's request for DNA was determined by evaluating the "totality of the circumstances" and balancing Kincade's needs for privacy and freedom from unreasonable search against the government needs that led to the demand for the DNA sample.

O'Scannlain cited cases to show that the courts considered convicts on parole, prohibition, or supervised release to have reduced rights, including reduced Fourth Amendment rights, as compared to ordinary citizens. Furthermore, he said, the bodily invasion involved in taking a blood sample was "not significant." The state, on the other hand, has an interest in supervising released convicts to make sure they fulfilled the conditions of their parole and did not commit further crimes, which they were more likely to do than nonconvicts.

Compared to Kincade's interests, O'Scannlain found the state's interests "undeniably compelling," indeed, "overwhelming." Those interests, he wrote, were more than great enough to permit warrantless searches, including demands for DNA samples, even when (as was true in Kincade's case) there was no reason to believe that a convict had committed a new crime. O'Scannlain therefore found that the DNA Act's demand for compulsory sample donation did not violate Kincade's Fourth Amendment rights. This court decision reversed the ruling of the three-judge panel and upheld the lower court's sentencing of Kincade to an additional prison term. Judge Ronald M. Gould filed a short concurring opinion, defending the DNA Act requirement as a "special needs" search.

Judge Stephen R. Reinhardt, known for his liberal rulings (and the author of the majority opinion in the October 2003 appeals court panel decision), filed a lengthy dissenting opinion, in which three of the other judges joined. (One of these judges, as well as the remaining judge who disagreed with the plurality's ruling, also filed shorter separate dissents.) Reinhardt felt that neither O'Scannlain's "totality of the circumstances" test nor the "special needs" exception, which does not cover suspicionless searches for standard law enforcement purposes, validated the DNA Act. He called O'Scannlain's standard "malleable and boundless" and said that the type of balance the plurality decision described would always tilt in favor of the government. He claimed that, whatever lower courts might have done, the Supreme Court had "never, ever, upheld a regime of suspicionless searches based on the government's desire to pursue ordinary law enforcement objectives."

Even more important in the long run, Reinhardt said, "the rationale employed by the plurality would set us on a dangerous path." He pointed out that CODIS, the FBI DNA database program, had already expanded considerably relative to its original coverage. In the years following passage of the DNA Act, for example, the rather limited number of federal crimes that the act named as qualifying for DNA collection had grown into a "laundry list" including forms of "civil disorder," interference with public officials, defacement of government property, and other activities that some people might consider legitimate protest. Reinhardt warned that the database was almost sure to expand further: Three states, for instance, were already collecting DNA from people merely arrested for qualifying crimes, whether or not they were convicted. (By 2007, the number of such states had risen to seven.) He feared that the range of groups from whom DNA was demanded would soon grow beyond those associated with crime in any way to encompass, for instance, people applying for driver's licenses or children attending public schools—indeed, ultimately, the entire U.S. population. (He noted that all members of the Armed Forces were already required to submit DNA samples, to aid in identifying the remains of those killed on duty.)

Reinhardt also pointed out that DNA profiles usually were kept in CODIS permanently. This fact, combined with the growing number of profiles, presented what he called a "catastrophic potential" for government misuse. In future, he said, the government might abuse this centralized source of information to "repress dissent or . . . to eliminate political opposition." Reinhardt was also concerned that, as science advanced, the supposedly "junk" DNA in the CODIS profiles would be found to contain revealing information about personal characteristics, such as susceptibility to disease or sexual preference. Thus, the dissenting judge wrote, the invasion of privacy involved in taking a DNA sample was potentially "substantial," amounting to much more than the minimal physical intrusion of a needle stick.

Finally, Reinhardt said that even if the state had a justifiable interest in invading the privacy of people like Kincade while they were on supervised release, that interest vanished when the supervised period ended and the convicts' debt to society had been fully paid. At that point, Reinhardt wrote, ex-convicts had just as much right to privacy as anyone else—yet their DNA would still be on file. "On balance," the judge concluded, "the government's desire to create a comprehensive DNA databank must give way when weighed against the privacy interests at issue and the extent of the intrusion involved." Following the reasoning of the plurality decision would result in "a fatally unwise and unconstitutional surrender to the government of our liberty for the sake of security." Reinhardt therefore held that the DNA Act was unconstitutional, and Kincade's claim that seizure of his DNA violated his Fourth Amendment rights was valid. "The erosion of conditional releasees' liberty," he wrote, "makes us all less free."

Impact

Replying to Judge Reinhardt's dissent in his main opinion, Judge O'Scannlain referred to Reinhardt's alarming pictures of the future, which included a reference to George Orwell's famous science fiction novel *1984*, as "Hollywood fantasies" and a "parade of horribles." The court was not called upon to rule on possible misuses of CODIS, O'Scannlain said, but only on the program "as currently structured and implemented." Nonetheless, commentators have pointed out, national forensic databases in the United States and some other countries, particularly Great Britain, have undeniably been expanding. Sooner or later, future courts will have to consider the constitutionality of this expanded coverage and new uses that may accompany it.

A report on the *Kincade* case in the December 2004 *Harvard Law Review* pointed out that, although this decision was by no means the first to support the constitutionality of the DNA Act, it expanded on earlier decisions by broadening the population subject to DNA testing and increasing the uses to which such testing could legally be put. This report also agreed with Judge Reinhardt that, although other courts had justified warrantless searches in which there were reasonable grounds for suspicion and had upheld certain types of suspicionless searches as justified by the "special needs" exception to the Fourth Amendment, judges had never supported a suspicionless search that was done strictly for law enforcement purposes—that is, detection of crime. Like Reinhardt, the Harvard writer criticized Judge O'Scannlain for having used the "totality of the circumstances" test rather than applying the much narrower "special needs" criterion. The writer said that the *Kincade* decision opened a dangerous gap in Fourth Amendment protection that the Supreme Court would be wise to close.

Rulings after *Kincade*, such as that by the U.S. Court of Appeals for the Third Circuit in *United States v. Sczubelek*, a 2005 case similar to Kincade's, have also supported the constitutionality of the DNA Act—but that decision also included a dissenting opinion that echoed many of the points Reinhardt had made. Courts surely will have to consider the constitutionality of collecting DNA for a national database again and again as the list of tested groups grows. A test case involving the constitutionality of collecting DNA from arrestees, for example, is almost sure to reach the courts—and quite possibly the Supreme Court—before the end of the 21st century's first decade.

Genetic Health Testing and Discrimination

BUCK V. BELL, 274 U.S. 200 (1927)

Background

In 1924, when she was 18 years old, Carrie Buck, a "feebleminded" (developmentally disabled) woman in the State Colony for Epileptics and Feeble Minded in Virginia, was ordered to be sexually sterilized under a newly passed state law that required such treatment for people living in state-supported institutions who were found to have hereditary forms of insanity or subnormal intelligence. Such sterilization was supposedly necessary to promote "the health of the patient and the welfare of society." Buck was deemed to be hereditarily feebleminded because her mother was of subnormal intelligence (she was confined in the same institution) and there were signs that Buck's illegitimate baby daughter was as well.

The Virginia law was typical of laws then existing, or soon to exist, in 34 states of the United States and several other countries, including Canada (some provinces), Britain, Germany, and the Scandinavian countries. These laws were based on the "scientific" doctrine of eugenics, which had been established in the late 19th century by British scientist Francis Galton and was widely accepted at the time of the Buck case. Galton and his followers believed that complex personality traits such as intelligence were inherited, and they claimed that the human race would be improved if groups such as the subnormally intelligent, habitual criminals, and the insane were prevented from reproducing—by force, if necessary. Such action, they said, would also save society considerable money by reducing the number of people who must be incarcerated and cared for at state expense.

Buck (or others acting on her behalf) sued the director of the institution to prevent her operation. The Virginia Supreme Court of Appeals supported the institution, but Buck's lawyers appealed, and the case came before the U.S. Supreme Court in 1927.

Legal Issues

Buck's suit alleged that she had been deprived of the right of due process of law guaranteed under the Fourteenth Amendment. She also claimed to have been denied equal protection of the laws because the Virginia law affected people inside institutions but not those outside. The underlying issue was whether the state had a right to forcibly prevent reproduction by people it deemed to suffer from inherited defects and therefore to be likely to produce undesirable offspring. "It seems to be contended that in no circumstances could such an order be justified," Supreme Court Justice Oliver Wendell Holmes noted in his majority opinion.

Decision

The Supreme Court ruled that the Virginia eugenics law did not violate either the due process clause or the equal protection clause of the Fourteenth Amendment, and it therefore denied Buck's right to sue. Justice Holmes wrote that Buck had been granted due process because the Virginia law contained "very careful provisions . . . [to] protect the patients from possible abuse," including requirement of a hearing attended by both the inmate and his or her guardian to determine whether the person was "the probable potential parent of socially inadequate offspring." "There can be no doubt that . . . the rights of the patient are most carefully considered" in this procedure, Holmes claimed, and all the steps of it had been followed with "scrupulous" care in Buck's case.

Holmes also wrote that Buck had not been denied equal protection, even though the law did not affect all citizens of subnormal intelligence equally. "It is the usual last resort of constitutional arguments to point out shortcomings of this sort," he complained. However, he said, "the law does all that is needed when it does all that it can, indicates a policy, applies it to all within the lines, and seeks to bring within the lines all similarly situated so far and so fast as its means allow."

Perhaps most important, Holmes defended the social as well as the legal validity of the eugenics law. He wrote:

We have seen more than once that the public welfare may call upon the best citizens for their lives. It would be strange if it could not call upon those who already sap the strength of the State for these lesser sacrifices, often not felt to be such by those concerned, in order to prevent our being swamped with incompetence. It is better for all the world if, instead of waiting to execute degenerate offspring for crime or to let them starve for their imbecility, society can prevent those who are manifestly unfit from continuing their kind. The principle that sustains compulsory vaccination is broad enough to cover cutting the Fallopian tubes. Three generations of imbeciles are enough.

Impact

The court's decision in *Buck v. Bell* not only upheld the constitutionality of at least some eugenics laws but demonstrated the prevalent thinking that found such laws both scientifically and morally justified. The *Buck* case was cited often in subsequent decisions about similar laws, such as *Skinner v. Oklahoma.* Ironically, however, Holmes and the others involved in the *Buck* case may have been wrong in the judgment that gave them jurisdiction over the family in the first place—the conclusion that the Bucks were "imbeciles." In a famous 1984 essay, scientist-writer Stephen Jay Gould reported that Carrie Buck had been reexamined in 1980 and was found to be of normal intelligence, and school records suggested that her daughter (who died in childhood) had been normal, too. The only "deficiencies" of the three generations of Bucks, Gould concluded, were that they were poor, uneducated, and violated contemporary sexual mores by giving birth to children out of wedlock.

SKINNER V. OKLAHOMA, 316 U.S. 535 (1942)

Background

Jack T. Skinner, the plaintiff in this case, had not led an exemplary life. He was convicted of stealing chickens in 1926, followed by convictions for robbery with firearms in 1929 and 1934. In 1935, after his second armed robbery conviction, he was sentenced to the state penitentiary. Oklahoma law considered all three of Skinner's crimes to be "felonies involving moral turpitude" and stated that anyone convicted of two or more such felonies and sentenced to prison was a "habitual criminal." As such, the Habitual Criminal Sterilization Act, a 1935 Oklahoma eugenics law based on the belief that criminal tendencies were inherited, made him subject to sexual sterilization.

Skinner sued to prevent the operation. A jury trial confirmed that he could be sterilized without endangering his health. When the Oklahoma Supreme Court supported this decision, Skinner appealed on constitutional grounds. His case came before the U.S. Supreme Court in May 1942 and was decided on June 1.

Legal Issues

Skinner, like Carrie Buck before him, claimed that he had been denied equal protection under the Fourteenth Amendment. He also said that the Oklahoma law violated the Eighth Amendment because sterilization was cruel and unusual punishment. Underlying the particulars of the suit, as in the *Buck* case, was the question of whether eugenics laws as a whole were constitutional. The suit alleged that "the act cannot be sustained as an exercise of the police power, in view of the state of scientific authorities respecting inheritability of criminal traits."

Decision

The Supreme Court ruled that Skinner had been denied equal protection because the Oklahoma law exempted embezzlers from the sterilization penalty but included those (like Skinner) who were convicted of grand larceny. The distinction between the two crimes was a very fine one, having to do with exactly when the convicted person had formed the intent of stealing. The state as a rule was entitled to make such fine distinctions, Justice William O. Douglas wrote in his majority opinion, but when a penalty as severe and permanent as sterilization was involved, they became much more dubious. Douglas explained:

Strict scrutiny of the classification which a state makes in a sterilization law is essential, lest unwittingly, or otherwise, invidious discriminations are made against groups or types of individuals in violation of the constitutional guaranty of just and equal laws. . . . When the law lays an unequal hand on those who have committed intrinsically the same quality of offense and sterilizes one and not the other, it has made as invidious a discrimination as if it had selected a particular race or nationality for oppressive treatment.

Douglas noted that there was no reason for assuming that a tendency to commit larceny was inheritable but a tendency to embezzle was not:

Oklahoma makes no attempt to say that he who commits larceny by trespass or trick or fraud has biologically inheritable traits which he who commits embezzlement lacks. . . . We have not the slightest basis for inferring that that line [between larceny and embezzlement] has any significance in eugenics.

In contrast to the decision in the *Buck* case, Chief Justice Harlan Fiske Stone concluded in a concurring opinion that Skinner had been denied due process because the Oklahoma law, unlike the Virginia one, did not provide for a hearing in which an individual could present evidence that he or she is not "the probable potential parent of socially undesirable offspring." Stone accepted that "science has found . . . that there are certain types of mental deficiency associated with delinquency which are inheritable" and affirmed the right of the state to "constitutionally interfere with the personal liberty of the individual to prevent the transmission by inheritance of his socially injurious tendencies." He insisted, however, that there was no proof that the traits of any entire legal category of criminals were inheritable, and individuals therefore had the right to a hearing to determine whether their particular "criminal tendencies are of an inheritable type." Skinner, he said, had been denied that right. "A law which condemns, without hearing, all the individuals of a class to so harsh a measure [as sterilization] . . . because some or even many merit condemnation, is lacking in the first principles of due process."

The most important difference between the *Buck* and *Skinner* cases lay in the court's comments about the underlying social issue of eugenics and the government's right to forcibly prevent certain people from reproducing. Justice Douglas wrote, "This case touches a sensitive and important area of human rights . . . a right which is basic to the perpetuation of a race—the right to have offspring." The case, he said, "raised grave and substantial constitutional questions."

We are dealing here with legislation which involves one of the basic civil rights of man. Marriage and procreation are fundamental to the very existence and survival of the race. The power to sterilize, if exercised, may have subtle, far reaching and devastating effects. In evil or reckless hands it can cause races or types which are inimical to the dominant group to wither and disappear. There is no redemption for the individual whom the law touches. Any experiment which the state conducts is to his irreparable injury. He is forever deprived of a basic liberty.

Justice Robert H. Jackson used equally strong words in a second concurring opinion.

I . . . think the present plan to sterilize the individual in pursuit of a eugenic plan to eliminate from the race characteristics that are only vaguely identified and which in our present state of knowledge are uncertain as to trans-

missibility presents . . . constitutional questions of gravity. . . . There are limits to the extent to which a legislatively represented majority may conduct biological experiments at the expense of the dignity and personality and natural powers of a minority—even those who have been guilty of what the majority define as crimes.

Impact

Even though the court found for the plaintiff in this case, *Skinner* did not reverse the effects of *Buck*. It did not declare eugenics laws to be unconstitutional, scientifically invalid, or morally reprehensible. It did, however, express the sort of doubts about such laws that many people were beginning to feel. Its claim that reproduction was a basic right would often be cited in later cases.

Eugenics laws remained on the books of many states and countries until the 1970s. After the 1940s, however, they were seldom enforced. A combination of better understanding of heredity, which suggested that complex personality traits such as intelligence and criminality were determined as much by environment as by genetics, and a revulsion for eugenics principles triggered by revelation of Nazi genocide following World War II helped to make such laws unpopular.

NORMAN-BLOODSAW V. LAWRENCE BERKELEY LABORATORY, 135 F.3D 1260 (9TH CIR. 1998)

Background

While examining her medical records in the process of applying for workers' compensation in January 1995, an employee of Lawrence Berkeley Laboratory (LBL), a California research facility managed by the University of California and the U.S. Department of Energy, discovered that blood and urine samples she had given during a preemployment medical examination had been tested in several ways without her knowledge. The same proved to be true of other LBL employees.

After receiving a letter from the Equal Employment Opportunity Commission saying that they had grounds for a suit, Marya S. Norman-Bloodsaw and six other administrative and clerical employees of LBL filed suit against the laboratory and others in September 1995. The suit alleged that the laboratory had tested employees' blood and urine for syphilis, sickle-cell trait (in the case of black employees), and pregnancy (in the case of women). It was filed on behalf of all present and past Lawrence employees who had been subjected to the tests in question.

The U.S. District Court for the Northern District of California dismissed all the employees' claims in June 1997. They appealed, and the case went to the Ninth Circuit Court of Appeals in February 1998.

Biotechnology and Genetic Engineering

Legal Issues

As Judge Stephen Reinhardt stated in the appeals court's written decision,

> *This appeal involves the question whether a clerical or administrative worker who undergoes a general employee health examination may, without his knowledge, be tested for highly private and sensitive medical and genetic information such as syphilis, sickle cell trait, and pregnancy.*

The LBL employees claimed that the medical tests in question had been administered without their knowledge or consent and that they were not notified later of the tests or their results. They said that their federal and state constitutional right to privacy had been violated by the conducting of the tests, the maintaining of the test results, and the lack of safeguards against disclosure of the results to others because of the "intimate" nature of the conditions tested. They claimed violations of Title VII of the Civil Rights Act of 1964 because only African-American employees had been tested for sickle-cell trait and only women had been tested for pregnancy; furthermore, they alleged, later blood samples from black and Hispanic, but not other, employees had been tested again for syphilis. Finally, the employees claimed violations under the Americans with Disabilities Act because the tests were not related to their job performance or business necessity. They did not claim that LBL had taken any negative action regarding their jobs because of the tests or that it had revealed the test information to others, but they said that the laboratory had provided no safeguards against dissemination of that information.

In addition to asking for damages for themselves, the employees were suing, according to the court record,

> *. . . to enjoin [forbid] future illegal testing, . . . to require defendants . . . to notify all employees who may have been tested illegally; to destroy the results of such illegal testing upon employee request; to describe any use to which the information was put, and any disclosures of the information that were made; and to submit Lawrence's medical department to "independent oversight and monitoring."*

The defendants denied that any of the employees' claims had merit. The tests, they said, represented only a minimal intrusion beyond that which the employees had consented to as part of taking the medical examination and giving blood and urine samples. They claimed that signs posted in examination rooms, furthermore, had announced the tests and that employees had been asked about some of the items tested on a questionnaire that they completed as part of their examination. The questionnaire asked if the employees had ever had medical conditions including sickle-cell anemia, venereal disease, or (in the case of women) menstrual disorders. They therefore should not have been surprised at being tested for such conditions, the defendants claimed. The defendants also said that the testing had occurred so long ago that the statute of limitations for complaints about it had expired and that, in any case, the laboratory had stopped doing the syphilis tests in 1993 (because such tests turned out to be an economically inefficient way of screening a healthy population), pregnancy tests in 1994, and sickle-cell tests in 1995 (because most blacks were by then tested for sickle-cell trait at birth).

Decision

The circuit court of appeals reversed the district court's ruling that the statute of limitations prevented the plaintiffs from suing. The time limit began to run, the court said, from the time when the plaintiffs learned about the tests—1995—not the time when the tests were made, as the district court had held. The circuit court said that the question of whether the plaintiffs knew or should have known that they were being tested would have to be settled at trial, but Judge Reinhardt maintained that the facts that the employees had consented to have a medical examination, give blood and urine samples, and answer written questions about certain medical conditions were "hardly sufficient" to establish an expectation of such testing. "There is a significant difference between answering [a questionnaire] on the basis of what you know about your health and consenting to let someone else investigate the most intimate aspects of your life," Reinhardt wrote. He also noted that "the record . . . contains considerable evidence that the manner in which the tests were performed was inconsistent with sound medical practice" in that the tests in question were not a routine or even an appropriate part of a standard occupational medical examination.

The appeals court upheld the district court's dismissal of the plaintiffs' claims under the Americans with Disabilities Act (ADA). First, Judge Reinhardt wrote, most of the testing at issue had occurred before January 26, 1992, the date on which the ADA began to apply to public entities. The employees tested after that date were tested as part of employee entrance examinations, which, unlike other examinations, are not required by the law to be limited to matters connected with a person's ability to perform job-related functions. The appeals court also disallowed claims under the ADA related to the way the employees' medical records were kept.

The appeals court supported the employees' right to sue on all the other grounds, however. It agreed that because of the tests' "highly sensitive" nature, they represented more than a minimal invasion of privacy beyond that involved in the medical examination that had been consented to. Judge Reinhardt wrote:

The constitutionally protected privacy interest in avoiding disclosure of personal matters clearly encompasses medical information and its confidentiality. . . . The most basic violation possible involves the performance of unauthorized tests—that is, the non-consensual retrieval of previously unrevealed medical information that may be unknown even to plaintiffs. These tests may also be viewed as searches in violation of Fourth Amendment rights. . . . The tests at issue . . . [also] implicate rights protected under . . . the Due Process Clause of the Fifth or Fourteenth Amendments. . . . One can think of few subject areas more personal and more likely to implicate privacy interests than that of one's health or genetic make-up.

The court ruled that discrimination in violation of Title VII of the Civil Rights Act of 1964 and the Pregnancy Discrimination Act was shown by the fact that certain tests were given to some employees but not to others as a condition of employment or were given more often to some employees during employment. The unauthorized obtaining of sensitive medical information on the basis of race or sex in itself constituted an "adverse effect" as defined by the act, even though no negative effects on employment occurred. The plaintiffs therefore had grounds to sue on this basis as well, the court decreed.

The fact that the tests had been discontinued did not make the plaintiffs' claims moot, the appeals court ruled, because the laboratory could reinstitute the tests at any time. Plaintiffs suffered ongoing injuries from the tests in that the test records were still in the employees' files and could potentially affect employment decisions or be given to others, even though this had not so far happened.

Impact

The decision confirmed employees' right to medical privacy and to not have tests, including tests for inherited conditions such as sickle-cell trait, run on them without their informed consent. In describing the decision, *U.S. News & World Report* writer Dana Hawkins said it represented "the first time a federal appeals court has recognized a constitutional right to genetic privacy."[7] The fact that Judge Reinhardt specifically mentioned genetic make-up in connection with privacy rights may prove particularly important to those concerned about privacy and discrimination related to genetic testing.

Partly because of this suit, Department of Energy contractors are now required to give employees a "clearly communicated" list of all medical examinations they will be expected to take, the purpose of the exams, and the results of the tests.

BRAGDON V. ABBOTT, 97 U.S. 156 (1998)

Background

In September 1994, Sidney Abbott visited her dentist, Randon Bragdon, in Bangor, Maine. In filling out a patient registration form, she indicated that she had been infected with HIV since 1986, although she had not yet developed any symptoms of AIDS. Bragdon examined Abbott, found that she had a cavity, and informed her that his policy was not to fill cavities of HIV-positive patients in his office. He then offered to do the work at a nearby hospital (where he felt he could take better precautions to protect himself from infection) if Abbott was willing to pay the extra cost of using the hospital's facilities. She declined.

Abbott sued Bragdon for violating her rights under Title III of the Americans with Disabilities Act (ADA) by not treating her, since places of "public accommodation" defined in that section include the "professional office of a health care provider." Bragdon's lawyer pointed out that the act stated that people could refuse to do something for an individual that was otherwise required by the act if they could show that "said individual poses a direct threat to the health and safety of others," and he claimed that Abbott fit that description. Abbott's attorney disagreed, pointing out that the Centers for Disease Control and Prevention (CDC) and others had written guidelines describing procedures by which dentists could treat people with HIV infection.

A district court ruled that Abbott's HIV infection satisfied the requirements of disability under the ADA and that Bragdon had not proved that treating her would put his health at risk. The Court of Appeals affirmed both of the district court's rulings. Bragdon then appealed to the U.S. Supreme Court, and the case came before the Court in June 1998.

Legal Issues

The Court agreed to rule on the following points: (1) whether Abbott, as an asymptomatic person with HIV infection, was disabled as defined by the ADA, and (2) whether sufficient evidence had been provided to show that Bragdon's health would have been endangered by treating her.

The underlying issue of the case for those concerned about genetic discrimination was whether a person who was presently healthy but likely to become ill later could be considered disabled under the ADA, since this description fitted healthy people whom tests revealed to have a genetic susceptibility to a disease. A factor likely to affect this issue was the section, or "prong," of the ADA under which Abbott claimed disability. The act defines disability as having to meet one of three criteria:

1. a physical or mental impairment that substantially limits one or more of the major life activities of an individual;
2. a record of such impairment; or
3. being regarded as having such impairment.

Abbott claimed disability under the first prong, saying that HIV infection limited her in the major life activity of reproduction and childbearing. By contrast, the Equal Employment Opportunity Commission (EEOC) had stated as policy in 1995 that healthy people with genetic predispositions were covered under the act's third prong.

Decision

In a 5-4 decision, the Supreme Court ruled that Abbott's HIV infection rendered her disabled according to the ADA's first criterion, that of limitation of a major life activity. Justice Anthony Kennedy in his majority opinion stated that reproduction was, without question, a major life activity and that Abbott was substantially limited in her pursuit of it, since the unprotected sex necessary to conceive a child would put her partner at significant risk of infection. Furthermore, if she did become pregnant, the child would also have a good chance of being infected. Kennedy also noted that, even though HIV infection did not produce obvious symptoms for years, it caused steady damage to the blood and immune systems and thus was "an impairment from the moment of infection."

In writing of the possible threat to Bragdon's health, Justice Kennedy noted that the ADA defined a direct threat to be "a significant risk to the health or safety of others that cannot be eliminated by a modification of policies, practices, or procedures or by the provision of auxiliary aids or services." The basic question, Kennedy wrote, was whether Bragdon's actions were reasonable in light of the medical evidence available to him at the time. Kennedy pointed out that, on the one hand, Bragdon had not produced medical evidence to show that he would have been any safer treating Abbott in a hospital than in his office. On the other hand, the CDC and other similar guidelines do not necessarily say that dentists will be safe while treating HIV-infected patients if they follow the recommended procedures. Some such guidelines do say that risk is minimal if proper precautions are followed, and medical testimony had been offered to this effect as well in the previous trials, but Kennedy noted that Bragdon may not have had this information at the time he treated Abbott. Kennedy ordered the Court of Appeals to reconsider the evidence supporting Bragdon's estimation of his health risk.

Impact

According to the summary of a February 1999 workshop held jointly by the National Human Genome Research Institute and the Hereditary Susceptibility Group of the National Action Plan for Breast Cancer to discuss the implications of the *Bragdon v. Abbott* decision for healthy people diagnosed with a genetic susceptibility to breast cancer (and, presumably, other gene-related illnesses), the Court's ruling "both excited and unnerved" those who hoped that the ADA could be interpreted in a way that would cover such people.[8] Chief Justice William Rehnquist, in fact, addressed this possibility when he wrote in his dissenting opinion that Abbott's argument, "taken to its logical extreme, would render every individual with a genetic marker for some debilitating disease 'disabled' here and now because of some possible future effects." On one hand, Rehnquist's words indicate that such reasoning is possible; on the other, he obviously was expressing disapproval of the idea.

In the workshop, law expert Paul Miller pointed out several hopeful signs in the *Bragdon* decision. The fact that the Court considered HIV infection an actual disability from the beginning because of its effects on immune cells, even though no obvious symptoms of illness appeared, suggested that genetic tendencies, which may also cause physical or chemical changes in cells without producing visible symptoms, might be similarly classified. Furthermore, since reproduction had been affirmed as being a major life activity covered by the ADA, people with inherited defects could argue, as Abbott did, that their ability to reproduce was limited because they risked passing their condition on to their offspring and thus endangering those offspring's health. Third, Miller said, the court had relied heavily on an EEOC policy ruling in determining that asymptomatic HIV infection qualified as a disability under the ADA. This added weight to other EEOC rulings, including the one about genetic predisposition.

On the other hand, it was not clear how broad the effect of the Court's ruling would prove to be. The ruling did not explicitly consider any life activity other than reproduction, for instance, so it might not cover, say, a postmenopausal woman or a gay man with asymptomatic HIV infection. More important, since Abbott claimed disability under the first prong of the ADA's definition, the Court decision did not illuminate the question of who would be included under the third prong, which many commentators feel is the one most likely to cover healthy people with genetic predispositions. Miller noted that many legislators are not supportive of the third prong, and a later speaker, Sharon Masling, said that the courts also have been interpreting it narrowly.

SUTTON V. UNITED AIR LINES, INC., 527 U.S. 471 (1999)

Background

In 1992, Karen Sutton and Kimberly Hinton, twin sisters, applied to United Air Lines for employment as global commercial airline pilots. They met all of the airline's criteria for this job except the minimum vision requirement, which called for uncorrected visual acuity of 20/100 or better. Both sisters were severely myopic (nearsighted), with uncorrected vision of 20/200 in their better eyes and 20/400 in their poorer ones. This degree of impairment, if uncorrected, would keep them from driving, shopping, and carrying out other everyday activities. With corrective lenses, however, both had vision of 20/20 or better. The airline refused to hire the women because they did not meet the minimum requirement for uncorrected vision. The sisters sued United, claiming that the airline had discriminated against them in a way that violated the Americans with Disabilities Act (ADA).

Legal Issues

The sisters stated that they had a disability that "substantially limits a major life activity," as required by the ADA, because of the limits that their uncorrected vision would place on working and other activities. They cited guidelines issued by the Equal Employment Opportunity Commission (EEOC) in support of their claim that their condition should be evaluated as a disability under ADA in its uncorrected rather than its corrected state. They also argued that United "regarded" them as having such a disability, whether they actually did or not, which would entitle them to protection under the third prong of the ADA.

United claimed that the women did not have a disability because glasses could easily restore their vision to normal levels. It also stated that the company did not regard them as being substantially limited in the activity of working because it did not exclude them from a large class of jobs, only from the specific job of global airline pilot.

This case, like *Bragdon v. Abbott*, had indirect implications for the question of whether currently healthy people diagnosed with a genetic susceptibility to disease are protected by the ADA. If the courts accepted the argument that the sisters' disability should be measured in terms of their uncorrected vision, this would suggest that healthy people with genetic susceptibilities might be protected by the ADA because they were likely to be disabled in the future, even though they were not so in the present, just as the sisters might be disabled without glasses even though their vision was normal when they wore corrective lenses. The question of whether being excluded from a specific job was sufficient to demonstrate that an employer regarded someone as disabled could also relate to employer decisions about whether to hire or retain people with flaws in their genetic makeup.

Decision

A district court heard the sisters' suit in 1996 and dismissed it, saying that they had failed to state a claim upon which relief could be granted. The court held that, because the women's vision could be corrected to normal, they were not actually disabled. It also ruled that, because United had denied them access only to one specific type of job, it had not demonstrated that it regarded them as substantially limited overall in the life activity of working. The 10th Circuit Court of Appeals affirmed the district court's judgment in 1997. The sisters' lawyers then appealed to the U.S. Supreme Court, which agreed to hear the case because the lower courts' decisions conflicted with other decisions holding that disabilities should be evaluated without consideration of whether they could be or were corrected.

The Supreme Court issued its ruling on June 22, 1999, with Justice Sandra Day O'Connor writing the majority opinion. The high court affirmed the ruling of the lower courts. O'Connor concluded, "we think the language [of the ADA] is properly read as requiring that a person be presently—not potentially or hypothetically—substantially limited [in a major life activity] in order to demonstrate a disability." By this standard, she said, the sisters were not disabled. She cited, for instance, a statistic on the number of disabled Americans included in the text of the ADA, saying that the figure would have been much larger if Congress had intended to include people with correctable as well as uncorrectable disabilities.

O'Connor also held that the sisters failed to qualify under the "regarded as" prong of the ADA because the position of global airline pilot did not represent a class of jobs. "The inability to perform a single, particular job does not constitute a substantial limitation in the major life activity of working," she wrote, and the fact that the airline refused to consider the sisters for this specific job therefore was not sufficient evidence that it regarded their (uncorrected) poor eyesight as substantially limiting their general ability to work. The women might, for example, have qualified as copilots, regional pilots, or pilot trainers. O'Connor pointed out that "the ADA allows employers to prefer some physical attributes over others and to . . . decide that some limiting, but not *substantially* limiting, impairments make individuals less than ideally suited for a job."

Justices John Paul Stevens and Stephen Breyer dissented from the majority opinion. They held that Congress had intended the ADA's coverage to be broad and, therefore, that the possibility of correction should be disregarded when determining whether a person is disabled as defined by the law. To Stevens, "it [was] quite clear that the threshold question whether an individual is 'disabled' within the meaning of the Act . . . focuses on her past or present physical condition without regard to mitigation." He claimed that eight of nine federal courts of appeals who had addressed the issue, all three of the executive agencies that had issued regulations or guidelines concerning the ADA, and members of congressional committees who had written reports during preparation of the ADA had taken this view. He also made the point that "if United regards petitioners [the sisters] as unqualified [to be commercial pilots] because they cannot see well without glasses, it seems eminently fair for a court also to use uncorrected vision as the basis for evaluating petitioners' life activity of seeing."

Impact

Unlike the decision in *Bragdon v. Abbott*, the decision in *Sutton v. United Air Lines* suggests that the ADA does not cover people who are presently healthy but have been shown by tests to have a genetic susceptibility that might cause them to become sick or disabled at a later time. The only exception might occur if such people could demonstrate that an employer had excluded them from an entire class of jobs or from any kind of employment because of their genetic predisposition.

---

[1] "Recombinant DNA Advisory Committee: Recombinant DNA and Gene Transfer," NIH fact sheet, p. 1. Available online. URL: http://www4.od.nih.gov/oba/rac/aboutrdagt.htm. Updated March 29, 2000.

[2] "Recombinant DNA Advisory Committee," p. 2.

[3] Mitchel Zoler, quoted in Martin Kenney, *Biotechnology: The University-Industrial Complex* (New Haven: Yale University Press, 1986), p. 256.

[4] Donald Dunner, quoted in Kenney, p. 256.

[5] Peter Farley, quoted in Edward J. Sylvester and Lynn C. Klotz, *The Gene Age: Genetic Engineering and the Next Industrial Revolution* (New York: Scribner, 1983), p. 118.

[6] Office of Technology Assessment (1987), quoted in *John Moore v. Regents of California*.

[7] Dana Hawkins, "Court Declares Right to Genetic Privacy," *U.S. News & World Report*, February 16, 1998, p. 4.

[8] National Action Plan on Breast Cancer, "*Bragdon v. Abbott*: Indications for Asymptomatic Conditions." Available online. URL: http://www.4women.gov/napbc/catalog.wci/napbc/heredita1.htm. Posted February 1999.

CHAPTER 3

CHRONOLOGY

This chapter presents a chronology of important events relevant to biotechnology, primarily the subset of biotechnology that involves genetic engineering. It also includes key events in the history of genetics, since genetic engineering would have been impossible without the knowledge gained from basic genetic research. It focuses primarily on the period following the invention of genetic engineering in the early 1970s and on events related to the ethical, legal, and social implications of biotechnology, genetic engineering, and human genetics.

ABOUT 10,000 YEARS AGO

- Biotechnology begins along with agriculture. It includes domestication and deliberate breeding of plants and animals, as well as (unknowing) use of microbial processes to make bread, cheese, alcoholic drinks, leather, and other products.

1665

- British scientist Robert Hooke discovers microscopic square bodies in a slice of cork and terms them cells.

1793

- U.S. Congress defines a patentable invention as "any new and useful art, machine, manufacture or composition of matter." Products of nature are not included.

1839

- German biologists Matthias Schleiden and Theodor Schwann propose that cells are the units of which all living things are made.

Biotechnology and Genetic Engineering

LATE 1850s–1860s

- Famed French chemist Louis Pasteur provides a scientific basis for part of traditional biotechnology when he shows that fermentation processes such as those used for millennia to produce wine, beer, buttermilk, and cheese depend on living microorganisms.

1859

- British biologist Charles Darwin publishes *On the Origin of Species*, in which he sets forth the theory of evolution by natural selection. The theory states that the members of a species with inherited characteristics that make them most suited to survive in a particular environment are the most likely to survive and bear young in that environment. Over generations, therefore, a change in the environment can produce a change in the predominant characteristics of a species.

1866

- Gregor Mendel, an Austrian monk, publishes a paper describing the basic mathematical rules by which characteristics are inherited. He had worked these out by breeding pea plants in his monastery garden.

1875

- German scientist Walther Flemming discovers that the nuclei (central bodies) of cells contain stringlike bodies that can be stained with dye; these are soon termed chromosomes, or "colored bodies."

1883

- British scientist Francis Galton coins the term *eugenics* (from Greek roots meaning "well born") in a book called *Inquiry into Human Faculty*; he and his followers believe that the human race can be improved by encouraging those with desirable traits to have children and discouraging or preventing those with undesirable traits from doing so.

1900

- Several scientists rediscover Mendel's work, which until this time has been virtually unknown; it now becomes the foundation of genetics.

1910

- American geneticist Thomas Hunt Morgan and coworkers at Columbia University prove that genes are located on chromosomes.

Chronology

1920s–1930s

- Thirty-four states in the United States pass eugenics laws requiring people with what are thought to be inherited defects (chiefly criminals, the insane, and the "feebleminded") to be forcibly sterilized.

1927

- In *Buck v. Bell*, the U.S. Supreme Court by an 8-1 vote upholds a Virginia eugenics law under which Carrie Buck, a woman thought to be developmentally disabled, was forcibly sterilized. Noting that Buck's mother and seven-month-old daughter both appear to be "feebleminded" as well, Justice Oliver Wendell Holmes writes, "Three generations of imbeciles are enough."

1930

- The U.S. Plant Patent Act allows breeders to protect plant varieties they have developed by refusing to allow others to reproduce such plants asexually.

1933

- The German government, recently seized by the Nazi Party, passes a eugenics law modeled on those of the United States.

1938

- U.S. Congress passes the Federal Food, Drug, and Cosmetic Act (FFDCA), which gives the Food and Drug Administration (FDA) the right to set tolerances for certain substances, including pesticides, on or in food and feed. Today the FDA shares this regulatory duty with the Environmental Protection Agency (EPA). Herbicide residues on genetically engineered herbicide-tolerant crops and pesticides (such as *Bacillus thuringensis* [Bt] toxin) produced in genetically altered food crops are regulated under the FFDCA.

1941

- George Beadle and Edward L. Tatum of Stanford University show that a single (structural) gene carries the instructions for making a single protein (enzyme).

1942

- In *Skinner v. Oklahoma*, the U.S. Supreme Court strikes down a state law requiring forced sterilization of convicted criminals, saying that procreation is "one of the basic civil rights of man."

Biotechnology and Genetic Engineering

1944

■ Oswald Avery and colleagues at the Rockefeller Institute in New York publish a paper demonstrating that DNA, not protein, carries inherited information.

1947

■ U.S. Congress passes the Federal Insecticide, Fungicide, and Rodenticide Act (FIFRA), which regulates the distribution, sale, use, and testing of chemical and biological pesticides. FIFRA has been amended to give the Environmental Protection Agency (EPA) control of these regulations, which cover plants and microorganisms that produce pesticidal substances, such as farm crops genetically modified to make *Bacillus thuringensis* (Bt) toxin.

1952

■ Scientists clone frogs by transplanting the nucleus of a frog cell into an unfertilized egg cell with the nucleus removed, a technique called nuclear transfer.
■ U.S. Congress revises patent law to state that patentable inventions must be novel, useful, and not obvious "to a person of ordinary skill in the art."

1953

■ *April 25:* James D. Watson and Francis Crick publish a paper in *Nature* in which they describe the structure of the DNA molecule.
■ *May:* Watson and Crick publish a second paper offering a theory that explains how DNA's structure allows the molecule to reproduce itself.

1958

■ Watson and Crick's theory of how the DNA molecule reproduces itself is confirmed.

1961

■ Francis Crick suggests that each "letter" of the genetic code—the unit specifying one amino acid in a protein—consists of three adjoining bases in a DNA molecule.

1961–1966

■ Molecular biologists decipher the genetic code, determining which amino acid each of the 64 possible combinations of three bases represents. They also work out the basic process by which the cell makes the proteins specified by the code.

1970

- U.S. Congress passes the Plant Variety Protection Act, which allows breeders to protect new plant varieties by forbidding others to reproduce the plants sexually.

1971

- Robert Pollack warns Paul Berg about the possible dangers of inserting genes from a cancer-causing virus into bacteria capable of infecting humans.

1972

- *November:* Herbert Boyer and Stanley Cohen meet in a Hawaiian delicatessen and begin planning recombinant DNA technology.

1972–1973

- Paul Berg, Herbert Boyer, and Stanley Cohen perform the first experiments in which pieces of DNA from one species are inserted into the genome of another species (recombinant DNA).

1973–1974

- Leading scientists in the field write two letters, published in *Science*, that warn of possible dangers of recombinant DNA experiments and call for a moratorium on some types of experiments.

1974

- Herbert Boyer and Stanley Cohen apply for a patent on their gene-splicing technique and assign all potential royalties from it to their respective universities (University of California–San Francisco and Stanford University).
- *October:* The National Institutes of Health establishes the Recombinant DNA Advisory Committee (RAC) to review the safety of recombinant DNA experiments.

1975

- North Carolina passes a law forbidding employers to discriminate against people with sickle-cell trait. This is the first American law to address genetic discrimination in the workplace.

- *February 24–27:* One hundred forty geneticists and molecular biologists meet at Asilomar, California, to draw up safety guidelines for recombinant DNA experiments.
- *April:* Congress holds its first hearings about safety of recombinant DNA research.

1976

- Robert Swanson and Herbert Boyer found Genentech, the first biotechnology company based on recombinant DNA technology.
- U.S. Congress passes the Toxic Substances Control Act (TSCA), which gives the Environmental Protection Agency (EPA) the authority to review new chemicals before they are introduced into commerce. TSCA is later amended to classify some genetically engineered organisms as "new chemicals" to be regulated under the act.
- *June:* National Institutes of Health (NIH) publishes safety guidelines for recombinant DNA experiments, based on the Asilomar guidelines. Britain and Europe adopt similar guidelines.

1977

- Sixteen bills regulating recombinant DNA research are introduced into Congress; none of them passes.

1978

- NIH guidelines are relaxed, reducing containment requirements for many recombinant DNA experiments.
- Genentech makes human insulin in genetically engineered bacteria.

1980

- NIH safety guidelines for recombinant DNA research are relaxed further.
- *June 16:* In *Diamond v. Chakrabarty*, a landmark case, the U.S. Supreme Court declares that living things can be patented if humans have altered them.
- *October:* Genentech stock is offered to the public for the first time and jumps from $35 to $89 a share in the first few minutes of trading, despite the fact that the company has not yet sold any products.

EARLY 1980s

- First transgenic plants and animals produced.

Chronology

1982

■ The FDA grants Genentech permission to sell genetically engineered human insulin.

1983

■ Kary Mullis discovers the polymerase chain reaction (PCR), which can be used to duplicate small amounts of DNA many times in a few hours.

1985

■ The first transgenic farm animals are created.
■ Alec Jeffreys of the University of Leicester, Great Britain, publishes a paper in *Nature* describing what he calls "DNA 'fingerprinting,'" a way to use genetic testing to identify individuals.
■ *October:* The first U.S. patent for a genetically altered plant is issued.

1986

■ First mammals (sheep) are cloned by nuclear transfer technique, using embryonic cells.
■ The Toxic Substances Control Act (TSCA) is amended to require an Environmental Protection Agency (EPA) permit for releasing genetically altered organisms into the environment.
■ *June:* U.S. Office of Science and Technology Policy issues framework for regulation of biotechnology, dividing such regulation among five government agencies.

1987

■ U.S. Congress passes the Plant Pest Act, which gives the U.S. Department of Agriculture's Animal and Plant Health Inspection Service (APHIS) the right to regulate any organisms, including those produced by genetic engineering, that are or might be plant pests.
■ Colin Pitchfork in Britain and Tommie Lee Andrews in the United States become the first people convicted of crimes on the basis of DNA identification testing.
■ *April:* The U.S. Patent and Trademark Office extends patentability to animals, cell lines, and genes, including human cells and genes.
■ *April 24:* Steven Lindow and coworkers oversee spraying of genetically altered bacteria onto strawberry plants in a field in Contra Costa County, California. This is the first deliberate release of genetically engineered organisms into the environment.

1988

- *April:* The Harvard Oncomouse, a mouse genetically altered to make it unusually susceptible to cancer and thus useful in cancer research and carcinogen testing, becomes the first genetically engineered animal to be patented.

1989

- Virginia opens the first state database of DNA identification profiles taken from convicted criminals.
- *May 22:* Steven Rosenberg and coworkers at the National Institutes of Health carry out the first successful insertion of foreign genes into a human being. The genes are intended only as markers and have no effect on health.

1990

- The Human Genome Project, which aims to sequence all human genes by a few years into the 21st century, begins.
- U.S. Congress passes the Americans with Disabilities Act (ADA), which might ban discrimination against healthy people with inherited defects revealed by genetic tests.
- *July:* The California Supreme Court rules in *John Moore v. Regents of University of California* that people do not retain ownership of their cells or tissues once these have been removed from their bodies.
- *September 14:* W. French Anderson and colleagues from the NIH administer the first successful gene therapy to four-year-old Ashanthi deSilva.

1991

- The FBI establishes guidelines for forensic DNA testing.
- James Watson, head of the Human Genome Project, earmarks 3 percent of the project's budget for investigation of its ethical, legal, and social implications.
- New York lawyers Barry Scheck and Peter Neufeld establish the Innocence Project, which seeks to obtain DNA testing for convicted criminals who say that they are innocent and that such testing could clear them.

1992

- U.S. Department of Defense begins requiring all military personnel to give DNA samples, which will be retained in a databank to help identify remains of soldiers killed in action.

- *May:* The FDA states that genetically altered foods do not have to receive special approval or be labeled as such if they are nutritionally the same as their natural equivalents and contain no new substances that might cause an allergic reaction.

1993

- The FDA approves recombinant bovine growth hormone (rBGH) for sale; it is the first genetically engineered animal hormone approved for sale in the United States.
- Kirk Noble Bloodsworth, falsely accused of the sex murder of a nine-year-old girl in Maryland in 1984, convicted, and sentenced to death, is cleared of the crime by DNA testing. He is the first death-row inmate to be exonerated by DNA evidence.
- *March:* The USDA streamlines its requirements for testing new varieties of genetically engineered corn, cotton, soybeans, potatoes, tomatoes, and tobacco to eliminate a former requirement for an environmental review before testing.
- *October:* Robert Stillman and Jerry Hall of George Washington University Medical Center in Washington, D.C., clone early-stage human embryos. The embryos, already due to be discarded by a fertility clinic, are not allowed to develop.

1994

- The Flavr Savr tomato becomes the first genetically engineered food to go on sale.
- The Equal Employment Opportunity Commission (EEOC) rules that denying employment to people who are healthy but have inherited defects revealed by genetic tests violates the Americans with Disabilities Act.
- The Environmental Protection Agency (EPA) decides to regulate plants genetically engineered to produce their own pesticides as if the plants were chemical pesticides.
- The USDA and W. R. Grace, a multinational corporation, obtain a patent for using the oil of the neem tree, native to India, against fungi on plants. People in India have used the oil in this way for millennia, and activists call the patent an example of biopiracy.
- *September 13:* Congress passes the DNA Identification Act, which gives the FBI national responsibility for training, funding, and proficiency testing of laboratories doing forensic DNA profiling and authorizes it to establish a national database of DNA information from crime scenes and convicted criminals.

Biotechnology and Genetic Engineering

1995

- The World Trade Organization's (WTO) Trade Related Intellectual Property Rights Agreement (TRIPS) goes into effect. This agreement requires WTO member countries to "harmonize" their patent laws with those of the United States and other industrialized countries, supposedly in order to facilitate the transfer of technology to those countries. Critics say its real purpose is to protect the rights and profits of multinational corporations.
- Britain establishes the Criminal Justice DNA Database, the world's first national database of DNA identity profiles of convicted criminals. It covers all crimes that result in imprisonment.
- *October:* Former football star O. J. Simpson is acquitted of the murder of his wife and an acquaintance, Ronald Goldman. The verdict comes despite DNA evidence linking Simpson to the crimes, after his defense lawyers show that samples from the crime scene were mishandled and may have been accidentally or deliberately contaminated with Simpson's blood before testing.

1996

- Genetically altered crops, including those that resist herbicides and those that produce their own insecticide, are sold commercially and planted on a large scale in the United States for the first time.
- *April:* U.S. Marines John C. Mayfield III and Joseph Vlakovsky are court-martialed and found guilty of disobeying a direct order because they refused to give DNA samples for storage in a military database, saying that doing so invaded their genetic privacy.
- *August 21:* U.S. Congress passes the Health Insurance Portability and Accountability Act, which includes a provision forbidding health insurers issuing group plans to deny coverage to healthy people because of preexisting genetic conditions.

1997

- The British government's Human Genetics Commission recommends a two-year moratorium on disclosure of genetic testing results to insurers. The moratorium is later extended for another five years.
- *January:* Deputy Attorney General Jamie Gorelick publicly admits that the FBI crime laboratory has a "serious set of problems" in its handling and testing of forensic DNA samples.
- *February 27:* Ian Wilmut and coworkers at the Roslin Institute in Scotland publish an article in *Nature* announcing that they have cloned a sheep (Dolly) from a mature udder cell taken from a six-year-old (adult) ewe.

Chronology

- ■ *March 4:* Reacting to fears stirred by the news about Dolly, President Bill Clinton announces a ban on the use of federal funds for research on cloning of human beings.
- ■ *April 11:* U.S. Congress amends the Toxic Substances Control Act (TSCA), tailoring the general screening program for microbial products of biotechnology to meet the special requirements of microorganisms used commercially for such purposes as production of industrial chemicals and breakdown of chemical pollutants in the environment (bioremediation).
- ■ *November:* The European Union begins requiring labeling of all genetically modified foods.
- ■ *November:* The General Conference of the United Nations Educational, Scientific and Cultural Organization (UNESCO) adopts the Universal Declaration on the Human Genome and Human Rights, which asserts, among other things, that "the human genome in its natural state shall not give rise to financial gains" and affirms individuals' rights to have genetic tests or research performed on them only with consent, to be free of discrimination based on genetics, and to have their genetic information kept private.

1998

- ■ The European Union places an informal moratorium on approval of new genetically modified (GM) crops and importation of any food products that might contain unapproved GM materials.
- ■ *January:* California becomes the first state to outlaw human reproductive cloning.
- ■ *February:* In *Norman-Bloodsaw v. Lawrence Berkeley Laboratory*, the U.S. Court of Appeals for the Ninth Circuit rules that Lawrence Berkeley Laboratory's genetic and other testing of employees' blood and urine samples without their informed consent was unconstitutional.
- ■ *June:* In *Bragdon v. Abbott*, the Supreme Court rules that the Americans with Disabilities Act (ADA) protects a healthy woman who is HIV positive, leading to speculation that it may also apply to healthy people with genetic problems revealed by testing.
- ■ *July 6:* After a decade of heated discussion and intense lobbying, the European Parliament and Council approve Directive 98/44, on the legal protection of biotechnological inventions. It permits patenting of plants but, on ethical grounds, bars patents on processes for cloning or modifying the genetic identity of human beings, use of human embryos for industrial purposes, and processes for modifying the genetic identity of animals that may cause them suffering unless such processes are likely to produce substantial medical benefits for humans. It allows patenting of

gene sequences only for particular industrial applications. It states that member countries may reject patents that they feel violate moral or ethical standards.

- *August:* Monsanto Corporation sues Canadian farmer Percy Schmeiser for patent infringement because he planted genetically altered ("Roundup Ready") canola without a license, even though the herbicide-resistant genes in Schmeiser's seeds apparently got there by accident.
- *October:* The FBI opens its National DNA Index System (NDIS), a national database that will store information from DNA samples taken from convicted murderers and sex offenders by the states, and CODIS, a program to coordinate information from state and national DNA databases.
- *November:* Geron Corp. and James A. Thomson of the University of Wisconsin announce that they have isolated and cultivated human embryonic stem cells, which potentially can provide tissues for transplantation. Early-stage embryos from fertility clinics are destroyed in the process of extracting the cells.

1999

- Ingo Potrykus, Peter Beyer, and other researchers create "golden rice," which contains added daffodil and bacterial genes that make it produce beta-carotene, a building block of vitamin A. Biotechnology supporters tout this genetically engineered crop as a cure or preventive for vitamin A deficiency, which affects up to 140 million children each year and makes about 500,000 of them go blind.
- *May:* A Cornell University study suggests that pollen from corn crops genetically altered to produce an insecticide may land on milkweed plants and poison monarch butterfly larvae.
- *June 22:* In a Supreme Court decision on *Sutton v. United Air Lines*, Justice Sandra Day O'Connor writes that the ADA protects only those who are presently disabled, dimming hopes that this law can be applied to healthy people with genetic predispositions to disease.
- *September 17:* Jesse Gelsinger, an 18-year-old Arizona man with a rare inherited disease, dies after taking part in a gene therapy experiment. His is the first death officially attributed to gene therapy.

2000

- The U.S. Department of Energy begins the Genomes to Life (later called Genomics:GTL) program, which will analyze the genomes of microorganisms in the hope of finding ways to use them in developing alternative energy sources.

Chronology

- *January 21:* Reporting on inspections conducted after the death of Jesse Gelsinger, the FDA sharply criticizes procedures in his and other gene therapy trials and suspends some of them.
- *January 24:* Congress passes the DNA Analysis Backlog Elimination Act, usually referred to as the DNA Act. In addition to authorizing grants to help states reduce backlogs of forensic DNA samples awaiting testing, this law authorizes collection of DNA samples from people convicted of "qualifying federal offenses," including murder and sexual abuse, who are on supervised release, parole, or probation.
- *January 29:* The compromise Cartagena Biosafety Protocol, which urges a "precautionary approach" to genetically modified crops, is completed and signed by representatives of more than 130 countries in Montreal, Quebec.
- *February:* Massachusetts-based Aqua Bounty Farms applies for federal approval for the first genetically modified animal intended for human consumption, a transgenic salmon that grows twice as fast as normal salmon.
- *February:* President Bill Clinton signs Executive Order 13145, which prohibits genetic discrimination in federal employment.
- *February:* Nearly 250 concerned environmentalists and other leaders sign an open letter warning that human germ-line genetic engineering "is an unneeded technology that poses horrific risks" and urging that it be banned.
- *April:* French scientists report that gene therapy has apparently cured several children with severe combined immune deficiency.
- *May 10:* The European Patent Office revokes the USDA–W. R. Grace patent on using neem oil against fungi on the grounds that it lacks novelty and inventiveness.
- *June 26:* Teams of scientists from the private company Celera Genomics and the government-funded National Human Genome Research Institute announce that they have finished a rough draft of the code of the complete human genome.
- *August 29:* A woman in Denver, Colorado, gives birth to a boy who was chosen by preimplantation genetic diagnosis not only to be free of an inherited disease that affected his older sister but to be of the same tissue type as the sister so that a stem cell transplant from him could cure her. Some people criticize this move because it seems too close to conceiving a child simply to produce "spare parts."
- *September 18:* The environmental group Friends of the Earth announces that Taco Bell taco shells contain traces of a type of genetically modified corn called StarLink, which had been approved for animal feed but not human food. StarLink is eventually found to have contaminated more than 430 million bushels of corn, and more than 300 types of corn-containing food products have to be recalled.

Biotechnology and Genetic Engineering

2001

- A fertility clinic announces that about 30 "normal, healthy" babies worldwide have been created by a technique that combines the nucleus of an egg from one woman with cytoplasm from the egg of another and sperm from a man, resulting in children whose cells, including their germ cells, contain DNA from all three people. This technique, done to avoid diseases caused by defective mitochondrial DNA in the nucleus donor, in effect alters germ-line genes.

- *January:* Scientists announce creation of the first genetically modified primate, a rhesus macaque monkey named ANDi (short for "inserted DNA" backward), whose cells contain a jellyfish gene used as a marker.

- *January:* Australian researchers say they accidentally found a way to modify the genes of a mouse virus that made it much more deadly, adding to fears that genetic engineering could be used to make exceptionally dangerous bioterrorism weapons.

- *January 5:* The U.S. Patent and Trademark Office issues final revised guidelines on the utility requirements for patenting human genetic sequences, affirming the "specific, substantial, and credible" standard it had established in 1999 but reemphasizing that an unaltered human gene could be patented as long as it had been isolated and purified.

- *January 22:* Britain legalizes cloning of human embryos for stem cell research.

- *February 9:* The federal Equal Employment Opportunity Commission (EEOC) files suit against the Burlington Northern Santa Fe Railroad, claiming that it violated the Americans with Disabilities Act (ADA) by making secret genetic tests of certain employees.

- *April 16:* Federal district court judge Albert V. Bryan, Jr., rules that denying convict James Harvey a chance to have DNA samples used in his trial retested, a move that Harvey claims would prove his innocence, violates the right to due process of law guaranteed by the Fourteenth and Fifteenth Amendments.

- *April 18:* The Burlington Northern Santa Fe Railroad agrees to an out-of-court settlement of the EEOC suit, in which it admits no wrongdoing but promises to stop all genetic tests of workers. It later also agrees to pay the tested employees $2.2 million.

- *August 9:* U.S. President George W. Bush announces that henceforth, federal funding can be used for research only on embryonic stem cell lines that already exist on this date. The ruling does not affect privately funded research.

- *fall:* New studies suggest that milkweed plants on which monarch butterfly caterpillars feed are unlikely to receive enough pollen from genetically modified, insecticide-containing corn to endanger the caterpillars.

- *fall:* Kenyan farmers harvest the first crop of yams, a food widely eaten in Africa, that have been genetically engineered to resist a virus that normally destroys up to 80 percent of the crop.
- *October 9:* The European Court of Justice rejects the claim of the Netherlands, Italy, and Norway that living things and human biological material should not be patentable and reaffirms the demand in European Union (EU) Directive 98/44 that the patent laws of all EU member countries be "harmonized."
- *November 29:* Ignacio Chapela and David Quist of the University of California–Berkeley publish an article in *Nature* claiming that they have found genes from bioengineered corn in Mexican corn crops, even though that country does not allow the planting of genetically engineered corn. This finding suggests that gene flow, which could lead to environmental damage, is taking place.
- *December:* CC (for Carbon Copy), the first cloned pet (a cat), is born at Texas A&M University. Even though she is genetically identical to her "mother," her fur coloration is quite different.

2002

- The British government's Human Fertilisation and Embryology Authority allows a couple to use preimplantation genetic diagnosis to select an embryo with a tissue type compatible with that of a sick older sibling because the embryo also needs to be selected to be free of a genetic disease, but it denies permission for another couple to do the same when determination of tissue type is the only reason for testing.
- A deaf lesbian couple deliberately chooses a sperm donor with a strong strain of inherited deafness in his family in an attempt to ensure that their child will be born deaf and thus will share fully in the deaf culture, which they value highly. Many people find this a controversial use of genetic selection.
- *January:* A three-judge panel of the Fourth Circuit Court of Appeals reverses district court judge Albert Bryan's decision in *Harvey v. Horan*, ruling that felons do not have a constitutional right to demand DNA testing after their conviction.
- *February:* James Harvey is granted access to the DNA sample he requested, thanks to a new state law in Virginia.
- *April 11:* *Nature* takes the unprecedented step of apologizing for publishing Ignacio Chapela and David Quist's paper claiming contamination of Mexican corn, saying that the authors' evidence is not sufficient to justify their conclusions.
- *May:* Reviewing *Alfaro v. Terhune*, the Third District Court of Appeals unanimously rejects the argument of California death-row inmates that requiring them to give blood and saliva samples for the state's forensic

DNA database violates their right to privacy and their Fourth Amendment right to be free of unreasonable searches and seizures.

- *July:* Scientists make polio viruses "from scratch," using gene sequences from a mail-order supplier and instructions downloaded from the Internet, suggesting that bioterrorists might also be able to make their own viruses.
- *August:* In spite of facing probable famine, the African countries of Zambia, Zimbabwe, Mozambique, and Malawi refuse to accept U.S. corn from the UN World Food Programme unless it is milled first. They fear that farmers might plant the corn, which could be genetically modified, and thereby ruin the countries' chances of selling corn to Europe, which forbids any importation of genetically modified food.
- *August:* On the basis of tests matching his DNA to that in semen found on the body of five-year-old Angela Bugay, a jury convicts Larry Graham, a California man, of raping and murdering the girl in 1983. His case is cited as an example of the solution of a long-past, "cold case" crime through DNA testing.
- *August:* The U.S. Department of Agriculture creates the Biotechnology Regulatory Services (BRS), a new unit within its Animal and Plant Health Inspection Services (APHIS) that will regulate and facilitate agricultural biotechnology.
- *September:* The FDA and USDA jointly issue guidelines for growing genetically altered crops that produce drugs.
- *September:* French scientists report that a child whom gene therapy had apparently cured of severe inherited immune deficiency has developed a leukemia-like blood condition. The overproduced cells in the child's blood contain the inserted gene, which suggests that the therapy somehow caused the cancer.
- *October:* The Food and Drug Administration (FDA) announces that it will regulate genetically altered animals and their products as drugs, providing a review that is strict but not open to the public.
- *November:* An ounce of corn genetically engineered to contain a vaccine, made by a company called ProdiGene, is found mixed with soybeans intended to become human food. The material is removed before it enters the food supply, and the USDA orders Prodigene to pay a $250,000 fine for not following safety procedures.
- *December:* In rejecting a patent application for a genetically engineered mouse, the Supreme Court of Canada declares that higher life-forms (multicellular differentiated organisms, including plants and animals) cannot be patented in that country.

2003

- The first conference on synthetic biology is held at the Massachusetts Institute of Technology (MIT).

Chronology

- The Supreme Court of Iceland rules in favor of plaintiffs in a suit against deCODE Genetics, which was attempting to discover disease genes by examining genetic and medical records from the country's entire population. The court held that deCODE had not taken sufficient steps to keep Icelanders' genetic information private and, therefore, that individuals could require the company to remove their or their relatives' information from its database.
- *January:* The FDA temporarily halts some gene therapy trials in the United States after learning that a second French child cured of severe inherited immune deficiency by a gene treatment has developed a form of leukemia.
- *February:* Dolly, the famous cloned sheep, is euthanized at the age of eight years because she has developed a severe lung disease.
- *March:* The British government announces that police will be able to keep DNA profiles indefinitely from everyone who has donated specimens, whether or not the people have been accused of a crime.
- *April:* Scientists clone a banteng, an endangered cattlelike animal from Java. This is the first clone of an endangered species to survive more than a few days beyond birth.
- *April 14:* The International Human Genome Sequencing Consortium announces that the Human Genome Project is completed, more than two years ahead of schedule.
- *May 13:* The United States, Argentina, and Canada ask the World Trade Organization to declare that the European Union's moratorium on importation of new genetically engineered food crops is illegally hampering trade.
- *July 2:* The European Parliament approves new regulations that require importing countries to provide traceability of food products "from farm to fork" and verify that foods not labeled as GM-containing have no more than 0.9 percent accidental GM contamination.
- *September 11:* The Cartagena Biosafety Protocol, the United Nations treaty governing international movement of genetically modified organisms, goes into effect.
- *October:* China's State Food and Drug Administration approves sale of Gendicine, a medication that inserts a tumor suppressor gene called p53 into cancer cells and thereby forces the cells to destroy themselves. Gendicine, used to treat certain cancers of the head and neck, is the world's first gene therapy medication approved for general marketing.
- *October 2:* Ruling in *United States v. Kincade*, a three-judge panel of the Ninth Circuit Court of Appeals votes 2-1 to find that the DNA Act's (2000) requirement of DNA samples from convicts on supervised release violates such people's Fourth Amendment rights.

Biotechnology and Genetic Engineering

2004

- The Office of the Inspector General, an independent Justice Department watchdog, investigates procedures in the DNA analysis unit of the Federal Bureau of Investigation's forensics laboratory and finds a number of problems.

- *January:* Panos Zavos, a retired professor of reproductive physiology at the University of Kentucky, announces at a press conference in London that he has implanted a cloned embryo into a woman's womb. No evidence of the subsequent birth of a cloned child ever appears, however.

- *February:* South Korean scientist Woo Suk Hwang makes headlines worldwide by announcing that he has succeeded in cloning a human embryo and extracting stem cells from it—the first time this has been done.

- *March:* A Mexican government study reports discovery of altered genes in 7.6 percent of sampled corn plants in Oaxaca state, supporting the disputed earlier claims of David Quist and Ignacio Chapela that gene flow from engineered to natural corn crops is occurring in that country.

- *March 29:* The Canadian Parliament passes the Assisted Human Reproduction Act, which prohibits human cloning (either reproductive or therapeutic), creation of embryos for research purposes (although embryos from fertility clinics may be used), or paying egg or embryo donors.

- *April 18:* New European Union rules on regulation of GM food go into effect. The rules establish strict new traceability and labeling requirements for foods containing genetically modified ingredients.

- *April 29:* Craig Harman, a British man, is convicted of killing a driver by dropping a brick from a freeway overpass. Harman was traced through a partial match of DNA on the brick with DNA from Harman's brother, whose profile was in the British forensic database. Harman is the first person to be convicted partly on the basis of a familial DNA search.

- *May:* The European Union ends its de facto moratorium on approval of genetically modified food products by allowing the importation of a strain of genetically modified sweet corn for human food, the first such product approved since 1998.

- *May:* Monsanto Corporation announces that it will discontinue its development of Roundup Ready spring wheat.

- *May 21:* The Supreme Court of Canada upholds lower court rulings against farmer Percy Schmeiser in the patent infringement suit filed by Monsanto Corporation.

- *June:* Robert N. Patton, Jr., is arrested in Columbus, Ohio, for committing 37 rapes—13 of which had occurred during two years in which

Patton's DNA, collected when he was first suspected of the crimes, lay untested in an overworked police laboratory.

■ *July:* Britain's Human Fertilisation and Embryology Authority begins to allow preimplantation genetic diagnosis (PGD) and embryo selection to be done for purposes other than preventing the birth of children with severe inherited diseases.

■ *August 11:* Britain's Human Fertilisation and Embryology Authority grants Alison Murdoch and other scientists at the University of Newcastle upon Tyne the country's first license to clone human embryos. The researchers want to clone embryos from cells of people with diabetes in order to extract embryonic stem cells and establish cell lines that will allow them to study how the disease develops and, they hope, eventually create treatments for it.

■ *August 18:* The Ninth Circuit Court of Appeals rules 6-5 in *United States v. Kincade* that requiring DNA samples from parolees, as mandated by the federal DNA Act (2000), does not violate the Fourth Amendment protection against unreasonable search and seizure.

■ *October:* One of two French children who developed a leukemia-like blood cancer after being given gene therapy dies.

■ *October 30:* The Justice for All Act becomes law. It expands grants for DNA testing and other forensics programs, including one for post-conviction DNA testing. It also makes DNA profiles of people convicted of any felony eligible for being uploaded to CODIS.

■ *November 2:* California voters pass Proposition 69, which will mandate (beginning in 2009) collection of DNA samples from all people, including juveniles, with past felony convictions and all adults arrested for felony crimes, whether or not they are later convicted.

■ *November 2:* California voters pass Proposition 71, which establishes the California Institute for Regenerative Medicine and earmarks $3 billion in funds to be spent on stem cell research, including research on embryonic stem cells, over the next 10 years.

2005

■ The government of Portugal announces its intention to create a forensic database containing DNA profiles of all Portuguese citizens.

■ A new British law mandates collection of DNA samples from people arrested for even minor offenses, such as speeding and littering.

■ A third French child treated with gene therapy is diagnosed with a leukemia-like cancer.

■ The Tillamook County Creamery Association, a cooperative based in Oregon that makes the famed Tillamook cheddar cheese, announces that it will no longer accept milk from cows treated with recombinant bovine growth hormone (rBGH).

- **February 8:** The British Human Fertilisation and Embryology Authority grants the country's second license to clone human embryos for research purposes to a research team coheaded by Ian Wilmut, creator of the cloned sheep Dolly. The group wants to establish embryonic stem cell lines from people with motor neuron disease (called amyotrophic lateral sclerosis, or Lou Gehrig's disease, in the United States) in order to learn how this fatal, incurable condition develops and possibly discover treatments for it.

- **February 25:** Dennis Rader is arrested and charged with murder in Wichita, Kansas. Known as the "BTK [bind, torture, kill] killer," Rader later confesses to killing at least 10 people. Some of the evidence that led to his arrest allegedly came from DNA analysis of a biopsy sample from his daughter, taken by police without her knowledge—an example of a familial DNA search.

- **March 14:** A new five-year moratorium on using information from genetic tests in setting most mortgage and health insurance rates, agreed to by the British government and the Association of British Insurers, goes into effect.

- **March 23:** By a vote of 84-32, the United Nations General Assembly adopts a declaration that calls on member states to ban all forms of human cloning.

- **April 26:** The National Academy of Sciences releases guidelines for human embryonic stem cell research, including recommendation of an oversight committee to review all research proposals. Compliance with the guidelines is strictly voluntary.

- **May:** The billionth acre of genetically modified crops (worldwide) is planted.

- **May:** Woo Suk Hwang announces that he has derived stem cell lines from 11 human embryos, cloned from patients with different diseases.

- **August:** Woo Suk Hwang and other South Korean scientists announce that they have produced the first cloned dog, an Afghan hound named Snuppy (for Seoul National University Puppy).

- **October 26:** Researchers announce completion of Phases 1 and 2 of the HapMap, a genetic atlas that focuses on stretches of DNA that differ from person to person and can potentially help to identify variations that predispose people to certain diseases.

- **December:** Seoul National University announces that an investigation has shown Woo Suk Hwang's embryonic stem cell research results to have been faked.

- **December:** A bill to authorize and fund the establishment of a national cord blood bank system is signed into law in the United States.

- **December:** The National Cancer Institute, part of the National Institutes of Health, announces the beginning of the Cancer Genome Atlas Project, a mapping project intended to identify all the important gene changes involved in human cancer.

2006

- **January 5:** The DNA Fingerprinting Act, passed by Congress in 2005, is signed into law. Among other things, it authorizes collection of DNA samples from people arrested for federal crimes, whether or not they are subsequently convicted, and from "non-United States persons who are detained under the authority of the United States."
- **April 10:** A German patient in a gene therapy trial dies; the death is blamed on a massive infection rather than on the treatment.
- **July:** Over protests from some member countries, the European Union adopts a measure that will fund embryonic stem cell research between 2007 and 2013. Only surplus embryos from fertility clinics may be used in the experiments.
- **July 19:** In the first veto of his presidential career, President George W. Bush refuses to approve the Stem Cell Research Enhancement Act, which would have allowed federal funding for stem cell research using cell lines from excess fertility clinic embryos.
- **August:** Robert Lanza and other researchers at Advanced Cell Technology claim to have proven that a researcher could take a single cell from an eight-cell embryo, much as is done for preimplantation genetic diagnosis, and derive stem cells from it, leaving the original embryo unharmed by the procedure.
- **September:** The European Commission reports that 20 percent of European rice is contaminated with an unapproved genetically modified rice strain that had been accidentally released in the United States and exported. The European Food Safety Authority states that the rice does not pose an imminent safety concern, but it has too little data to provide a full risk assessment.
- **October:** In a final ruling, the World Trade Organization holds that the European Union's moratorium on approvals for new genetically modified food crops, in place from 1999 to 2004, violated international trade laws.
- **December 28:** The U.S. Food and Drug Administration tentatively concludes that milk and meat from cloned farm animals is safe to eat.

2007

- Food and Water Watch, an advocacy group, launches a national campaign against dairy products from cows given recombinant bovine growth hormone (rBGH). Shortly afterward, California Dairies Inc., the second-largest dairy cooperative in the United States, asks its members to stop using the hormone. The coffee chain Starbucks also announces that its outlets in parts of several states, amounting to 37 percent of its total production, will use only dairy products from rBGH-free cows.

- *January 7:* Scientists report isolating stem cells from amniotic fluid, the fluid surrounding an unborn child in the uterus, that seem similar to embryonic stem cells. If confirmed, this finding could lead to a way to harvest embryonic stem cells without destroying embryos.
- *February:* Genetic Savings and Clone, a company that had offered to clone customers' pet cats for $50,000 each (later reduced to $32,000), closes due to lack of demand.
- *April 11:* The Senate passes a bill to allow unrestricted federal funding of embryonic stem cell research by a vote of 63-34.
- *April 30:* A unanimous Supreme Court ruling in *KSR International Co. v. Teleflex, Inc.* makes it easier for the U.S. Patent Office to declare an invention "obvious" and therefore not patentable. Representatives of several industries, including the biotechnology and drug industries, express concern about the decision because it will make patents harder to obtain.
- *May:* Arcadia Biosciences, a biotechnology company in Davis, California, announces that it has made an agreement with the government of China's main rice-growing region to adapt the company's new strain of genetically engineered rice to that region. The rice absorbs nitrogen more efficiently than natural rice and therefore needs less nitrogen-based fertilizer. Arcadia says that the rice could help to reduce global warming because soil bacteria convert such fertilizer to nitrous oxide, a greenhouse gas that has 300 times more power to induce warming than carbon dioxide.
- *May 3:* The public comment period on the federal Food and Drug Administration's (FDA) tentative decision to allow meat and milk from cloned animals and their offspring to enter the food supply ends. Thousands of consumers protested the FDA's decision during the comment period, but most scientists who made comments supported it.
- *May 14:* By a vote of 420 to 3, the House of Representatives passes the Genetic Information Nondiscriminating Act, which would prohibit employers from using genetic information in making employment or health benefit decisions.
- *May 16:* The California Supreme Court declines to hear an appeal of a lower court decision finding that an initiative passed by the state's voters in November 2004, which established an institute for embryonic stem cell research, was constitutional. This ruling allows the lower court's decision

to stand, disappointing right-to-life groups and others who had brought suit to stop the initiative.

■ *June:* A researcher in Japan announces that he has engineered a strain of rice to produce a protein vaccine against cholera, a digestive illness caused by bacteria that is often fatal. Unlike standard vaccines, the rice vaccine can be taken by mouth and does not require refrigeration, making it potentially useful in developing countries.

■ *June 5:* California's state embryonic stem cell research institute announces distribution of more than $50 million in grants to finance construction of new laboratories.

■ *June 6:* A bill to ban human reproductive cloning is defeated by a vote of 213-204 in the House of Representatives. Many Republican representatives opposed the bill because it did not ban therapeutic or research cloning (cloning of embryos to produce stem cells).

■ *June 7:* The House of Representatives votes 247-176 to approve the Senate bill to permit unrestricted funding of embryonic stem cell research.

■ *June 13:* Scientists say that they have reprogrammed adult mouse skin cells to show the properties of embryonic stem cells, providing a potential way to obtain the stem cells without killing embryos.

■ *June 14:* An international research project called the Encyclopedia of DNA Elements (ENCODE), which studied the details of genetic interactions in 1 percent of the human genome, announces that the operation of genes and other DNA sequences is much more complex than had been thought. For example, different genes may share sequences on a DNA molecule, or, conversely, parts of the same gene may be widely separated on the molecule.

■ *June 20:* President George W. Bush vetoes the recently passed funding bill for embryonic stem cell research. Congress does not have the votes needed to override the veto.

■ *July 24:* Jolee Mohr, a young Illinois woman with rheumatoid arthritis, dies after taking part in a trial of a gene therapy treatment for her illness.

■ *September:* Scientists announce that they have fully sequenced the genetic code of an individual (genetic sequencing pioneer Craig Venter) for the first time (the Human Genome Project studied a composite genome from several individuals).

■ *September 17:* Investigators advise the National Institutes of Health that Jolee Mohr probably died from a fungus infection, to which she was susceptible because drugs she took for her rheumatoid arthritis suppressed her immune system, rather than from the effects of experimental gene therapy.

- *October 8:* Three researchers win the Nobel Prize in physiology or medicine for creating techniques for altering genes in mice that improve use of the animals in medical research.
- *November 1:* New rules for filing patents go into effect in the United States; the rules streamline the application process for the patent office but put more burden on applicants, irritating (among others) the biotechnology industry.

CHAPTER 4

BIOGRAPHICAL LISTING

This chapter offers brief biographical information on people who have played major roles in the development of biotechnology and genetic engineering or in events relating to the ethical, legal, and social implications of these fields. Some pioneers of genetics and molecular biology are also included, but the list focuses primarily on people who have been active since the development of genetic engineering in the early 1970s.

Sidney Abbott, an HIV-positive woman with no symptoms of AIDS. She sued her dentist, Randon Bragdon, for refusing to treat her, saying that his refusal violated her rights under the Americans with Disabilities Act (ADA). In June 1998 the Supreme Court confirmed her right to sue, leading to hope that the ADA might also cover healthy people with a genetic predisposition to disease.

W. French Anderson, pioneer in gene therapy. He headed the National Institutes of Health (NIH) team that made the first successful use of altered genes to treat a human disease in September 1990. He is currently at the University of Southern California.

Tommie Lee Andrews, the first American convicted of a criminal charge primarily through DNA identification testing. Andrews, a 24-year-old warehouse worker, was convicted of the rape and beating of several women in Orlando, Florida, in late 1987 and early 1988 after the DNA profile from a sample of his blood matched that of semen found on one of the victims.

Oswald Avery, Canadian-born bacteriologist who proved that DNA rather than protein carries inherited information. In 1944, Avery and his colleagues at the Rockefeller Institute in New York published a paper showing that a form of *Pneumococcus* bacteria incapable of causing disease became able to cause illness after it took up DNA from a disease-causing form of the same bacteria. The change was passed on to the bacteria's descendants.

Paul Berg, Stanford University biochemist, sometimes called "the father of genetic engineering." Berg did the first recombinant DNA experiments, transferring genes between two types of viruses, in the winter of 1972–73. Berg was also one of the first scientists to question the safety of recombinant DNA experiments.

Peter Beyer, scientist at the Swiss Federal Institute of Technology. He and Ingo Potrykus led the team that created "golden rice" in 1999. This rice contains daffodil and bacterial genes that allow it to make beta-carotene, a building block of vitamin A. Biotechnology supporters claim that eating the rice can help to prevent vitamin A deficiency, which affects 100 to 140 million children a year and makes about 500,000 of them go blind.

Kirk Noble Bloodsworth, falsely accused of the sex murder of a nine-year-old girl in Maryland in 1984, convicted, and sentenced to death. In 1993, DNA evidence cleared him, and the real killer was also identified. Bloodsworth was the first death-row inmate to be exonerated by DNA evidence.

Herbert Boyer, a biochemist. In 1973, while at the University of California at San Francisco, he worked with Stanley Cohen of Stanford to develop the first practical technique for transferring genes between organisms. Boyer and venture capitalist Robert Swanson cofounded Genentech, the first business built on recombinant DNA technology, in 1976.

Albert V. Bryan, Jr., U.S. District Court judge in Virginia. In April 2001 he ruled in *Harvey v. Horan* that felons had a constitutional due process right to obtain DNA samples for testing that might prove their innocence. The U.S. Fourth Circuit Court of Appeals reversed his ruling in January 2002.

Carrie Buck, an institutionalized woman thought to be developmentally disabled, who appealed a decision to forcibly sterilize her under a Virginia eugenics law. The U.S. Supreme Court upheld the law by an 8-1 vote in 1927. Noting that Buck's mother and seven-month-old daughter as well as Buck herself seemed to be "feebleminded," Justice Oliver Wendell Holmes wrote in the majority opinion on the case that "three generations of imbeciles are enough." Buck may in fact have had normal intelligence.

Ananda Chakrabarty, General Electric Corporation scientist who obtained the first U.S. patent on a living thing. His application for a patent on a genetically engineered bacterium that digested petroleum was rejected at first by the Patent and Trademark Office, but in 1980 the Supreme Court ruled by a 5-4 vote that Congress had intended "anything under the sun that is made by man" to be patentable, including altered organisms such as the one Chakrabarty had invented.

Ignacio Chapela, researcher at the University of California–Berkeley. His paper, coauthored with David Quist, in the November 29, 2001, issue of *Nature* caused widespread controversy by claiming that engineered genes had appeared in native corn crops in Mexico. On April 11, 2002, *Nature* took the unprecedented step of apologizing for publishing the paper, saying that evidence did not justify the authors' conclusions. Chapela, Quist, and some other scientists stand by their report, however, and later reports by other scientists seem to confirm it.

Stanley Cohen, Stanford University geneticist. With Herbert Boyer, he invented the first practical method of transferring genes between organisms in 1973. Cohen had studied plasmids, small rings of DNA that bacteria sometimes use to transfer genetic information, and he and Boyer worked out a method of combining plasmids from two types of bacteria.

Francis Collins, a molecular biologist. He currently directs the National Institutes of Health's National Human Genome Research Institute and was a leader in the Human Genome Project.

Francis Crick, British molecular biologist. With James Watson, he worked out the structure of the DNA molecule at Cambridge University and showed how this structure allowed the molecule to reproduce itself. Watson, Crick, and Maurice Wilkins shared a Nobel Prize for this work in 1962. In the early 1960s, Crick set out a theory explaining how inherited information is coded in DNA and how the cell translates this information into proteins. His ideas were later shown to be basically correct.

Charles Darwin, author of the theory of evolution by natural selection, which he set forth in *On the Origin of Species* (1859). The theory states that the members of a species with inherited characteristics that make them most suited to survive in a particular environment will be the most likely to survive and bear young. Over generations, therefore, a change in the environment can produce a change in the predominant characteristics of a species.

Ashanthi deSilva was four years old in September 1990, when she made history as the first person to receive altered genes as a treatment for disease. Ashanthi had inherited a rare condition called ADA (adenosine deaminase) deficiency, which left her without a functioning immune system. After several treatments in which her own blood cells were reinjected after being given normal ADA genes in the laboratory, she became healthy enough to lead an essentially normal life.

Drew Endy, professor at the Massachusetts Institute of Technology (MIT) and a founder of the new discipline of synthetic biology. Endy has developed what fellow MIT synthetic biologist Tom Knight calls BioBricks: strands of DNA (usually comprising several genes) that perform specific functions and can be joined together in any order. Endy and other synthetic biologists hope to use BioBricks as standardized components that can be assembled into switches, circuits, and other "devices" that control chemical reactions in cells.

Walther Flemming, a German biologist. In 1875, he discovered chromosomes ("colored bodies") in the nuclei, or central bodies, of cells and showed how these change as cells reproduce.

Francis Galton, a cousin of Charles Darwin. He founded the pseudoscience of eugenics, which held that personality characteristics such as intelligence or laziness are inherited and that the human race can be improved by encouraging people with desirable characteristics to have children and discouraging reproduction in those with undesirable characteristics. Some of Galton's supporters advocated sterilization by force. Eugenics has fallen into disrepute, but Galton should also be remembered as a founder of biostatistics (application of statistics to animal populations) and of identification by fingerprinting.

Jesse Gelsinger, an Arizona man with a rare genetic disease who died at age 18 on September 17, 1999, when his immune system reacted violently to viruses injected into him as part of an experimental gene therapy treatment. His death was the first attributed to gene therapy. It resulted in suspension or cancellation of a number of gene therapy trials in the United States and a considerable tightening of the regulations governing human trials.

David Golde, physician at the University of California–Los Angeles (UCLA) Medical Center, who was a defendant in a suit concerning ownership rights to tissue that was made into a commercial cell line. John Moore, whom Golde had treated for leukemia, sued Golde and others in 1984 because they had developed a lucrative cell line from Moore's spleen (removed as part of a cancer treatment) without obtaining Moore's permission or offering him any recompense. The California Supreme Court ruled in 1990 that Moore had no ownership rights to his tissue but also stated that Golde had violated his "fiduciary duty" to Moore by not telling him about his commercial plans.

Larry Graham, a California man, who was convicted in August 2002 of raping and murdering a five-year-old girl 19 years before. Testing that matched his DNA with that in semen found on the girl's body was crucial in the conviction, and his case was cited as an example of the "cold cases" that could be solved, years after the crimes were committed, through DNA testing and databases.

Craig Harman, the first person to be convicted partly as a result of a familial DNA search. Harman killed a driver by dropping a brick from a freeway overpass in May 2003, and police traced him after finding that DNA on the brick was a close partial match to the DNA of Harman's brother, whose genetic profile was in the British forensic database. Harman was convicted of the murder on April 19, 2004.

James Harvey, a Virginia man convicted of rape and forcible sodomy in 1990. He protested his innocence and, with the help of the New York–based Innocence Project, demanded access to the rape kit used as evidence in his trial so that improved DNA testing methods could be applied to it. When the prosecutor, Robert F. Horan, Jr., refused to hand over the kit,

Harvey and his lawyers sued Horan in July 2000, claiming violation of 42 U.S.C. 1983, a civil rights statute. A District Court judge stated in April 2001 that Horan's refusal violated Harvey's due process rights under the Fourteenth and Fifteenth Amendments, but in January 2002 the U.S. Fourth Circuit Court of Appeals reversed the lower court's decision, saying that convicts did not have a constitutional right to demand DNA testing after their conviction. In February, another judge, citing a new Virginia state law, ordered that Harvey be given the rape kit. The DNA test, when finally performed, confirmed Harvey's guilt.

Kimberly Hinton, one of a pair of twin sisters who applied for the job of global airline pilot with United Air Lines in the 1990s (the other was Karen Sutton). The airline rejected Hinton because she failed to meet its requirement for uncorrected vision, since she was extremely nearsighted without glasses, although her vision with corrective lenses was 20/20 or better. She and her sister sued the airline for violation of the Americans with Disabilities Act (ADA), but the U.S. Supreme Court rejected their suit in June 1999 because, as Sandra Day O'Connor wrote in the Court's majority opinion, the ADA requires someone to be "presently—not potentially or hypothetically—substantially limited" in a major life activity. The court did not consider the women disabled because their disability could be corrected. This decision suggested that people with a genetic predisposition to disease but no actual illness also would not be covered by the ADA.

Mae-Wan Ho, a major opponent of genetically engineered foods and biotechnology in general. She heads the Institute of Science in Society, a British (London) organization devoted to "science, society, and sustainability."

Oliver Wendell Holmes, Supreme Court Justice from 1902 to 1932. In the 1927 case *Buck v. Bell,* Holmes wrote the majority opinion that supported a Virginia eugenics law under which a woman thought to be developmentally disabled, Carrie Buck, was forcibly sterilized. Noting that not only Buck but her mother and her seven-month-old daughter appeared to be of subnormal intelligence, Holmes wrote that "three generations of imbeciles are enough."

Leroy Hood, a pioneer inventor of machines that handle DNA automatically. In 1983 he and coworkers at the California Institute of Technology invented a machine that could assemble short stretches of DNA of known sequence. They invented one that could automatically determine the sequence of stretches of DNA in 1986. Hood's machines have helped make gargantuan tasks like the Human Genome Project possible.

Robert Hooke, a 17th-century British scientist, inventor, and philosopher. In addition to many achievements in chemistry and physics, Hooke invented or greatly improved the compound microscope. In 1665, using this invention, he observed tiny, square structures in a slice of cork and dubbed them cells. In fact, what he saw were merely the walls that were all that remained of these basic units of living matter.

Woo Suk Hwang, South Korean cloning and stem cell researcher. Hwang appeared to have achieved major breakthroughs in embryonic stem cell research in February 2004, when he announced that his laboratory had cloned human embryos and extracted embryonic stem cells from them, and again in May 2005, when he said that he had derived 11 stem cell lines from embryos cloned from different patients. Late in 2005, however, doubts began to arise about his achievement. Seoul National University, where Hwang worked, publicly stated in December that an investigation had shown the scientist's results to have been faked. Hwang's claim to have cloned a dog in August 2005, however, apparently stood up to examination.

Alec Jeffreys, of the University of Leicester in Great Britain, invented the technique of "DNA 'fingerprinting,'" as he called it, in 1985. Based on the observation that certain stretches of DNA contain short repeated base sequences that vary considerably in length from person to person, Jeffreys's technique was used first in immigration and paternity cases. It began to be used for identification of criminals in 1987.

Leon R. Kass, University of Chicago bioethicist and author, headed the President's Council on Bioethics from 2002 to 2005. (As of 2007, he was still a member of the council.) Kass disapproves strongly of human cloning for any purpose and of alteration of human germ-line genes, both of which he considers an affront to human dignity.

Jay Keasling, professor at the University of California–Berkeley, and *Discover* magazine's scientist of the year for 2006. Keasling, a pioneer in the new field of synthetic biology, is using genes from several different species to reengineer *E. coli* bacteria so that they can inexpensively produce artemisinin, a valuable antimalaria drug that currently can be made only from certain plants.

Thomas Kincade, a convicted bank robber. In March 2002, while Kincade was on supervised release after serving time in prison, his probation officer ordered him to provide a DNA sample as ordered by the DNA Act, a federal law that went into effect in 2000. Kincade refused and challenged the act's constitutionality in a federal district court. The court ruled that the law was constitutional. A three-judge panel of the Ninth Circuit Court of Appeals reversed that decision in October 2003, but when the entire court reheard the case, it ruled 6-5 on August 18, 2004, that the law did not violate the Fourth Amendment.

Robert P. Lanza, vice president for medical and scientific development at Advanced Cell Technology (Alameda, California, and Worcester, Massachusetts) and a leader in U.S. research on cloning and embryonic stem cells. Lanza claimed in August 2006 that his team had harvested embryonic stem cells grown from a single cell taken out of an embryo at the eight-cell stage (much as is done for preimplantation genetic diagnosis) without harming the original embryo. This procedure for acquiring the stem cells may avoid the objections of those who say killing embryos in order to derive stem cells from them is unethical.

Steven Lindow, a plant pathologist at the University of California–Berkeley. He oversaw the first release of genetically altered organisms (bacteria) into the environment in 1987. His "ice-minus" bacteria protected crops against frost damage.

John C. Mayfield III, a 21-year-old lance corporal in the U.S. Marines in 1995. He and Corporal Joseph Vlakovsky refused to give samples for DNA testing and archiving, which had been required of all U.S. military personnel since 1992. Mayfield and Vlakovsky said that being forced to give samples for the DNA database, which is supposed to help identify the remains of soldiers killed in action, violated their right to privacy. A court-martial in April 1996 convicted the two of refusing to obey a direct order.

Gregor Mendel, an Austrian monk. He bred pea plants in his monastery garden to work out the basic rules by which characteristics are inherited. He published a paper describing his work in 1866, but it remained virtually unknown until several scientists rediscovered it in 1900. Mendel's work became the foundation of genetics and provided the mechanism for evolution by natural selection, first described by Mendel's contemporary Charles Darwin.

Pat R. Mooney, head of the ETC Group, a Canadian watchdog group. Mooney opposes a number of forms of biotechnology, including agricultural biotechnology and nanobiotechnology. He fears that these technologies will damage the environment, human health, or both and will be used in ways that take advantage of poor people in the developing world. In 2004 he called for a moratorium on nanoscience research until risks of the technology could be evaluated.

John Moore, a Seattle businessman. He sued his doctor (David Golde), the University of California, and several drug companies in 1984 for having made a profitable laboratory cell line from his spleen, which had been removed as a cancer treatment, without consulting him. A lower court supported him, but in 1990 the California Supreme Court ruled that Moore had no ownership claims on his cells once they had been removed from his body.

Thomas Hunt Morgan, pioneer geneticist, worked at Columbia University in New York (1904–28) and later at the California Institute of Technology (1928–45). At Columbia, he and his coworkers and students drew on experiments with fruit flies to relate changes in chromosomes to the inheritance of particular characteristics, thus providing a physical basis for the patterns of trait transmission that Gregor Mendel had observed. Morgan won a Nobel Prize for his work in 1933.

Alison Murdoch, scientist at the University of Newcastle upon Tyne in England. On August 11, 2004, Britain's Human Fertilisation and Embryology Authority granted Murdoch's research team the country's first license to clone human embryos. Murdoch wants to clone embryos from cells of people with diabetes in order to extract embryonic stem cells and establish cell lines that will allow her to study how the disease develops and, she hopes, eventually create treatments for it.

Peter Neufeld, a New York defense attorney. He and his partner, Barry Scheck, have become famous for acting in criminal cases involving identification by DNA testing and have appeared in a number of trials, including the O. J. Simpson murder trial in 1995. In 1991 they founded the Innocence Project, which has used DNA testing to show that certain convicted criminals actually were not guilty.

Marya Norman-Bloodsaw, a worker at Lawrence Berkeley Laboratory, a California research facility run by the Department of Energy. In 1995, she and some of her coworkers found out that the laboratory had run several tests, including a genetic test, on samples of their blood and urine without their knowledge or consent. They sued for violation of their rights under the Fourth Amendment and the 1964 Civil Rights Act, and the U.S. Court of Appeals for the Ninth Circuit upheld them.

Louis Pasteur, famed 19th-century French chemist. He provided a scientific basis for part of traditional biotechnology when he showed, starting in the 1850s, that fermentation processes such as those used for millennia to produce wine, beer, buttermilk, and cheese depended on living microorganisms.

Robert N. Patton, Jr., convicted rapist. Patton was arrested in Columbus, Ohio, in June 2004 for committing 37 rapes—13 of which had occurred during the previous two years, while Patton's DNA, collected when he was first suspected of the crimes, had remained untested in a police laboratory. The Patton case drew attention to the immense backlog of forensic DNA samples collected by police that state and federal laboratories lack the money and personnel to analyze.

Zhaohui Peng, Chinese researcher and entrepreneur. Peng is the cofounder (in 1998) and head of SiBono, a Chinese company that manufactures Gendicine, the first gene therapy medication approved for sale anywhere in the world. Gendicine inserts copies of a tumor suppressor gene called p53 into cancer cells, forcing the cells to kill themselves. China's State Food and Drug Administration approved Gendicine for treatment of certain types of head and neck cancer in October 2003.

Biographical Listing

Colin Pitchfork, a 27-year-old baker in a village in Leicestershire, England, the first person to be convicted of a crime partly on the basis of DNA "fingerprinting." When police asked for an unprecedented mass "blooding" (blood sample donation for DNA testing) of young men in several villages in an attempt to solve the rape and murder of two teenaged girls, Pitchfork tried to evade the process by having a coworker give blood in his place. He was caught, however, and his sample proved to match the semen on one of the victims. He confessed and was found guilty of the crimes in 1987.

Robert Pollack, a geneticist. He was perhaps the first to express concern about the safety of recombinant DNA experiments. Then working at Cold Spring Harbor Laboratory in New York, Pollack learned in 1971 that Paul Berg of Stanford University was planning to insert a gene from a cancer-causing virus into a type of bacterium that could infect human beings. Pollack persuaded Berg that this might be dangerous, and Berg agreed to transfer the gene between viruses but not go on to infect the bacterium.

Ingo Potrykus, scientist at the Swiss Federal Institute of Technology. He and Peter Beyer led the team that created "golden rice" in 1999. This rice contains daffodil and bacterial genes that allow it to make beta-carotene, a building block of vitamin A. Biotechnology supporters claim that eating the rice can help to prevent vitamin A deficiency, which affects 100 to 140 million children a year and makes about 500,000 of them go blind.

David Quist, researcher at the University of California–Berkeley. His paper, coauthored with Ignacio Chapela, in the November 29, 2001, issue of *Nature* caused widespread controversy when it claimed that engineered genes had appeared in native corn crops in Mexico. On April 11, 2002, *Nature* took the unprecedented step of apologizing for publishing the paper, saying that evidence did not justify the authors' conclusions. Quist, Chapela, and some other scientists stand by their report, however, and later reports by other scientists seem to confirm it.

Dennis Rader, a serial killer from Wichita, Kansas, known as the "BTK [bind, torture, kill] killer." Allegedly, when police began to suspect that Rader was the perpetrator of a series of brutal murders in the Wichita area, they obtained a warrant to take a biopsy sample from a physician treating Rader's daughter. The daughter was not told that the sample had been seized. DNA from the biopsy material was tested and found to be a close match to that found at the scenes of some of the killings. The crime scene profile matched what paternity analysis would predict for the father of the biopsy's owner—that is, Dennis Rader. This match provided some of the evidence that led to Rader's arrest on February 25, 2005. He pleaded guilty and confessed, during his trial in June, to killing at least 10 people.

Jeremy Rifkin, founder and president of the Foundation on Economic Trends. He has been known as a critic of biotechnology and genetic engineering since the mid-1970s. In addition to filing lawsuits and organizing protest demonstrations, Rifkin has written numerous books on the impact of scientific and technological change on society and the environment, including *The Biotech Century* (1998).

Barry Scheck, a New York defense attorney. With his partner, Peter Neufeld, he has become famous for acting in criminal cases involving identification by DNA testing and has appeared in a number of trials, including the O. J. Simpson murder trial in 1995. In 1991, Scheck and Neufeld founded the Innocence Project, which has used DNA testing to show that certain convicted criminals actually were not guilty.

Matthias Schleiden, 19th-century German biologist, coauthor of the cell theory with Theodor Schwann. In 1839, the two proposed that the microscopic membrane-bound bodies called cells were the units of which all living things are made.

Percy Schmeiser, Canadian farmer. In 1997, Schmeiser sprayed the Monsanto herbicide Roundup on his canola crop and discovered that 60 percent of the plants survived the spraying. Their resistance was probably due to herbicide-resistant genes accidentally transferred to his plants from nearby farms that raised genetically altered, "Roundup Ready" canola. Schmeiser kept seeds from some of the resistant plants and planted them the following year. In August 1998, Monsanto sued Schmeiser for patent infringement for using the seeds without a license from the company. Several courts, including the Supreme Court of Canada in May 2004, upheld the suit. Opponents of agricultural biotechnology say that farmers, including those in the developing world, who save and reuse genetically altered seeds might face similar suits.

Theodor Schwann, 19th-century German biologist, coauthor of the cell theory in 1839 with Matthias Schleiden.

Vandana Shiva, Indian physicist, author, and environmental activist. She is the director of Navdanya (Research Foundation for Science, Technology, and Ecology), which works to conserve biodiversity, protect the rights of indigenous peoples, and oppose centralized systems of monoculture. Shiva strongly opposes promotion of genetically engineered crops in the developing world and large biotechnology companies' patenting of natural materials and traditional knowledge from tropical countries.

Lee J. Silver, Princeton University geneticist. Generally a supporter of human gene alteration, Silver predicted in *Remaking Eden* (1997) and other books that, because only the wealthy will be able to afford to alter their children's genes, the altered "Gen-Rich" and unaltered "Naturals" may become so different that they will evolve into separate species.

Scott Simplot, head of J. R. Simplot Co., a large beef-producing company based in Boise, Idaho. Simplot is one of the strongest advocates of cloning cattle as a method of producing better and more dependable quality in beef and milk. In early 2007, he had 22 cloned cattle and at least 26 offspring of these clones (produced in the natural way).

O[renthal]. J. Simpson, a famous African-American former football star. In 1995, he went on trial for the murder of his ex-wife, Nicole Brown Simpson, and an acquaintance of hers, Ronald Goldman. Despite DNA evidence that pointed to his guilt, Simpson was acquitted, probably at least partly because his defense lawyers, who included Barry Scheck and Peter Neufeld, demonstrated that the Los Angeles police could have accidentally or deliberately contaminated evidence from the crime scene with samples of Simpson's blood.

Gregory Stock, head of the Program on Science, Technology, and Society at the University of California–Los Angeles. He supports alteration of germ-line genes with the aim of improving the human species.

Karen Sutton, one of twin sisters who applied for the job of global airline pilot with United Air Lines in the 1990s (the other was Kimberly Hinton). The airline rejected Sutton because she failed to meet its requirement for uncorrected vision, since she was extremely nearsighted without glasses, although her vision with corrective lenses was 20/20 or better. She and her sister sued the airline for violation of the Americans with Disabilities Act (ADA), but the U.S. Supreme Court rejected their suit in June 1999 because, as Sandra Day O'Connor wrote in the Court's majority opinion, the ADA requires someone to be "presently—not potentially or hypothetically—substantially limited" in a major life activity. The Court did not consider the women disabled because their disability could be corrected. This decision suggested that people with a genetic predisposition to disease but no actual illness also would not be covered by the ADA.

Robert Swanson, a venture capitalist, persuaded Herbert Boyer to join him in exploring the business possibilities of Boyer and Stanley Cohen's newly developed recombinant DNA technology. Swanson and Boyer founded Genentech (from GENetic ENgineering TECHnology) in California in 1976. When the company offered its stock to the public for the first time in 1980, the price per share jumped from $35 to $89 in the first few minutes of trading, even though Genentech had yet to sell a product. Swanson died in December 1999.

Edward L. Tatum, a Stanford University molecular biologist, showed in 1941 that each gene in a DNA molecule coded for one enzyme (that is, a protein molecule). He and his colleague, George Beadle, won a Nobel Prize for this work in 1958.

James A. Thomson, the first person to isolate stem cells from human embryos. He did so in 1998, while at the University of Wisconsin–Madison.

Harold Varmus, a pioneer researcher on the genetics of cancer, shared a Nobel Prize with J. Michael Bishop, his co-researcher at the University of

California–San Francisco, in 1989. He was head of the National Institutes of Health in Bethesda, Maryland, from 1993 to 1999 and later became chief executive officer of the Memorial Sloan-Kettering Cancer Center in New York City. His positions on biotechnology and genetic engineering issues include a defense of research on embryonic stem cells.

Craig Venter, molecular biologist. While at the National Institutes of Health (NIH) in the early 1990s, he developed important new methods of identifying and sequencing genes. He left NIH in 1992 and subsequently founded several biotechnology companies, including the Institute for Genomic Research (TIGR) and Celera Genomics. Celera "raced" the government-funded Human Genome Project to be the first to sequence the entire human genome; the competition is said to have ended in a tie. In the mid-2000s, Venter traveled around the world to collect samples of microorganisms for genetic sequencing, investigated possible uses of microbes to destroy pollution and generate energy, and worked on a project to create an artificial microorganism with a "minimal genome."

Joseph Vlakovsky, a 25-year-old U.S. Marine corporal in 1995, joined Lance Corporal John C. Mayfield III in refusing to give samples of DNA for testing and archiving, as had been required of all U.S. military personnel since 1992. The men said that keeping their DNA in a database violated their right to privacy. A court-martial in April 1996 convicted the two of refusing to obey a direct order.

Florence Wambugu, Kenyan molecular biologist. She speaks frequently about the power of genetically engineered crops to help developing-world farmers. She also helped to create a genetically engineered yam able to resist a virus that destroys much of the yam crop in Africa.

James Dewey Watson, a young American studying at Cambridge University, worked with British molecular biologist Francis Crick to discover the structure of DNA in 1953. For this groundbreaking work, which ultimately showed how inherited information is coded and passed on, Watson, Crick, and Maurice Wilkins received a Nobel Prize in 1962. Watson went on to direct Cold Spring Harbor Laboratory in New York and was also the first head of the Human Genome Project. He resigned from the project in 1992.

Ian Wilmut, a scientist at the Roslin Institute in Scotland. He startled the world in February 1997 by announcing that he and his coworkers had cloned a sheep (which they named Dolly) from a mature body cell, something many scientists had thought impossible. On February 8, 2005, a research team co-headed by Wilmut was granted Britain's second license to clone human embryos. They want to establish embryonic stem cell lines from people with motor neuron disease (called amyotrophic lateral sclerosis or Lou Gehrig's disease in the United States) in order to learn how this fatal, incurable condition develops and possibly discover treatments for it. Wilmut is opposed to human reproductive cloning.

Panos Zavos, retired professor of reproductive physiology at the University of Kentucky. In mid-January 2004, he announced at a press conference in London that he had implanted a cloned embryo into a woman's womb. No evidence of the subsequent birth of a cloned child ever appeared, however.

CHAPTER 5

GLOSSARY

Biotechnology, genetic engineering, and genetics are complex fields with highly technical vocabularies. This chapter presents some of the terms that the general reader is most likely to encounter while researching these fields and their ethical, legal, and social implications. Several web sites also offer online glossaries (see chapter 6, "How to Research Biotechnology and Genetic Engineering").

ADA (adenosine deaminase) deficiency A rare inherited disorder in which lack of a certain enzyme in white blood cells causes an essentially complete failure of the immune system. It was the first illness to be treated successfully by gene therapy (in 1990).

adenine One of the four bases in DNA and RNA. It pairs with thymine in DNA and uracil in RNA.

adult stem cells Stem cells taken from an adult, fetus, or newborn (including cells from the umbilical cord and placenta). Adult stem cells probably cannot differentiate into as many different kinds of tissue as embryonic stem cells, but their use is less controversial because they can be harvested without killing living things. *See also* **embryonic stem cells; stem cells.**

adverse selection The tendency of people who know they are likely to need life or health insurance, while insurers do not know of this risk, to take out large amounts of insurance, thus throwing off the insurance industry's statistical methods and raising premiums for everyone. Allowing policyholders but not insurance companies to know the results of genetic tests could increase adverse selection.

altered nuclear transfer A process for therapeutic cloning that disables genes required for full embryonic development before harvesting cells from what the process's developers say can no longer be considered an embryo.

Americans with Disabilities Act (ADA) Passed in 1990, this law protects the disabled from discrimination. It covers healthy people who are perceived as disabled, which may include those with inherited defects revealed by genetic tests.

amino acids   The small molecules of which proteins are made. There are 20 different kinds.

amniotic fluid   The fluid surrounding an unborn child in the uterus. In January 2007, researchers claimed that they had found cells in amniotic fluid that had many of the properties of embryonic stem cells.

amyotrophic lateral sclerosis (ALS)   *See motor neuron disease.*

Animal and Plant Health Inspection Service (APHIS)   The part of the U.S. Department of Agriculture (USDA) responsible for regulating genetically altered agricultural plants and animals. In August 2002, the USDA created a new unit within APHIS, the Biotechnology Regulatory Services (BRS), to focus on regulating and facilitating agricultural biotechnology.

Asilomar conference   A conference held at a California retreat center in February 1975, in which 140 geneticists and molecular biologists worked out safety standards for recombinant DNA experiments.

Bacillus thuringensis **(Bt)**   A bacterium that produces a toxin that kills a variety of pest insects but is harmless to most nonpest insects and other living things, including humans. Sprays containing the bacteria have been used as short-lived, "organic" insecticides. The gene that produces the Bt toxin also has been inserted into several kinds of crop plants.

bases   Small molecules that usually exist in pairs in DNA and RNA. In DNA, the bases are adenine, thymine, guanine, and cytosine; in RNA, uracil replaces thymine. The sequence of bases in a nucleic acid molecule contains the genetic code.

BioBricks   Sequences of DNA, usually comprising several genes, that code for particular cell functions and can be attached to one another in any order, providing standardized components that scientists in the new discipline of synthetic biology hope to use to turn cells into controllable chemical reactors or even perhaps create new organisms. BioBricks were named by Tom Knight and developed by Drew Endy, both professors at the Massachusetts Institute of Technology (MIT).

bioinformatics   The application of computer and statistical techniques, especially database management, to the organization of biological information, such as DNA or protein sequences.

bioprospecting   Searching an environment for living things or parts of living things (including genetic material) that may have commercial use. Critics call this process "biopiracy."

bioremediation   A branch of biotechnology that uses living things, usually microorganisms, to repair environmental damage, for instance by breaking down oil or other toxic chemicals.

biotechnology   Using or altering living things in processes that benefit humankind; now frequently applied to commercial processes that use organisms with altered genes.

blastocyst A stage of early embryonic development (about five days after fertilization in humans) in which the embryo consists of a ball of about 200 cells. This is the stage at which embryonic stem cells are harvested, destroying the embryo.

bovine growth hormone (bovine somatotropin, BGH, BST) A cattle hormone that can be made by bacteria containing recombinant DNA (the recombinant hormone is known as rBGH). The hormone is given to dairy cattle to increase milk production. Its use is controversial.

BRCA1, BRCA2 Genes found in the early 1990s to be inherited in some families in which breast and ovarian cancer are unusually common and occur at an early age. Five to 10 percent of women with breast cancer have a mutated form of one of these genes, which can be detected by a test.

Bt *See* ***Bacillus thuringensis.***

Cancer Genome Atlas Project A genome mapping project, launched by the National Cancer Institute (part of the National Institutes of Health) in December 2005, that intends to identify all the important genes involved in human cancer.

carrier An organism that has inherited a gene related to a disease and therefore can pass it on to offspring but does not suffer from the disease.

Cartagena Biosafety Protocol An international treaty (actually worked out in Montreal in January 2000 after a meeting in Cartagena, Colombia, in 1999 had deadlocked) that regulates international movement of genetically modified crops in ways that minimize their threat to biodiversity; it is an addendum to the United Nations Convention on Biological Diversity (1992). The treaty went into effect on September 11, 2003.

cell The basic unit of which all organisms are composed, made up of a microscopic piece of living material surrounded by a membrane. It is the simplest living system that can exist independently.

cell line A group of cells altered so that they will multiply indefinitely in culture in a laboratory. All cells in a cell line are descended from a single cell and usually are genetically identical.

chimera An organism that contains cells from two or more different species.

chromosome One of a group of threadlike bodies containing the DNA that carries a cell's basic genetic information (they also contain protein). In cells with nuclei, they are found in pairs inside the nucleus. Chromosomes ("colored bodies") reproduce themselves during cell division.

Clinical Laboratory Improvement Amendments (CLIA) The law, established in 1988, under which the Centers for Medicare and Medicaid Services, a federal agency, regulates clinical laboratories.

clone A gene, cell, or organism that is the exact genetic duplicate of another gene, cell, or organism. Genes or other stretches of DNA may be cloned in bacteria for study or for biotechnology processes. Plant and (to a more limited extent) animal clones are used in biotechnology and research. The ethics of cloning humans has been hotly debated.

cold cases   Crimes committed years or even decades previously, for which there is insufficient evidence to justify an arrest (and often no leads at all). Many of these cases are now being solved by comparing DNA in samples taken from the crime scenes to DNA profiles of convicted criminals stored in local, state, or national databases.

Combined DNA Index System (CODIS)   Software developed by the Federal Bureau of Investigation (FBI) to coordinate information from local, state, and national databases of DNA profiles from criminals and crime scenes.

cytoplasm   The living material in the main body of a cell (outside the nucleus in nucleated cells).

cytosine   One of the four bases in DNA and RNA. It always pairs with guanine.

differentiation   The process of maturation in which a cell takes on characteristics associated with a particular type of tissue, such as nerve tissue or muscle tissue. Differentiation is usually, but not always, irreversible.

Directive 98/44   A European Union (EU) directive on the legal protection of biotechnological inventions, passed by the EU Parliament and Council on July 6, 1998, which excludes from patenting (on ethical grounds) processes for cloning or modifying the genetic identity of human beings, uses of human embryos for industrial purposes, and processes for modifying the genetic identity of animals that may cause them suffering (unless such processes are likely to produce substantial medical benefits for humans). The directive also states that individual patents may be denied if they are deemed to violate moral or ethical standards. It permits patenting of gene sequences, but only for particular industrial applications.

DNA (deoxyribonucleic acid)   The chemical in which inherited information is encoded, except in some viruses. Each DNA molecule consists of two phosphate-sugar backbones twined around each other in the shape of a double helix. Pairs of bases (adenine, thymine, guanine, and cytosine) joined by weak hydrogen bonds stretch between the backbones like rungs on a twisted ladder.

DNA Act (DNA Analysis Backlog Elimination Act)   A federal law, passed on January 24, 2000, that, among other things, requires submission of a DNA sample as a condition of a convict's supervised release (parole). *See also* **DNA Fingerprinting Act; Justice for All Act.**

DNA chip (gene chip, biochip)   A small piece of glass or other material embossed with microarrays of fragments of known genes (often hundreds of them on a single chip). When a drop of sample is placed on the chip, the active genes in the sample are bound by matching fragments on the chip. The matches are revealed when the chip is placed in an electronic detector, thereby showing which genes in the sample are active. Affymetrix invented this device, which it called the GeneChip, in 1996.

DNA "fingerprinting" (profiling, identification testing)   A technique invented by Alec Jeffreys of the University of Leicester in Great Britain in the early 1980s, which uses stretches of DNA that differ considerably from person to person as a means of determining whether a sample of DNA came from a particular person. It is most often used to determine family relationships or to find out whether a sample of DNA from a crime scene could have come from a certain suspect.

DNA Fingerprinting Act   Signed into law on January 5, 2006, this act (P.L. 109-162), among other things, authorizes expansion of the federal DNA database to include DNA collected from "individuals arrested, and from non-United States persons who are detained under the authority of the United States." *See also* **DNA Act; Justice for All Act.**

DNA Identification Act   Passed in 1994, this law gives the Federal Bureau of Investigation (FBI) responsibility for overseeing training, funding, and proficiency testing of all laboratories doing forensic DNA profiling. It also authorizes establishment of a national database of DNA profiles.

DNA profile   Material stored in a forensic database that describes only the 13 loci used in DNA identification testing. Unlike an actual sample of blood or tissue, a profile reveals no health-related genetic information.

Dolly   A sheep (named after singer-actress Dolly Parton) cloned from an udder cell of an adult ewe by Ian Wilmut and his colleagues at the Roslin Institute in Scotland. She was the first mammal to be cloned from a mature adult cell. Her existence was announced in an article in *Nature* on February 27, 1997. Dolly was euthanized in February 2003 because she had developed a virus-induced lung tumor.

dominant gene   A gene that is expressed even if only one copy of it is inherited. *Compare with* **recessive gene.**

double helix   The shape of a DNA molecule, in which two parallel strands twist or coil like a corkscrew.

embryonic stem cells   Stem cells taken from an embryo, usually at the blastocyst stage. Embryonic stem cells have greater capacity to differentiate into different types of tissue than adult stem cells; they probably can form any tissue in the body. Many researchers therefore would rather work with embryonic than adult stem cells, but their use is more controversial because embryos must be destroyed in order to harvest them. *See also* **adult stem cells; stem cells.**

enzyme   A protein that catalyzes a chemical reaction, speeding the rate at which it occurs.

epigenetics   The study of the influence of the environment on genes. Scientists are learning that environmental cues can considerably modify the way particular genes are expressed, having major effects on the body and even on an organism's offspring without changing the genetic code.

Escherichia coli (E. coli)   A common type of bacterium that usually lives harmlessly in the human intestine (though some strains of it can cause serious illness). It grows easily in the laboratory and was frequently used in early recombinant DNA and other genetics experiments.

ethanol   A fuel made from agricultural products that can be substituted for gasoline in vehicles. In the future, industrial biotechnology may allow it to be made from plant waste.

eugenics   A pseudoscience founded in the late 19th century by Francis Galton (he coined the word, from Greek words meaning "well born," in 1883). It holds that complex behaviors and characteristics such as intelligence are inherited and that the human race can be improved by encouraging people with desirable traits to reproduce and discouraging or forcibly preventing those with undesirable traits from doing so.

eukaryote   An organism made up of one or more cells with nuclei. All living things except viruses, bacteria, and blue-green algae are eukaryotes. *Compare with* **prokaryote.**

exon   The part of a gene that carries the code for making a protein. *Compare with* **intron.**

familial search   A search of a forensic DNA database to identify close partial matches with DNA from crime scenes (which suggest that the perpetrator of the crime is related to a person whose profile is in the database), or obtaining (often without consent) DNA samples from relatives of a suspect who refuses to donate DNA, with the intention of testing those samples in place of the suspect's DNA.

Federal Food, Drug, and Cosmetic Act (FFDCA)   First passed in 1938, this act gave the Food and Drug Administration (FDA) and, later, the Environmental Protection Agency (EPA) the right to set tolerances for certain substances on and in food and feed. Herbicide residues on genetically engineered herbicide-tolerant food crops and pesticides (such as Bt toxin) produced by genetically engineered food crops are regulated under the FFDCA.

Federal Insecticide, Fungicide, and Rodenticide Act (FIFRA)   A law, passed in 1947, that governs the testing and use of pesticides. After the Environmental Protection Agency (EPA) was established in 1970, it took over the administration of regulations required by FIFRA. FIFRA has been amended to include genetically engineered crops that produce Bt *(Bacillus thuringensis)* toxin in its definition of pesticides.

fermentation   A group of processes in which microorganisms such as bacteria or yeasts cause chemical changes that include the production of gas. Fermentation is part of many traditional biotechnology processes, such as those used to make alcoholic beverages. It was thought to be strictly a chemical reaction until the mid-19th century, when Louis Pasteur proved that it required living microorganisms.

Flavr Savr A type of tomato that a company called Calgene genetically altered to delay its breakdown after ripening. In 1994, Flavr Savr became the first genetically modified food to go on sale. It quickly failed in the marketplace because consumers thought the added gene might be dangerous.

"Frankenfoods" Disparaging term used for foods containing genetically modified material, especially in Europe. It refers to the monster created by a scientist in Mary Shelley's novel *Frankenstein* (1818).

Frye rule A rule (first stated in a 1923 case, *Frye v. United States*) that judges often use to decide whether evidence based on new scientific techniques will be admitted in court. The rule recommends accepting such evidence only if the technique on which it is based has "gained general acceptance" among scientists in its field.

fullerene ("buckyball") A hollow sphere made of 60 carbon atoms, one common kind of nanoparticle. It was named after American inventor Buckminster Fuller (1895–1983) because its shape reminded scientists of Fuller's geodesic domes.

gene The basic unit of inherited information, consisting of a sequence of bases in a DNA molecule that carries the code for production of a specific protein or RNA molecule or for control of another gene.

gene chip *See* **DNA chip.**

gene flow Transfer of genes from one population to another, such as the transfer of genes from bioengineered crops to wild relatives through the spread of pollen.

Genentech The first biotechnology company to be based on recombinant DNA technology. Herbert Boyer and Robert Swanson founded it in 1976.

gene splicing Common term for insertion of one or more genes from one species into the genome of another; also called recombinant DNA technology.

gene therapy Treatment of a disease by altering genes. Gene therapy was first successfully used on a human being in September 1990.

genetic code The sequence of bases in a DNA or RNA molecule that determines the sequence of amino acids in a protein. Each "letter" of the code consists of a group of three bases.

genetic determinism The belief that most physical and mental characteristics, including complex behaviors, are determined primarily or exclusively by genetics (heredity).

genetic engineering Direct manipulation of genetic information or transfer of genes from one type of organism to another to produce new biological structures or functions useful to human beings. Genetic engineering includes, but is not limited to, recombination of DNA.

Glossary

genetics The study of the patterns and mechanisms by which traits are inherited (passed from parents to offspring).

genetic screening Testing of a population to identify people at risk for suffering from a genetic disease or passing such a disease to their children.

genome An organism's complete collection of genetic information.

genomics The science of identifying genes and their functions, including building maps and databases of genes. Genomics studies large numbers of genes, or whole genomes, at once, including ways that groups of genes interact with each other and the environment.

genotype The nature of an organism or group as determined by its genes.

germ-line genes Genes that are contained in the sex cells (the cells that will become sperm and eggs) and therefore can be passed on to offspring.

glyphosate A common type of herbicide (weed killer). Monsanto Corporation markets it under the name Roundup and has genetically engineered certain crops to be resistant to it. *See also* **"Roundup Ready" crops.**

GM (genetically modified) foods Foods that contain, or may contain, material from transgenic organisms.

golden rice Rice engineered in 1999 to contain daffodil and bacterial genes that allow it to make beta-carotene, a building block of vitamin A. Biotechnology supporters have hailed it as a possible cure or preventive for vitamin A deficiency, a widespread health problem for children in developing countries.

guanine One of the four bases in DNA and RNA. It always pairs with cytosine.

haplotype A segment of DNA, containing multiple areas of variation, that tends to be inherited as a block. *See also* **HapMap Project.**

HapMap Project A genetic atlas, the first draft of which was completed in 2005, that focuses on stretches of DNA (haplotypes) that differ from person to person and potentially can help to identify variations that predispose people to certain diseases.

Harvard Oncomouse A type of mouse genetically engineered to be unusually susceptible to cancer. Intended for use in medical research and testing of possible carcinogens, it was patented by Harvard in 1988. It was the first genetically altered animal to be patented.

Health Insurance Portability and Accountability Act (HIPAA) Passed in 1996, this act includes a provision forbidding health insurers that issue group plans to employers from denying insurance on the basis of preexisting genetic conditions. It does not apply to individual plans or to employers who ensure their own workers.

hemoglobin The iron-containing pigment in red blood cells that gives the cells (and the blood) their color and carries oxygen throughout the body. Several inherited blood diseases, including sickle-cell disease, produce abnormal forms of hemoglobin.

Human Genome Project   An international project, launched in 1990, that aimed to sequence all the genes in the human genome. It announced a draft version of the human genome in June 2000 and a complete sequence of the genome in April 2003.

Huntington's disease   An inherited form of incurable brain degeneration caused by a single dominant gene. Affected people usually show no symptoms until middle age. A person who inherits even one copy of the gene will develop the disease, and the child of someone with the disease has a 50 percent chance of developing it. The gene that causes Huntington's was identified in 1993 and can be detected by a test.

hybrid   An offspring of parents that differ significantly in genetic nature.

industrial biotechnology ("white" biotechnology)   Biotechnology that applies living things or chemicals derived from them to manufacturing processes and products.

Innocence Project   A project established in 1991 by New York City lawyers Barry Scheck and Peter Neufeld to obtain DNA testing for convicted felons who declared their innocence. As of October 2007, the project had freed 208 people, some of whom had been sentenced to death.

intron   A DNA base sequence that interrupts the protein-coding portion of a gene (exon). The function of introns is currently unknown, so they are sometimes called "junk DNA."

in vitro fertilization   Combination of an egg and a sperm in a laboratory to create an embryo that is then implanted in a uterus; often called test-tube fertilization. ("In vitro" is Latin for "in glass.")

"junk DNA"   Popular name for introns. *See also* **intron.**

Justice for All Act   A federal law (P.L. 108-405) that took effect on October 30, 2004. It expanded a grant program, begun by the DNA Act (2000), to help states reduce their backlogs of samples awaiting forensic DNA testing, as well as establishing other programs to train personnel in DNA collection and to provide access to postconviction DNA testing in federal cases. It also expanded the national DNA database to include samples from everyone convicted of a federal felony offense. *See also* **DNA Act; DNA Fingerprinting Act.**

ligase   A type of enzyme that permanently links complementary single strands of DNA.

locus (pl. *loci*)   The place on a chromosome occupied by a gene or other specified sequence of DNA used as a marker.

macular degeneration   An eye disease, usually related to age, that is the most common cause of blindness in older people. Its exact cause is unknown, but certain genetic variations are associated with increased susceptibility to it.

maize Alternate name for the crop commonly called corn in the United States.

mitochondrial DNA DNA contained within mitochondria, organelles that help a cell use energy. Unlike nuclear DNA, mitochondrial DNA is inherited only through the female line. It has been used to trace family relationships and track missing persons. Some diseases can be carried in mitochondrial DNA, and special in vitro fertilization techniques have been developed to avoid transmission of these illnesses. Such techniques may inadvertently change germ-line genes.

molecular biology A branch of biology dealing with the molecular basis of biological processes such as protein synthesis and transmission of inherited characteristics.

moratorium A temporary halt to an activity.

motor neuron(e) disease The British name for one of a group of diseases that causes slowly developing loss of control of the muscles that produce movement. (In the United States, the illness is called **amyotrophic lateral sclerosis [ALS]** or Lou Gehrig's disease, after a famous baseball player who had it.) Its cause is presently unknown, and it is incurable and usually fatal. British researcher Ian Wilmut hopes to learn more about it by creating and studying embryonic stem cells from cells of people with the disease.

mutation An inheritable change in a DNA sequence.

nanobiotechnology The part of nanotechnology that deals with biological materials and applications that affect the human body.

nanotechnology Technology that deals with particles one to 100 nanometers (billionths of a meter) in size.

National DNA Index System (NDIS) A national database of DNA identity profiles of convicted criminals, established by the FBI in October 1998.

natural selection The mechanism of evolution described by Charles Darwin in *On the Origin of Species* (1859) and independently by Alfred Russel Wallace. It states that the members of a species with inherited characteristics that make them most suited to survive in a particular environment will be most likely to survive and bear young in that environment. Over generations, therefore, a change in the environment can produce a change in the predominant characteristics of a species.

neem A tree native to India, products of which have been used for millennia to fight bacteria, fungi, and pest insects. Multinational giant W. R. Grace and the U.S. Department of Agriculture (USDA) obtained a European patent for using neem oil against fungi on plants in 1994, which activists considered a "biopiracy" theft of traditional people's knowledge. The European Patent Office revoked the patent in 2000.

nuclear transfer   A technique for cloning in which the nucleus of a cell (embryonic or, more recently, mature) is transferred into an unfertilized egg from which the nucleus has been removed. The two are then fused by electricity. The resulting single cell develops into a clone of the organism that provided the nucleus.

nucleic acid   A large molecule composed of nucleotides. The most common nucleic acids are DNA and RNA.

nucleotide   A subunit of DNA or RNA composed of a base and an attached "backbone" of phosphate and sugar. Each DNA or RNA molecule contains thousands of nucleotides linked together.

nucleus   The membrane-bound central body in eukaryotic cells that contains the cell's main genetic material.

parthenote   An embryolike structure that sometimes develops from an unfertilized egg. It does not go on to form a fetus. Some researchers have derived embryonic stem cells from nonhuman parthenotes.

pharmacogenomics   An applied science that studies the relationship between particular genetic variations and the actions of drugs.

pharming   Use of genetically altered farm animals or plants to produce human body chemicals and drugs.

Plant Patent Act   Passed in 1930, this act stops short of allowing conventional patents on plants but allows plant breeders to forbid others to clone (reproduce asexually) hybrid varieties that they have developed.

Plant Pest Act   Passed in 1987, this act gives the U.S. Department of Agriculture's Animal and Plant Health Inspection Service (APHIS) the right to regulate any organisms, including those produced by genetic engineering, that are or might be plant pests.

Plant Variety Protection Act   This act, passed in 1970, extends the Plant Patent Act to forbid sexual reproduction of plant varieties without their developers' permission.

plasmid   A circular DNA molecule, used by some bacteria to transfer genes from one microorganism to another. It is separate from the main bacterial genome and can reproduce on its own. Herbert Boyer and Stanley Cohen used plasmids in some of the first recombinant DNA experiments.

polylactide (PLA)   A clear plastic made from corn that can be shaped into biodegradable containers, clothing, or other materials. It is a product of industrial biotechnology.

polymerase chain reaction (PCR)   A method for rapidly multiplying copies of a DNA sequence, developed in the early 1980s by Kary Mullis. It can be used to increase tiny samples of DNA to a size usable in DNA identification testing or gene sequencing.

precautionary principle   The idea that a new technology should be assumed to be, and regulated as, potentially dangerous until it has been proved safe, even if the dangers have not been proved. The Cartagena

Glossary

Biosafety Protocol and numerous national and international laws and treaties, especially in Europe, apply this principle to genetically modified foods.

preimplantation genetic diagnosis (PGD) A technique developed in the early 1990s, in which a group of embryos are created from a couple's eggs and sperm in a fertility clinic and then analyzed genetically. Only embryos that meet specified criteria, such as absence of a disease-causing gene or presence of genes producing tissue compatibility with a sibling, are implanted and allowed to develop into babies.

prokaryote A cell or organism that lacks a separate nucleus, such as a bacterium. *Compare with* **eukaryote.**

protein One of a large family of substances that are composed of amino acids arranged in a certain order. Proteins carry out most functions in cells, including acting as enzymes, structural components, and signaling molecules. They are assembled according to instructions in the genetic code contained in DNA and RNA. Instructions for each type of protein are carried on a separate gene.

proteomics The study of the structure and function of the proteins that are produced by genes.

rBGH *See* **bovine growth hormone.**

reach-through rights Rights demanded by some holders of biotechnology patents as part of licensing agreements. If a company has reach-through rights to a technology, it can demand royalties on all products developed with that technology, even if the development work is done by others.

recessive gene A gene that does not produce a detectable characteristic unless copies of that gene have been inherited from both parents. *Compare with* **dominant gene.**

recombinant DNA DNA from different types of organisms that is combined directly in a laboratory. *See also* **gene splicing.**

Recombinant DNA Advisory Committee (RAC) A group of experts formed by the National Institutes of Health in October 1974 to judge the safety of recombinant DNA experiments. It still exists, but it now reviews only experiments that differ from previous ones in a substantial way or involve treatment of humans (gene therapy).

reproductive cloning Cloning of human embryos or cells for the purpose of producing a child. *Compare with* **therapeutic cloning.**

restriction enzyme (restriction endonuclease) An enzyme produced by some bacteria to break up the DNA of invading viruses. A restriction enzyme cuts a DNA molecule at any spot where a particular sequence of bases occurs, leaving a piece of single-stranded DNA at each end. There are many types of restriction enzymes, each of which cuts at a different sequence. Restriction enzymes have been used in recombinant DNA technology since its start.

RNA (ribonucleic acid) A single-stranded nucleic acid with a molecular structure similar to that of DNA but with uracil substituting for thymine among the bases. Unlike DNA, RNA is found in the cytoplasm of the cell as well as the nucleus. It plays a vital role in turning the DNA code into protein molecules and in other chemical activities of the cell. It exists in several forms, including messenger RNA and transfer RNA.

RNA interference (RNAi) A phenomenon in which short stretches of double-stranded RNA are used to turn off particular genes. New drugs applying RNAi may soon be used against cancer and other diseases.

"Roundup Ready" crops Crops genetically engineered by Monsanto Company to be resistant to glyphosate, a type of herbicide that the company markets under the name Roundup.

short tandem repeats (STRs) Areas of DNA containing many short, identical sequences in a row. The number of repeats varies considerably from person to person. A test using STRs at 13 loci has become the standard form of DNA profiling in the United States.

sickle-cell disease (sickle-cell anemia) An inherited blood disease, fairly common among people of African descent, that is caused by abnormal hemoglobin produced by a defective gene. The mutant hemoglobin deforms round red blood cells to a sickle shape and causes them to block tiny blood vessels, starving the body of oxygen and causing pain, illness, and sometimes an early death. Because the gene that causes the disease is recessive, people develop it only if they inherit the mutant gene from both parents. A person with only one mutant gene is called a sickle-cell carrier or a possessor of sickle-cell trait.

sickle-cell trait A trait possessed by people, usually of African descent, who inherit one normal and one mutant copy of the gene that, when a person inherits two copies, causes sickle-cell anemia. People with sickle-cell trait are also called sickle-cell carriers because they can pass the defective gene to their children. They themselves, however, are perfectly healthy. Indeed, they seem to have unusual resistance to malaria, a serious blood disease caused by a parasite that is widespread in Africa. People with sickle-cell trait have sometimes been discriminated against in employment or insurance because of the mistaken impression that they are or will become ill.

somatic cell Any cell in the body except the sex cells (sperm and eggs) and their precursors.

somatic cell nuclear transfer A cloning technique in which the cloned cell is a mature body cell. This technique was used to create Dolly the sheep. *See also* **nuclear transfer.**

StarLink A type of genetically engineered corn approved for animal feed but not for human food because of the possibility that it might cause allergic reactions. Small amounts of it were found to have tainted corn-containing foods in fall 2000, causing a wide outcry and a recall of 430 million bushels of corn and 300 types of corn products.

stem cell A cell that has not yet differentiated and has the potential to produce cells of a wide range of types, in some cases any type. The study of stem cells may eventually produce tissues for transplantation or treatment of degenerative diseases. Stem cells from early embryos (embryonic stem cells) are thought to have the ability to develop into more different types of tissues than stem cells taken from adults, newborns, or fetuses (adult stem cells). Many scientists therefore prefer them for research, but their use is more controversial because embryos must be destroyed in order to harvest them.

"sticky ends" Pieces of single-stranded DNA at the ends of a double-stranded segment left after cutting by restriction enzymes or certain other treatments. They attract and bind to other pieces of single-stranded DNA with a complementary sequence. This fact was used in the creation of recombinant DNA in the early 1970s.

structural gene A gene that contains coded instructions for making a particular protein or RNA molecule.

sustainability The ability to meet the needs of the present without damaging the power of future generations to meet their own needs.

SV40 (Simian Virus 40) A monkey virus that can cause cancer in mice. Paul Berg used it in the first recombinant DNA experiments.

synthetic biology A new discipline that integrates engineering with molecular biology in order to create artificial biological systems, even perhaps whole organisms, from standardized component parts.

technology transfer The process of converting the results of basic scientific research into commercially useful products.

"Terminator" seeds Seeds containing a desirable genetically engineered trait that have also been modified to make them unable to reproduce, thus forcing farmers to buy new seeds each year from the company that owns the patented seeds. Such seeds have been developed, but, as of 2007, they have never been sold.

therapeutic cloning Cloning for the purpose of producing embryos from which embryonic stem cells can be harvested; the embryos are never allowed to develop into babies. It is also called research cloning. *Compare with* **reproductive cloning**.

thymine One of the four bases in DNA. It always pairs with adenine. In RNA it is replaced by uracil.

Toxic Substances Control Act (TSCA) An act passed in 1976 that gives the Environmental Protection Agency (EPA) the right to regulate release and sale of "new" chemicals. TSCA was amended in 1986 to classify genetically engineered organisms expressing new traits or containing genes from two different genera as equivalent to new chemicals and thus susceptible to regulation under the act.

Trade Related Intellectual Property Rights Agreement (TRIPS) Agreement of the World Trade Organization (WTO), which went into effect in 1995. This agreement requires WTO member countries to "harmonize" their patent laws with those of the United States and other industrialized countries, supposedly in order to facilitate the transfer of technology to those countries. Critics say its real purpose is to protect the rights and profits of multinational corporations.

transgenic organism A living thing that has been changed by adding genes from another species, removing genes, or changing the activity (expression) of existing genes.

umbilical cord blood Blood taken, just after a baby's birth, from the cord that connects the baby to the placenta (which nourished it in the womb). This blood contains stem cells that, although not as capable of differentiating into many tissues as embryonic cells, are more capable than other adult stem cells. Cord blood is used to treat some diseases, and some parents store it in case it is needed to treat their children or other family members. A bill authorizing and providing funding for a national cord blood bank was signed into law in December 2005.

uracil One of the four bases in RNA. It is the equivalent of thymine in DNA and pairs with adenine.

vector A tool for carrying genes from one organism into the genome of another. Plasmids and certain viruses are examples of vectors.

xenotransplantation Transplantation of tissues or organs from one species into the body of another species.

PART II

GUIDE TO FURTHER RESEARCH

CHAPTER 6

How to Research Biotechnology and Genetic Engineering

Although students, teachers, journalists, and other investigators may ultimately have different objectives in doing research on biotechnology and genetic engineering, all are likely to begin with the same basic steps. The following approach should be suitable for most purposes:

- Gain a general orientation by reading the first part of this book. Chapter 1 can be read as a narrative, while chapters 2–5 are best skimmed to get an idea of what is covered. They can then be used as a reference source for helping make sense of the events and issues encountered in subsequent reading.

- Skim some of the general books listed in the first section of the bibliography (chapter 7). Neutral overviews and books that provide pro and con essays on various issues in the field are particularly recommended.

- Browse the many web sites provided by organizations involved in various aspects of biotechnology and genetics issues (see chapter 8). Whether they favor, oppose, or are neutral toward the field, their pages are often rich in news, articles, and links to other organizations as well as describing particular cases and discussing the pros and cons of various practices involving biotechnology.

- Use the relevant sections of chapter 7 to find more books, articles, and online publications on particular topics of interest.

- Find more (and more recent) materials by using the bibliographic tools such as the library catalogs and periodical indexes discussed later.

- To keep up with current events and breaking news, check back periodically with media and organization web sites and periodically search the catalogs and indexes for recent material.

173

The rest of this chapter is organized according to the various types of resources and tools. The three major categories are online resources, bibliographic resources, and legal research, including both laws and court cases.

ONLINE RESOURCES

The tremendous growth in the resources and services available through the Internet (and particularly the World Wide Web) is providing powerful new tools for researchers. Mastery of a few basic online techniques enables today's researcher to accomplish in a few minutes what used to require hours in the library poring through card catalogs, bound indexes, and printed or microfilmed periodicals.

Not everything is to be found on the Internet, of course. While a few books are available in electronic versions, most must still be obtained as printed text. Many periodical articles, particularly those more than 10 years old, must still be obtained in hard-copy form from libraries. Nevertheless, the Internet has now reached critical mass in the scope, variety, and quality of material available. Thus, it makes sense to make the Internet the starting point for most research projects. This is particularly true regarding recent events in biotechnology and genetic engineering. Web/Internet links can lead the researcher not only to companies, professional organizations, and journals in the field but even to complex databases of DNA sequences and other highly technical material.

For someone not used to it, searching the Internet and the World Wide Web can feel like spending hours trapped inside a pinball machine. The shortest distance between a researcher and what he or she wants to know is seldom a straight line, at least not a single straight line. These things are called nets and webs for good reason: Everything is connected by links, and often a researcher must travel through a number of them to find the desired information.

Web searching is best approached with a combination of patience, alertness, and, preferably, humor. A given search often will not reveal the desired information but will unearth at least three things, or groups of things, that are even more interesting. The information sought on the initial search, meanwhile, will be uncovered by chance at a later time when the researcher is looking for something else entirely. The sooner one accepts this, the sooner searching is likely to become rewarding rather than painful. In addition to specific files related to particular areas of research, it is a good idea for researchers to save promising URLs (universal resource locators or web addresses) in their web browser's Favorites list.

It is easy to feel lost on the Web, but it is also easy to find one's way back. During any given search, the Back button on the browser is the Ariadne's thread that will guide the researcher back through the labyrinth to the beginning of the adventure on the browser's home page, passing en route through all the sites visited (so that one can stop for another look or, if desired, jump off to somewhere else). Alternatively, Home (the house icon) takes users directly back to their home page, like Dorothy clicking her red shoes. The History button provides a list of all the sites visited on recent previous sessions.

Finally, a word of caution about the Internet. It is important to critically evaluate all online materials. Many sites have been established by well-known, reputable organizations or individuals. Others may come from unknown individuals or groups. Their material may be equally valuable, but it should be checked against reliable sources. Different groups' biases may be obvious or subtle. Gradually, each researcher will develop a feel for the quality of information sources as well as a trusty tool-kit of techniques for dealing with them.

TOOLS FOR ORGANIZING RESEARCH

Several techniques and tools can help the researcher keep materials organized and accessible:

Use the web browser's "Favorites" or "Bookmarks" menu to create a folder for each major research topic (and optionally, subfolders). For example, folders used in researching this book included: organizations, laws, cases, current news, reference materials, and bibliographical sources.

Use favorites or bookmark links rather than downloading a copy of the actual web page or web site, which can take up a large amount of both time and disk space. Exception: if the site has material that will definitely be needed in the future, download it to guard against its disappearance from the Web.

Use a simple database program (such as Microsoft Works) or, perhaps better, a free-form note-taking program. This makes it easy to take notes (or paste text from web sites) and organize them for later retrieval.

WEB INDEXES

A web index is a site that offers what amounts to a structured, hierarchical outline of subject areas. This organization enables the researcher to zero in on a particular aspect of a subject and find links to web sites for further exploration. A staff of researchers constantly adds to and updates the links.

The best known (and largest) web index is Yahoo! (http://www. yahoo. com). Its directory page (http://dir.yahoo.com) gives the top-level list of topics. Five of these are of particular use for researching biotechnology and genetic engineering:

Numerous topics under the major heading *Science* are relevant. In addition to general subjects such as journals and news/media, specific secondary subject headings likely to be helpful are Agriculture; Animals, Insects, and Pets; Biology; Ecology; Forensic Science; Life Sciences; Medicine; and Nanotechnology. Selecting the secondary heading Biology produces a further host of useful subheads, including Bioethics, Biochemistry, Biodiversity, Biological Safety, Biotechnology, Cell Biology, Companies, Genetics, Institutes, Journals, Molecular Biology, Organizations, Pharmacology, and Reproduction.

Government provides listings including documents, ethics, law, and research labs. Categories under the *Law* subhead that are likely to be useful include Cases, Countries, Criminal Justice, Disability, Employment Law, Environmental, Health, Indigenous Peoples, Intellectual Property, Legal Research, Privacy, and Technology.

Health produces subheads dealing with medicine, health care, reproductive health, and diseases and conditions, which includes genetic disorders.

Headings under *Business and Economy* that may be useful include Ethics and Responsibility, Global Economy, Intellectual Property, and Law.

Society and Culture may have biotechnology information under the headings of Disabilities, Environment and Nature, and Issues and Causes, as well as under more general headings such as Cultural Policy.

In addition to following Yahoo!'s outline-like structure, there is also a search box into which the researcher can type one or more keywords and receive a list of matching categories and sites. (The box is rather confusingly labeled "Search the Web," but it also searches Yahoo!'s directories, and the results of this search appear at the top of the page.)

Web indexes such as Yahoo! have two major advantages over undirected surfing. First, the structured hierarchy of topics makes it easy to find a particular topic or subtopic and then explore its links. Second, Yahoo! does not make an attempt to compile every possible link on the Internet (a task that is virtually impossible, given the size of the Web). Rather, sites are evaluated for usefulness and quality by Yahoo!'s indexers. This means that the researcher has a better chance of finding more substantial and accurate information. The disadvantage of web indexes is the flip side of their selectivity: The researcher is dependent on the indexer's judgment for determining what sites are worth exploring.

Two other web indexes are LookSmart (http://search.looksmart.com) and About.com (http://www.about.com), the latter of which is run by About, a company formerly named The Mining Company. About.com features overviews or guides prepared by self-declared experts.

SEARCH ENGINES

Search engines take a very different approach to finding materials on the Web. Instead of organizing topically in a "top down" fashion, search engines work their way "from the bottom up," scanning through Web documents and indexing them. There are hundreds of search engines, but some of the most widely used are the following:

AltaVista (http://www.altavista.com)
Excite (http://www.excite.com)
Go.com (http://go.com)
Google (http://www.google.com)
Hotbot (http://www.hotbot.com)
Lycos (http://www.lycos.com)
WebCrawler (http://www.webcrawler.com)

Search engines are generally easy to use by employing the same sorts of keywords that work in library catalogs. There are a variety of Web search tutorials available online (try "Web search tutorial" in a search engine).

Here are a few basic rules for using search engines:

When looking for something specific, use the most precise term or phrase. For example, when looking for information about DNA fingerprinting, use "DNA fingerprinting" or "forensic DNA testing," not "DNA." (When using phrases as search specifications, enclose them in quotation marks.)

When looking for a more general topic, use several descriptive words (nouns are more reliable than verbs), such as privacy genetic information. (Most engines will automatically put pages that match all three terms first on the results list.)

Use "wildcards" when a desired word may have more than one ending. For example, gene* matches gene, genetics, genetic engineering, and so on.

Most search engines support Boolean (AND, OR, NOT) operators, which can be used to broaden or narrow a search.

Use AND to narrow a search. For example, agriculture AND biotechnology will match only pages that have both terms.

Use OR to broaden a search. Agriculture OR biotechnology will match any page that has *either* term.

Use NOT to exclude unwanted results. Biotechnology NOT agriculture finds articles about biotechnology other than agricultural biotechnology.

Since each search engine indexes somewhat differently and offers somewhat different ways of searching, it is a good idea to use several different search engines, especially for a general query. Several "metasearch" programs automate the process of submitting a query to multiple search engines. These include:

Dogpile (http://www.dogpile.com)
Metacrawler (http://www.metacrawler.com)
Search (http://www.search.com)
Surfwax (http://www.surfwax.com)

Metasearch engines may overwhelm researchers with results (and may insufficiently prune duplicates), however, and they often do not use some of the more popular search engines, such as Google.

MEDIA SITES

News (wire) services, most newspapers, and many magazines have web sites that include breaking news stories and links to additional information. The following media sites have substantial listings for stories on biotechnology and genetic engineering:

- ABC News: URL: http://abcnews.go.com
- Cable News Network (CNN): URL: http://www.cnn.com
- *New York Times:* URL: http://www.nytimes.com (offers only abstracts unless readers pay)
- Reuters: URL: http://www.reuters.com
- *Time magazine:* URL: http://www.time.com/time
- *Washington Post:* URL: http://www.washingtonpost.com

Yahoo! maintains a large set of links to many newspapers that have web sites or online editions: see http://dir.yahoo.com/News_and_Media/Newspapers/ Web_Directories.

METASITES: EVERYTHING ABOUT BIOTECH AND THEN SOME

One basic principle of research is to take advantage of the fact that other people may have already found and organized much of the most useful information about a particular topic. For biotechnology and genetic engineering, there are several web sites that can serve as excellent starting points for research because they provide links to vast numbers of other resources.

- Access Excellence, http://www.accessexcellence.org, is aimed at biology teachers and students. It has an extensive "About Biotech" page at http:// www.accessexcellence.org/RC/AB. Subsections of this page are Issues and Ethics, Biotech Applied, and Biotech Chronicles (history). It also has links to biotechnology web sites at http://www.accessexcellence.org/ RC/biotech.html and genetics web sites at http://www.accessexcellence. org/RC/genetics.html.
- Bioethics Resources on the Web, http://bioethics.od.nih.gov, is sponsored by the National Institutes of Health and has links to sites about various bioethical issues. Its genetics section covers genetics and genom-

ics, gene patenting, genetic testing and counseling, and gene therapy and gene transfer.

- Two sites sponsored by biotechnology trade organizations present highly favorable views of the technology. The U.S. site, from the Biotechnology Industry Organization, is http://www.bio.org; the European one, from Europabio, is http://www.europabio.org/index.htm.

- The J. Craig Ventar Institute provides a links page at http://www.tigr.org/ links.shtml. The links cover news, genome databases, DNA educational sites, electronic journals, and more.

- Indiana University and the University of Texas have a combined site, Biotech: Life Sciences Resources and Reference Tools, at http://biotech.icmb. utexas.edu. It includes an illustrated dictionary, an extensive list of Science Resources links, and BioMedLink, a mega-database of biomedical sites.

- The U.S. Food and Drug Administration (FDA) web site has a page on genetically engineered crops, http://www.cfsan.fda.gov/~lrd/biotechm.html, and one on cellular and gene therapy, http://www.fda.gov/cber/gene.htm. The FDA's Center for Veterinary Medicine, which regulates transgenic and cloned animals and animal feeds, has a site on genetically engineered animals and animal drugs at http://www.fda.gov/cvm/bio_drugs.html. They have a page on animal cloning at http://www.fda.gov/cvm/cloning.htm.

- The international Organisation for Economic Co-operation and Development (OECD) has considerable material about biotechnology at http:// www.oecd.org/topic/0,3373,en_2649_37437_1_1_1_1_37437,00.html.

- GeneWatch UK, http://www.genewatch.org, a British organization that is generally negative toward genetic modification, provides access to a wide variety of news stories and reports on the subject. Topics covered include genetically modified crops and foods, genetically modified animals, human genetics, laboratory use, biological weapons, and patenting. Some reports are not available online and must be ordered.

- The site of the Council for Responsible Genetics, http://www.gene-watch. org, presents an (at best) cautious view of biotechnology and human genetic alteration. It discusses genetic determinism, discrimination and privacy in connection with genetic testing, human genetic manipulation and cloning, and biotechnology and agriculture.

- Foundations of Classical Genetics, http://www.esp.org/foundations/ genetics/classical, offers original manuscripts of some of the most important papers in the history of genetics research. It also has a chronology and links to related sites.

- The National Center for Biotechnology Information, http://www.ncbi. nlm.nih.gov, provides access to a variety of human genome and other gene sequence databases.

SPECIFIC INTERNET/WEB RESOURCES

In addition to the metasites, there are many web pages devoted to particular aspects of biotechnology and genetic engineering. Here is a small sampling of the most interesting and extensive ones:

Agricultural Biotechnology

- For a neutral, international perspective, see the United Nations Food and Agriculture Organization (FAO) site on biotechnology in food and agriculture, http://www.fao.org/biotech/index.asp?lang=en.

- The British point of view on genetically modified food can be found at a site sponsored by the University of Reading, http://www.ncbe.reading.ac.uk/ncbe/gmfood/menu.html.

- The Pew Initiative on Food and Biotechnology, http://pewagbiotech.org, presents a neutral view of GM food technology from the United States. Its site contains numerous reports, papers, and news items.

- Two sites strongly supportive of agricultural biotechnology, sponsored by or linked to leading biotechnology trade organizations, are AgBioWorld (http://www.agbioworld.org) and Council for Biotechnology Information (http://www.whybiotech.com).

- For a negative view of this technology, stressing its potential risks to environment and health, see a set of pages on genetic engineering in agriculture sponsored by the Union of Concerned Scientists (http://www.ucsusa.org/food_and_environment/genetic_engineering.)

- The Third World Network's Biosafety Information Centre, http://www.biosafety-info.net/biosafety.php, presents similar information from the point of view of developing countries.

Human Genetics and Gene Modification

- The National Human Genome Research Institute, which carries out the Human Genome Project in the United States, has a variety of resources related to the project on its site at http://www.genome.gov. Its Policy and Ethics section has material on privacy of genetic information, genetic discrimination in employment and insurance, commercialization and patenting, DNA forensics, and genetics and the law. It includes a set of links to online resources on genetics at http://www.genome.gov/10000464 and to educational materials at http://www.genome.gov/Education. The project's Ethical, Legal, and Social Implications (ELSI) Research Program has its own site at http://www.genome.gov/10001618, as does the more recent HapMap, http://www.hapmap.org.

180

- Another government site, http://www.ornl.gov/sci/techresources/Human_ Genome/education/education.shtml, also has links to educational resources on the Human Genome Project, although dated items are not recent.
- For a British view of the Human Genome Project and genetics gener- ally, see the Sanger Institute's site, http://www.yourgenome.org/hgp/ hgp_1.shtml.
- The American Museum of Natural History has a series of linked pages related to an exhibition on advances in human genetics, "The Genomic Revolution," at http://www.amnh.org/exhibitions/genomics/0_home/ index.html.
- The University of Utah's Genetic Science Learning Center, http://learn. genetics.utah.edu, has educational materials about genetics, including "do-it-yourself genetics research" and material on stem cells, cloning, gene therapy, and personalized medicine.
- The High School Human Genome Program of Washington University has a "virtual sequencing" student exercise at http://hshgp.genome. washington.edu/student_activities/virtseq1.htm.
- The Center for Genetics and Society, http://geneticsandsociety.org, has accounts of various human genetic technologies (cloning, genetic en- hancement, and the like) and regulation of these technologies in different countries. It has recently expanded its site. Although the center presents views on both sides of these issues, it seems to be mostly opposed to al- teration of the human genome.
- The same is true of the Council for Responsible Genetics, which has a site on genetics and the law at http://www.genelaw.info. Topics covered include newborn screening, criminal justice, privacy and confidentiality, medical malpractice, employment, and health and life insurance.
- On the other hand, the Reproductive Cloning Network, http://www.re- productivecloning.net, and the Human Cloning Foundation, http://www. humancloning.org, contain essays explaining and supporting human re- productive cloning (cloning to produce children), as well as lists of books and other resources on the subject.

Other Subjects

- Information about the FBI's CODIS (Combined DNA Index System) program can be found at http://www.fbi.gov/hq/lab/codis/index1.htm. NDIS, the U.S. national forensic DNA database, is described more briefly at http://foia.fbi.gov/dna552.htm.
- The National Institutes of Health's stem cell research site, http://stem- cells.nih.gov, has links to basic descriptions, answers to frequently asked questions, descriptions of U.S. federal policy regarding stem cell research, reports, a glossary, and more.

- The International Society for Stem Cell Research, http://www.isscr.org, has numerous resources on that topic.
- The U.S. government's National Nanotechnology Initiative site, http://www.nano.gov, has resources and links on that subject.
- To get a feel for synthetic biology and a look at BioBricks and some of the other "parts" from which synthetic biologists are trying to compile biological circuits and possibly whole organisms, see the Massachusetts Institute of Technology (MIT)'s Synthetic Biology Working Group site, http://parts.mit.edu/registry/index.php/Main_Page.

FINDING ORGANIZATIONS AND PEOPLE

Lists of organizations connected with biotechnology can be found on archive sites (see Specific Internet/Web Resources, p. 155) and index sites such as Yahoo!, as well as in Chapter 8 of this book. If such sites do not yield the name of a specific organization, the name can be given to a search engine. Put the name of the organization in quotation marks.

Another approach is to take a guess at the organization's likely Web address. For example, the American Civil Liberties Union (which includes genetic discrimination among its concerns) is commonly known by the acronym ACLU, so it is not a surprise that the organization's web site is at http://www.aclu.org. (Note that noncommercial organization sites normally use the .org suffix, government agencies use .gov, educational institutions have .edu, and businesses use .com.) This technique can save time, but it does not always work.

There are several ways to find a person on the Internet. Put the person's name (in quotes) in a search engine and possibly find that person's home page on the Internet. Contact the person's employer (such as a university for an academic, or a corporation for a technical professional). Most such organizations have web pages that include a searchable faculty or employee directory. Try one of the people-finder services such as Yahoo! People Search (http://people.yahoo.com) or BigFoot (http://www.bigfoot.com). This may yield contact information such as an e-mail address, regular address, and/or phone number.

BIBLIOGRAPHIC RESOURCES

Bibliographic resources generally include catalogs, indexes, bibliographies, and other guides that identify the books, periodical articles, and other printed resources that deal with a particular subject. They are essential tools for the researcher.

LIBRARY CATALOGS

Most public and academic libraries have replaced their card catalogs with online catalogs, and many institutions now offer remote access to their catalog, either through dialing a phone number with terminal software or connecting via the Internet.

Access to the largest library catalog, that of the Library of Congress, is available at http://catalog.loc.gov. This page explains the different kinds of catalogs and searching techniques available.

Yahoo! offers a categorized listing of libraries at http://dir.yahoo.com/ reference/libraries. Of course, one's local public library (and for students, the high school or college library) is also a good source for help in using online catalogs.

With traditional catalogs, lack of knowledge of appropriate subject headings can make it difficult to make sure the researcher finds all relevant materials. Online catalogs, however, can be searched not only by author, title, and subject but also by matching keywords in the title. Thus a title search for "biotechnology" will retrieve all books that have that word somewhere in their title. (Of course, a book about biotechnology may not have the word *biotechnology* in the title, so it is still necessary to use subject headings to get the most comprehensive results.)

Once the record for a book or other item is found, it is a good idea to see what additional subject headings and name headings have been assigned. These in turn can be used for further searching.

BOOKSTORE CATALOGS

Many people have discovered that online bookstores such as Amazon.com (http://www.amazon.com) and Barnes & Noble (http://www.barnesandnoble. com) are convenient ways to shop for books. A lesser-known benefit of online bookstore catalogs is that they often include publisher's information, book reviews, and readers' comments about a given title. They can thus serve as a form of annotated bibliography.

On the other hand, a visit to one's local bookstore also has its benefits. While the selection of titles available is likely to be smaller than that of an online bookstore, the ability to physically browse through books before buying them can be very useful.

PERIODICAL DATABASES

Most public libraries subscribe to database services such as InfoTrac, which index articles from hundreds of periodicals, including some specialized ones. The database can be searched by author or by words in the title, subject headings, and sometimes words found anywhere in the article text. Depending on the database used, "hits" in the database can result in just a bibliographical description (author, title, pages, periodical name, issue date, and so on), a description plus an abstract (a paragraph summarizing the contents of the article), or the full text of the article itself.

Many libraries provide dial-in or Internet access to their periodical databases as an option in their catalog menu. However, licensing restrictions usually mean that only researchers who have a library card for that particular library can access the database (by typing in their name and card number). Check with local public or school libraries to see what databases are available.

A somewhat more time-consuming alternative is to find the web sites for magazines likely to cover a topic of interest. Some scholarly publications are putting all or most of their articles online. Popular publications tend to offer only a limited selection. Some publications of both types offer archives of several years' back issues that can be searched by author or keyword.

Nearly all newspapers now have web sites with current news and features. Generally a newspaper offers recent articles (perhaps from the last 30 days) for free online access. Earlier material can often be found in an archive section. A citation and perhaps an abstract are frequently available for free, but a fee of a few dollars may be charged for the complete article. One can sometimes buy a "pack" of articles at a discount as long as the articles are retrieved within a specified time. Of course, back issues of newspapers and magazines may also be available in hard copy, bound, or microfilm form at local libraries.

LEGAL RESEARCH

As issues related to biotechnology, genetic engineering, and human genetics continue to capture the attention of legislators and the public, a growing body of legislation and court cases concerning these issues is emerging. Because of the specialized terminology of the law, legal research can be more difficult to master than bibliographical or general research. Fortunately, the Internet has also come to the rescue in this area, offering a variety of ways to look up laws and court cases without having to pore through huge bound volumes in law libraries (which may not be accessible to the general public, anyway). To begin with, simply entering the name of a law, bill, or court case into a search engine will often lead the researcher directly to both text and commentary.

FINDING LAWS

When federal legislation passes, it becomes part of the U.S. Code, a massive legal compendium. Laws can be referred to either by their popular name or by a formal citation. For example, the DNA Identification Act is cited as 42 U.S.C. 14131, meaning title 42 of the U.S. Code, section 14131.

The U.S. Code can be searched online in several locations, but the easiest site to use is probably the one from Cornell Law School (a major provider of free online legal reference material), at http://www4.law.cornell.edu/uscode. The fastest way to retrieve a law is by its title and section citation, but phrases and keywords can also be used.

Federal laws are generally implemented by a designated agency that writes detailed rules, which become part of the Code of Federal Regulations (C.F.R.). A regulatory citation looks like a U.S. Code citation and takes the form *vol. C.F.R. sec. number*, where *vol.* is the volume number and *number* is the section number. Regulations can be found at the web site for the relevant government agency, such as the Food and Drug Administration or the Environmental Protection Agency.

Many state agencies have home pages that can be accessed through the Findlaw state resources web site (http://www.findlaw.com/11stategov). This site also has links to state law codes. These links may or may not provide access to the text of specific regulations, however.

KEEPING UP WITH LEGISLATIVE DEVELOPMENTS

Pending legislation is often tracked by advocacy groups, both national and those based in particular states. See chapter 8, "Organizations and Agencies," for contact information.

The Library of Congress Thomas web site (http://thomas.loc.gov) includes files summarizing legislation by the number of the Congress. Each two-year session of Congress has a consecutive number; for example, the 110th Congress was in session in 2007 and 2008. Legislation can be searched for by the name of its sponsor(s), the bill number, or by topical keywords. (Laws that have been passed can be looked up under their Public Law number.) For instance, selecting the 110th Congress and typing the phrase "human cloning" into the search box at the time of writing retrieved three bills containing that phrase. Further details retrievable by clicking on the bill number and then the link to the bill summary and status file include text, sponsors, committee action, and amendments.

A second extremely useful site is maintained by the Government Printing Office (http://www.gpoaccess.gov/index.html). This site has links to the Code of Federal Regulations, the Federal Register (which contains announcements of new federal agency regulations), the Congressional

Record, the U.S. Code, congressional bills, a catalog of U.S. government publications, and other databases. It also provides links to individual agencies, grouped under government branch (legislative, executive, or judicial), and to regulatory agencies, administrative decisions, core documents of U.S. democracy such as the Constitution, and hosted federal web sites.

FINDING COURT DECISIONS

Like laws, legal decisions are organized using a system of citations. The general form is: *Party 1 v. Party 2 volume court reports (year).*

Here are some examples from chapter 2:

Bragdon v. Abbott, 97 U.S. 156 (1998)

Here the parties are Bragdon (plaintiff) and Abbott (defendant), the case is in volume 97 of the *U.S. Supreme Court Reports,* and the year the case was decided is 1998. (For the Supreme Court, the name of the court is omitted).

John Moore v. Regents of California, 51 Cal. 3d 120 (1990)

Here the parties are John Moore (plaintiff) and the Regents of the University of California (defendant), the decision is in volume 51 of the California Supreme Court records, and the case was decided in 1990.

To find a federal court decision, first ascertain the level of court involved: district (the lowest level, where trials are normally held), circuit (the main court of appeals), or the Supreme Court. The researcher can then go to a number of places on the Internet to find cases by citation and, often, the names of the parties or subject keywords. Two of the most useful sites are the following:

The Legal Information Institute (http://supct.law.cornell.edu/supct/index.html) has all Supreme Court decisions since 1990 plus 610 of the most important older decisions. It also links to other databases with early court decisions.

Washlaw Web (http://www.washlaw.edu) has a variety of courts (including state and international courts) and legal topics listed, making it a good jumping-off place for many sorts of legal research. However, the actual accessibility of state court opinions varies widely.

For more information on conducting legal research, see the "Legal Research FAQ" at http://www.faqs.org/faqs/law/research. After a certain point, however, the researcher who lacks formal legal training may need to consult with or rely on the efforts of professional researchers or academics in the field.

CHAPTER 7

ANNOTATED BIBLIOGRAPHY

Hundreds of books and thousands of articles and Internet documents relating to biotechnology, genetic engineering, and genetics have appeared in recent years as these fields have grown in complexity and importance. They range from extremely technical "how-to" articles and descriptions of particular advances to reviews and opinion pieces aimed at the general public. This bibliography lists a representative sample of serious nonfiction sources selected for clarity and usefulness to the general reader, recent publication (all dated citations are from 2004 or later), and variety of points of view.

Listings in this bibliography are grouped by subject under the following headings:

- General Material on Biotechnology, Genetic Engineering, and Genetics
- Agricultural Biotechnology
- Patenting Life
- Human Genetics
- DNA "Fingerprinting" and Databases
- Genetic Health Testing and Discrimination
- Medical Biotechnology, Gene Therapy, and Human Gene Alteration
- Stem Cell Research and Human Cloning
- The Future of Biotechnology and New Forms of Biotechnology

Within each subject, entries are grouped by type (books, articles, and Web documents).

GENERAL MATERIAL ON BIOTECHNOLOGY, GENETIC ENGINEERING, AND GENETICS

BOOKS

Aldridge, Susan. *The Thread of Life*. New York: Cambridge University Press, 2004. Aldridge delves behind the headlines to explore the exciting world of molecular biology and its present and future applications.

Alexander, Brian. *Rapture: How Biotechnology Became the New Religion*. New York: Basic Books, 2004. Focusing on William Haseltine, a former Harvard professor who went on to run one of the world's largest biotechnology companies, Alexander explores the utopian dreams of entrepreneurs who helped to develop the biotechnology industry.

Augen, Jeff. *Bioinformatics in the Post-Genome Era: Genome, Transcriptome, Proteome, and Information-Based Medicine*. Boston: Addison-Wesley Professional, 2004. Covering both theoretical and practical aspects of bioinformatics, Augen shows how tools developed by combining information technology and new genetic knowledge have greatly accelerated the development of drugs and diagnostic techniques.

Avise, John C. *The Hope, Hype, and Reality of Genetic Engineering: Remarkable Stories from Agriculture, Industry, Medicine, and the Environment*. New York: Oxford University Press, 2004. Avise shows how researchers are reshaping organisms to meet human needs.

Bailey, Ronald. *Liberation Biology: The Scientific and Moral Case for the Biotech Revolution*. Amherst, N.Y.: Prometheus Books, 2005. Bailey provides an optimistic view of biotechnology, stressing its benefits and claiming that the fears of critics are misplaced.

Bains, William. *Biotechnology from A to Z*. 3d ed. New York: Oxford University Press, 2004. Three hundred entries provide clear descriptions of the concepts and terminology of biotechnology.

Barnum, Susan R. *Biotechnology: An Introduction*. 2d ed. Pacific Grove, Calif.: Brooks Cole, 2006. Barnum's textbook presents a broad view of biotechnology, including its medical, economic, social, and ethical implications.

Beyond Borders: Global Biotechnology Report 2006. New York: Ernst & Young, 2006. Annual report from an international business advice company presents statistics defining the biotechnology industry in four different regions of the world.

Bunton, Robin, and Alan Petersen, eds. *Genetic Governance: Health, Risk and Ethics in the Biotech Era*. New York: Routledge, 2005. These essays consider ways that new developments in genetics may affect contemporary liberal governance and risk assessment, particularly in the area of health care.

Annotated Bibliography

Burley, Justine, and John Harris, eds. *A Companion to Genethics*. Malden, Mass.: Blackwell, 2004. These essays consider the ethical, philosophical, political, and social significance of science's revolutionary new understanding of human genetics.

Canini, Mikko, ed. *Genetic Engineering*. San Diego, Calif.: Greenhaven Press, 2005. This anthology, part of the History of Issues series, presents the history of controversies concerning genetically modified crops, human cloning, and gene patenting.

Colson, Charles W., and Nigel M. de S. Cameron, eds. *Human Dignity in the Biotech Century: A Christian Vision for Public Policy*. Downers Grove, Ill.: InterVarsity Press, 2004. This essay collection presents conservative Christian thought (chiefly negative) about biotechnology and its ethical and social implications.

Dewar, Elaine. *The Second Tree: Stem Cells, Clones, Chimeras, and Quests for Immortality*. New York: Carroll & Graf, 2005. Dewar chronicles the lives, discoveries, and personal conflicts among the creators of modern genetics and biotechnology, offering a lively portrait of the way biologists make breakthroughs and showing how they could alter human evolution.

Duncan, David Ewing. *The Geneticist Who Played Hoops with My DNA, and Other Masterminds from the Frontiers of Biotech*. New York: Morrow, 2005. This collection presents biographical profiles of seven men and women on the frontiers of biotechnology research.

Engdahl, Sylvia Louise, ed. *Genetic Engineering*. Farmington Hills, Mich.: Greenhaven Press, 2006. Part of the Contemporary Issues Companion series, this anthology provides overviews and pro and con pieces on different aspects of gene alteration.

Finegold, David L., et al. *Bioindustry Ethics*. New York: Elsevier Academic, 2005. Finegold describes the ethical questions raised by the biotechnology industry, the industry's possible effects on global health, and the approaches to ethical issues taken by biotech and pharmaceutical companies.

Francioni, Francesco, and Tullio Scovazzi, eds. *Biotechnology and International Law*. Oxford, U.K.: Hart Publishing, 2006. The authors discuss legal questions raised by international biotechnology concerning ownership of genetic resources, environmental protection, trade, and human rights.

Fumento, Michael. *BioEvolution: How Biotechnology Is Changing Our World*. New York: Encounter Books, 2004. Fumento shows how biotechnology is changing people's lives and will do so to an even greater extent in the near future.

Gehring, Verna V., ed. *Genetic Prospects: Essays on Biotechnology, Ethics, and Public Policy*. Lanham, Md.: Rowman & Littlefield, 2004. The authors in this essay collection apply philosophical analysis to three kinds of questions about biotechnology and its implications.

Gerdes, Louise I. *Opposing Viewpoints: Genetic Engineering*. San Diego, Calif.: Greenhaven Press, 2004. This essay collection for high school students presents pro and con views of aspects of genetic engineering and its regulation, including genetically modified foods and "designer babies."

Glasner, Peter, ed. *Reconfiguring Nature: Issues and Debates in the New Genetics*. Burlington, Vt.: Ashgate, 2004. Controversial topics considered in this anthology include public understanding of genetic issues, health and commercialization of genetic knowledge, ethical issues raised by genetic databanks, genetic screening, human cloning, and xenotransplantation.

———, Paul Atkinson, and Helen Greenslade, eds. *New Genetics, New Social Formations*. New York: Routledge, 2006. This collection of case studies, with an international emphasis, focuses on the way new genetic developments in health and agriculture have contributed to new social formations such as genetic data banks, nongovernmental organizations, and national research laboratories.

———, and Harry Rothman. *Splicing Life? The New Genetics and Society*. Burlington, Vt.: Ashgate, 2004. The authors discuss how advances such as mapping the human genome challenge people's sense of self, culture, and society.

Grace, Eric S. *Biotechnology Unzipped: Promises and Realities*. 2d ed. Washington, D.C.: Joseph Henry Press, 2006. Grace offers the lay reader a clear explanation of biotechnology's nature, history, and ethical repercussions.

Herring, Mark Y. *Historical Guides to Controversial Issues in America: Genetic Engineering*. Westport, Conn.: Greenwood Press, 2005. Herring provides a balanced history of public perceptions of genetics and genetic engineering, including such topics as genetically altered foods, cloning, and the use of DNA profiling in law enforcement.

Hindmarsh, Richard, and Geoffrey Lawrence, eds. *Recoding Nature: Critical Perspectives on Genetic Engineering*. Sydney, Australia: University of New South Wales Press, 2004. This anthology presents varied criticisms of biotechnology and genetic engineering.

Jasanoff, Sheila. *Designs on Nature: Science and Democracy in Europe and the United States*. Rev. ed. Princeton, N.J.: Princeton University Press, 2007. Jasanoff describes the evolution of public policy on biotechnology from the 1980s forward, comparing policy decisions about medical and agricultural biotechnology. The patenting of higher life forms is among the topics discussed.

Krimsky, Sheldon, and Peter Shorett. *Rights and Liberties in the Biotech Age*. Lanham, Md.: Rowman & Littlefield, 2005. The authors say that citizens need a "genetic bill of rights" to keep developments such as genetically modified foods and patents on living things under control.

Annotated Bibliography

Lee, Keekok. *Philosophy and Revolutions in Genetics: Deep Science and Deep Technology*. New York: Palgrave Macmillan, 2005. Lee provides a philosophical discussion of the implications of genetic research and biotechnology, focusing on the establishment of classical genetics at the start of the 20th century and more recent discoveries about DNA.

McGiffen, Steven P. *Biotechnology: Corporate Power versus the Public Interest*. London: Pluto Press, 2005. McGiffen claims that the United States and the European Union, by shaping the laws that govern technologies based on genetics, are removing decision-making power from citizens and their elected representatives throughout the world. He calls for more democratic decision-making procedures.

McKelvey, Maureen, and Luigi Orsenigo, eds. *The Economics of Biotechnology*. 2 vols. Northampton, Mass.: Edward Elgar, 2006. Thirty-nine articles introduce readers to the distinctive economic features of the biotechnology industry.

Mehta, Michael C., ed. *Biotechnology Unglued: Science, Society, and Social Cohesion*. Vancouver, B.C.: University of British Columbia Press, 2006. These eight case studies show how applications of biotechnology affect the social cohesiveness of different groups.

Melendez-Ortiz, Ricardo, and Vicente Sanchez, eds. *Trading in Genes: Development Perspectives in Biotechnology, Trade, and Sustainability*. London: Earthscan, 2005. This book provides varied views from the developing world on the relationships between biotechnology, trade, safety, and sustainable development.

Moody, Glyn. *Digital Code of Life: How Bioinformatics Is Revolutionizing Science, Medicine, and Business*. Hoboken, N.J.: Wiley, 2004. Moody examines bioinformatics, viewed as the most lucrative discipline within biotechnology, from the point of view of commercial applications and investment opportunities.

Neumann-Held, Eva M., and Christoph Rehmann-Sutter, eds. *Genes in Development: Rereading the Molecular Paradigm*. Durham, N.C.: Duke University Press, 2005. This essay collection presents new views of the way genes work, stressing the influence of other factors on genes and opposing the idea that genes provide unalterable, controlling blueprints for development.

Nill, Kimball. *Glossary of Biotechnology and Nanobiotechnology Terms*. 4th ed. Washington, D.C.: CRC Press, 2005. The new edition includes terms related to nanotechnology as well as to traditional biotechnology.

Omoto, Charlotte K., and Paul F. Lurquin. *Genes and DNA: A Beginner's Guide to Genetics and Its Applications*. New York: Columbia University Press, 2004. This textbook for nonbiologists describes classical, molecular, and population genetics and their many applications, from genetic testing to genetically modified foods.

Rabinow, Paul, and Talia Dan-Cohen. *A Machine to Make a Future: Biotech Chronicles.* Princeton, N.J.: Princeton University Press, 2004. The authors focus on the leaders and activities of California biotechnology company Celera Diagnostics in 2003 as the company helped to develop new knowledge of the human genome into a powerful diagnostic tool that could reshape health care.

Rajan, Kaushik Sunder. *Biocapital: The Constitution of Postgenomic Life.* Durham, N.C.: Duke University Press, 2006. Rajan compares drug development in the United States and India to illustrate his argument that contemporary biotechnologies can be underestood only in relation to the economic environment within which they emerge.

Shanks, Pete. *Human Genetic Engineering: A Guide for Activists, Skeptics, and the Very Perplexed.* New York: Nation Books, 2005. Shanks provides a highly critical view of many types of genetic engineering. He focuses on forms that manipulate humans, but he also discusses the genetic modification of food crops.

Shmaefsky, Brian. *Biotechnology on the Farm and in the Factory.* New York: Chelsea House, 2005. Shmaefsky offers an overview of the technology for young adults.

Smith, Gina. *The Genomics Age: How DNA Technology Is Transforming the Way We Live and Who We Are.* New York: AMACOM, 2004. Smith, a science writer, interviews 23 experts in genomics and covers the major sources of controversy in the field, including gene-based discrimination and "designer babies."

Suzuki, David, and Holly Dressel. *From Naked Ape to Superspecies: Humanity and the Global Ecocrisis.* Rev. ed. Vancouver, B.C.: GreyStone Books, 2005. The authors focus on what they see as humans' disregard for other species, especially small and microscopic ones, and reckless use of technical innovations such as biotechnology. They also introduce people who are resisting these trends.

Thacker, Eugene. *The Global Genome: Biotechnology, Politics, and Culture.* Cambridge, Mass.: MIT Press, 2006. Thacker asks readers to consider the interrelationships between DNA as biological material, as information, and as intellectual property defined in a patent. He stresses the importance of globalism for understanding the biotechnology industry.

Vettel, Eric James. *Biotech: The Countercultural Origins of an Industry.* Philadelphia: University of Pennsylvania Press, 2006. This history of the biotechnology industry focuses on the 1960s cultural background of the people who established modern biotechnology in the San Francisco Bay Area in the 1970s.

Walker, Sharon. *Biotechnology Demystified.* New York: McGraw-Hill Professional, 2006. This book is a self-teaching guide to the fundamentals in all major subtopics of biotechnology.

Annotated Bibliography

ARTICLES

Adams, Nathan A., IV. "Creating Clones, Kids, and Chimera: Liberal Democratic Compromise at the Crossroads." *Issues in Law and Medicine*, vol. 20, Summer 2004, pp. 3–69. Adams recommends applying to human genetic engineering the same "liberal democratic compromise" that has been used in regard to, for instance, reproductive rights and experimentation on human subjects, banning or keeping tight control of some types of technologies while allowing others to flourish.

Bucchi, Hassimiano, and Federico Neresini. "Why Are People Hostile to Biotechnologies?" *Science*, vol. 304, June 18, 2004, p. 1749. The authors conclude that the negative attitudes toward biotechnologies revealed by polls among Italians are not part of a general public prejudice against science. Instead, citizens are concerned about the procedures for connecting scientific expertise with political decision-making and want a larger role in the process.

Ghadar, Fariborz, and Heather Spindler. "The Growth of Biotechnology." *Industrial Management*, vol. 47, March 1, 2005, pp. 19–26. The authors summarize present and predicted future biotechnology advances in agriculture, health care, industry, and environmental remediation. They warn that regulation of these technologies needs to be upgraded.

Kogan, Lawrence A. "Exporting Europe's Protectionism." *The National Interest*, vol. 77, Fall 2004, pp. 91–99. Kogan strongly disapproves of Europe's "precautionary" approach to biotechnology, which he considers to be protectionism in disguise.

Montpetit, Eric. "A Policy Network Explanation of Biotechnology Policy Differences Between the United States and Canada." *Journal of Public Policy*, vol. 25, September-December 2005, pp. 339–366. Montpetit believes that policy networks can explain why Canada has a more restrictive biotechnology policy than the United States.

Moore, Patrick. "Battle for Biotech Progress." *Review—Institute of Public Affairs*, vol. 56, March 31, 2004, pp. 10–13. Moore stresses the benefits of biotechnology and says that fear-mongering by the technology's opponents deny these benefits to some parts of the world.

Sapolsky, Robert M. "Of Mice, Men, and Genes." *Natural History*, vol. 113, May 2004, pp. 21–24, 31. Beginning before birth, environment can have major influences on the way genes express themselves.

Scheitle, Christopher P. "In God We Trust: Religion and Optimism Toward Biotechnology." *Social Science Quarterly*, vol. 86, December 2005, pp. 846–856. Analyzing the 1997–1988 U.S. Biotechnology Study, the author finds that belief in a personal God is associated with relative optimism about biotechnology. He speculates that this is so because such a belief makes people feel that they have a divine "safety net" protecting them from harm.

Silver, Lee M. "The Clash of Biotechnology and Post–Christian Spirituality." *Skeptical Inquirer*, vol. 31, March-April 2007, pp. 32-37. The strongest opponents of biotechnology are not fundamentalist Christians but people who, sometimes without being aware of it, are driven by a "post–Christian" religiosity in which Mother Nature substitutes for God. Holders of this view tend to see everything natural as good and everything human-made as harmful.

"Surviving the Treacherous Transgenics Path." *Genetic Engineering News*, vol. 26, May 1, 2006, pp. 40-41. The financial and regulatory issues affecting genetic engineering research are discussed.

Watters, Ethan. "DNA Is Not Destiny." *Discover*, vol. 27, November 2006, pp. 33–37, 75. Epigenetic factors—proteins that turn genes on and off—are inheritable and can be influenced by the environment. Thus, people's behavior and environment can affect not only their own future health but possibly even that of their descendants.

WEB DOCUMENTS

"Biotechnology for Sustainable Growth and Development." Organisation for Economic Co-operation and Development. Available online. URL: http://www.oecd.org/dataoecd/60/63/23536372.pdf. Posted on January 8, 2004. This extended executive summary of a longer document concludes that the OECD should strengthen its contribution to development of biotechnology as a driver for sustainable growth.

"DNA from the Beginning." Josiah Macy, Jr. Foundation. Available online. URL: http://www.dnaftb.org/dnaftb/. Accessed on March 28, 2007. This animated primer on the basics of DNA, genes, and heredity is divided into classical genetics, molecules of genetics, and genetic organization and control.

Shaw, Sabrina, and Risa Schwartz. "Trading Precaution: The Precautionary Principle and the WTO." Institute for Advanced Studies, United Nations University. Available online. URL: http://www.ias.unu.edu/binaries2/Precautionary%20Principle%20and%20WTO.pdf. Posted in November 2005. The authors call for international agreement on approaches to risk assessment, including a precise definition of the "precautionary principle," in order to avoid conflicts within the World Trade Organization. They conclude that the WTO dispute settlement system is not well suited to solving disagreements about risk assessment, which are often culturally based.

Strickland, Debbie, ed. "Guide to Biotechnology 2007." Biotechnology Industry Organization. Available online. URL: http://bio.org/speeches/pubs/er. Posted in 2007. This guide provides statistics about the biotechnology industry and descriptions of different types of biotechnology and their applications, covering medicine and health care, agriculture and aquaculture, industrial and environmental applications, national security, and other applications. It also includes statements of this trade organization's ethical principles.

van Beuzekom, Brigitte, and Anthony Arundel. "OECD Biotechnology Statistics—2006." Available online. Organisation of Economic Co-operation and Development. URL: http://www.oecd.org/dataoecd/51/59/ 36760212.pdf. Posted in 2006. This report includes data for 23 OECD countries and two observer countries, plus China (Shanghai). It covers biotechnology firms, research, employment, sales, applications, patents, agriculture, alliances, and venture capital, as well as providing profiles of the industry in major countries.

AGRICULTURAL BIOTECHNOLOGY

BOOKS

Altieri, Miguel A. *Genetic Engineering in Agriculture: The Myths, Environmental Risks, and Alternatives.* 2d ed. Oakland, Calif.: Food First, 2004. Altieri discusses recent developments in the study of the biological, environmental, social, and economic impacts of transgenic crops and concludes that more sustainable alternatives exist and are preferable.

Bhalgat, Mahesh K., et al., eds. *Agricultural Biotechnology: Challenges and Prospects.* Washington, D.C.: American Chemical Society, 2004. Experts in industrial, academic, and government research examine current challenges and future prospects of agricultural biotechnology.

Bruce, Donald, and Ann Bruce, eds. *Engineering Genesis: Ethics and the GM Revolution.* London: Earthscan Publications, 2004. Experts brought together by the Society, Religion and Technology Project of the Church of Scotland present case studies and consider the ethical and social implications of genetic engineering in plants and animals, including genetically modified foods, drugs produced by animals, and animal cloning.

Cook, Guy. *Genetically Modified Language: The Discourse of the GM Debate.* New York: Routledge, 2004. Cook carefully analyzes the language of the debate over genetically modified foods and considers how spokespeople's choice of words influences policy and public opinion.

Cummins, Ronnie, and Ben Lilliston. *Genetically Engineered Food: A Self-Defense Guide for Consumers.* 2d ed. New York: Marlowe & Co., 2004. The authors detail what they see as the scientific, political, economic, and health threats of genetically modified foods and tell consumers what they can do to avoid these foods.

Entine, Jon. *Let Them Eat Precaution: How Politics Is Undermining the Genetic Revolution in Agriculture.* Washington, D.C.: AEI Press, 2006. Entine feels that cultural politics and trade disputes are blocking development and use of beneficial genetically enhanced foods such as golden rice, which can prevent blindness resulting from lack of vitamin A.

Evenson, R. E., and V. Santaniello, eds. *Consumer Acceptance of Genetically Modified Foods.* Cambridge, Mass.: CABI Publishing, 2004. The authors of the papers in this book, presented at meetings of the International Consortium on Agricultural Biotechnology Research, studied consumer attitudes toward genetically modified foods in a variety of countries.

————. *International Trade and Policies for Genetically Modified Products.* Cambridge, Mass.: CABI Publishing, 2006. Papers presented at the 2004 meeting of the International Consortium on Agricultural Biotechnology discuss controversies concerning international trade and government policies about genetically modified foods.

————. *The Regulation of Agricultural Biotechnology.* Cambridge, Mass.: CABI Publishing, 2004. This essay collection reviews changes that the expansion of agricultural biotechnology has produced in regulatory systems and considers the relationship between these changes and innovation, market development, and international trade.

Falkner, Robert, ed. *The International Politics of Genetically Modified Food: Diplomacy, Trade, and Law.* New York: Palgrave Macmillan, 2006. These analyses of the international politics of regulating genetically modified food explore disputes between Europe and North America, growing differences between North and South, and the GM food battle's implications for environmental law and international trade.

Fedoroff, Nina, and Nancy Marie Brown. *Mendel in the Kitchen: A Scientist's View of Genetically Modified Foods.* Washington, D.C.: Joseph Henry Press, 2006. The authors contend that risks from these foods are minimal and the potential benefits are great.

Fox, Michael W. *Killer Foods: When Scientists Manipulate Genes, Better Is Not Always Best.* Guilford, Conn.: Lyons Press, 2004. Fox concludes that the creation of transgenic plants and animals is both unethical and unnecessary; traditional husbandry practices and sustainable organic farming are better solutions to the developing world's food needs.

Hilbeck, A., and D. A. Andrew. *Environmental Risk Assessment of Genetically Modified Organisms, Vol. I.: A Case Study of Bt Maize in Kenya.* Cambridge, Mass.: CABI Publishing, 2004. This case study is a product of an attempt by the International Organization for Biological Control to establish biosafety guidelines for genetically modified crops. It considers effects of such crops on the environment and agriculture but not on human health or society.

Ho, Mae-Wan, and Lim Li Ching. *GMO-Free: Exposing the Hazards of Biotechnology to Ensure the Integrity of Our Food Supply.* Ridgefield, Conn.: Vital Health Publishing, 2004. Ho and Ching urge a worldwide ban on genetically altered crops. They recommend replacing them with sustainable agriculture and organic farming.

Inui, Akio, ed. *Epigenetic Risks of Cloning*. Washington, D.C.: CRC Press, 2005. Essays in this anthology survey the techniques and risks of animal cloning, stressing that only a tiny percentage of cloned embryos survive to adulthood.

Just, Richard E., Julian M. Alston, and David Zilberman, eds. *Regulating Agricultural Biotechnology: Economics and Policy*. New York: Springer, 2006. This book offers a comprehensive economic analysis of current agricultural biotechnology regulation, including costs of approving new products, liability, and implications for social welfare and biosafety.

Kloppenberg, Jack Ralph. *First the Seed: The Political Economy of Plant Biotechnology*. 2d ed. Madison: University of Wisconsin Press, 2005. Kloppenberg presents the history of plant breeding and shows how efforts to control ownership of seeds have shaped the agricultural biotechnology industry.

Lurquin, Paul. *High Tech Harvest*. Boulder, Colo.: Westview Press, 2004. Lurquin presents the scientific background for an optimistic view of genetically modified food.

Maclean, Norman, ed. *Animals with Novel Genes*. New York: Cambridge University Press, 2006. The authors in this collection conclude that the potential positive impact of transgenic animals on agriculture and medicine is very great.

Miller, Henry I., and Gregory Conko. *The Frankenfood Myth: How Protest and Politics Threaten the Biotech Revolution*. Westport, Conn.: Praeger Publishers, 2004. Miller and Conko argue that overregulation is strangling food biotechnology research in universities and small businesses and depriving people, especially in developing countries, of crops that could increase food yield and benefit health.

National Research Council. *Biological Confinement of Genetically Engineered Organisms*. Washington, D.C.: National Academies Press, 2004. This book examines biological techniques, such as induced sterility, that can be used to confine genetically altered microorganisms, plants, and animals from escaping into the environment and breeding there. The authors recommend a systematic approach that uses multiple methods of bioconfinement, since no single technique is completely effective.

——— and Institute of Medicine. *Safety of Genetically Engineered Foods*. Washington, D.C.: National Academies Press, 2004. This book presents methods for evaluating unintended changes in food composition and harmful effects on human health.

Nestle, Marion. *Safe Food: Bacteria, Biotechnology, and Bioterrorism*. Berkeley, Calif.: University of California Press, 2004. Nestle describes the political aspects of food safety and threats to the food production system, including threats from biotechnology.

Panno, Joseph. *Animal Cloning: The Science of Nuclear Transfer*. New York: Facts On File, 2004. Panno presents the basic concepts and history of animal cloning and offers a neutral discussion of the controversies surrounding it, including the possibility of human cloning.

Parekh, Sarad R., ed. *The GMO Handbook: Genetically Modified Animals, Microbes, and Plants in Biotechnology*. Totowa, N.J.: Humana Press, 2004. This book surveys technologies for creating genetically modified organisms, current accomplishments of GMO research, and social, legal, and political issues raised by the technology.

Paul, Helena, and Ricarda Steinbrecher with Derek Kuyek and Lucy Michaels. *Hungry Corporations: Transnational Biotech Companies Colonise the Food Chain*. New York: Zed Books, 2004. The authors describe the growing control that multinational corporations, including agricultural biotechnology companies, have over people's lives.

Pechan, Paul, and Gert E. de Vries. *Genes on the Menu: Facts for Knowledge-Based Decisions*. New York: Springer, 2005. The authors present facts intended to help the public make intelligent decisions about agricultural biotechnology and genetically modified food. An accompanying film appears on an enclosed CD-ROM.

Pringle, Peter. *Food, Inc.—Mendel to Monsanto: The Promises and Perils of the Biotech Harvest*. New York: Simon & Schuster, 2005. Pringle believes that biotech crops have great potential for feeding the hungry and reducing use of chemical pesticides, but he says that large corporations' attempts to force GM foods on people have destroyed the public's good will toward the technology.

Rees, Andy. *Genetically Modified Food: A Short Guide for the Confused*. London: Pluto Press, 2006. A leading activist with GM Watch, an organization that opposes genetically modified food, is alarmed by the worldwide spread of this technology and its control by multinational corporations.

Roseboro, Ken. *Food at Risk: Genetically Altered Foods and Your Health*. Laguna Beach, Calif.: Basic Health Publications, 2004. Roseboro stresses the risks of genetically modified foods to health and environment and recommends organic food instead.

Smith, Jeffrey M. *Genetic Roulette: The Documented Health Risks of Genetically Engineered Foods*. White River Junction, Vt.: Chelsea Green, 2007. In a clear, accessible style, Smith describes almost 40 health risks that consumers may face from genetically modified food. He lists scientific studies that document each claim.

Smyth, Stuart, et al. *Regulating the Liabilities of Agricultural Biotechnology*. Cambridge, Mass.: CABI Publishing, 2004. With an international focus, this book examines how society, industry, and government interact to evaluate the risks of and regulate agricultural biotechnology.

Annotated Bibliography

Stewart, C. Neal. *Genetically Modified Planet: Environmental Impacts of Genetically Modified Plants*. New York: Oxford University Press, 2004. Aimed at concerned consumers, this book argues that the environmental benefits of genetically modified crops far outweigh such crops' risks.

Strauss, Steven H., and H. D. Bradshaw, eds. *The Bioengineered Forest: Challenges for Science and Society*. Washington, D.C.: RFF Press, 2004. The papers in this book describe the many opportunities that genetic engineering offers to forestry. It also lists harmful environmental and social effects that could come from misapplication of genetic modification.

Thomson, Jennifer A. *Genes for Africa: Genetically Modified Crops in the Developing World*. Lansdowne, Zambia: Juta Academic Publishers, 2004. Thomson shows how, in her opinion, genetically modified crops can help to fight starvation, poverty, and disease in developing African countries.

Toke, Dave. *The Politics of GM Food*. New York: Routledge, 2004. Toke uses discourse analysis to compare the interplay between science and politics regarding genetically modified food in the United States, Britain, and the European Union. He considers how this issue relates to globalization.

Wambugu, Florence. *Modifying Africa: How Biotechnology Can Benefit the Poor and Hungry*. Self-published, 2005. Wambugu, a Kenyan scientist who has used biotechnology to improve local food crops, offers her country as a case study to show how this technology can help poor farmers.

Wesseler, J. H., ed. *Environmental Costs and Benefits of Transgenic Crops*. New York: Springer, 2005. These essays explore the complex and far-reaching economic implications of possible environmental effects of transgenic crops.

Winston, Mark L. *Traveling in the Genetically Modified Zone*. Rev. ed. Cambridge, Mass.: Harvard University Press, 2005. The author seeks a middle ground between opposing views in the politicized arena of agricultural biotechnology, presenting interviews with a variety of participants and discussing possible solutions to problems.

Wu, Felicia, and William P. Butz. *The Future of Genetically Modified Crops: Lessons from the Green Revolution*. Santa Monica, Calif.: RAND Corporation, 2004. Wu and Butz compare the 1960s Green Revolution with the current "Gene Revolution." They consider the cultural, political, and economic factors that determine whether governments, farmers, and consumers will accept a new agricultural technology.

Zerbe, Noah. *Agricultural Biotechnology Reconsidered: Western Narratives and African Alternatives*. Trenton, N.J.: Africa World Press, 2005. Illustrating the principle that technology is mediated by social factors, Zerbe shows how very different cultural beliefs have shaped the development of biotechnology in the United States and in Africa. He concludes that if Africa is to benefit from this technology, it must adapt the technology to the African social model.

Biotechnology and Genetic Engineering

ARTICLES

Anderson, Clifton E. "Biotech on the Farm." *The Futurist*, vol. 39, September 1, 2005, pp. 38-42. So far, genetically modified crops have not fulfilled their promise of high yields that will feed the hungry, nor have consumers been convinced that their benefits outweigh their risks.

"Animal Cloning Spurs Controversy, Too." *CQ Researcher* vol. 14, October 22, 2004, pp. 888–889. Animals have been cloned for a variety of reasons, and animal cloning has inspired ethical controversies.

Bernauer, Thomas. "Causes and Consequences of International Trade Conflict over Agricultural Biotechnology." *International Journal of Biotechnology*, vol. 7, March 10, 2005, pp.7 ff. This article shows why and how differences in the way countries regulate biotechnology create tensions that often develop into serious trade conflicts. It provides a fairly pessimistic view of the effects of such conflicts on biotechnology and developing countries.

Berwald, Derek, Colin A. Carter, and Guillaume P. Gruere. "Rejecting New Technology: The Case of Genetically Modified Wheat." *Agricultural Economics*, vol. 88, May 2006, pp. 432–447. According to these authors, Monsanto Co. halted its development of genetically modified spring wheat in 2004 because Canada's stringent regulations prevented release of the wheat in that country. Monsanto decided that if it could not release the technology simultaneously in both Canada and the United States, it would not introduce the wheat at all.

Bokor, Raymond K. "Resistance to Genetic Engineering in Africa." *Synthesis/Regeneration*, vol. 34, Spring 2004, pp. 11–13. Bokor, who opposes genetically modified foods, surveys resistance to use of such foods in Africa.

Brookes, Graham, and Peter Barfoot. "GM Crops: The Global Economic and Environmental Impact—The First Nine Years, 1996–2004." *AgBioForum*, vol. 8, 2005, pp. 187–196. The authors provide evidence that genetically modified crops have given substantial net economic benefits to farmers and have also significantly reduced pesticide use and greenhouse gas emissions.

Clapp, Jennifer. "The Political Economy of Food Aid in an Era of Agricultural Biotechnology." *Global Governance*, vol. 11, October 1, 2005, pp. 467–485. The advent of genetically modified crops has repoliticized the giving and acceptance of food aid. Hungry countries have refused food donations that might contain GM material for fear of contaminating their own crops and ruining their export market.

Cohen, Joel I., and Robert Paarlberg. "Unlocking Crop Biotechnology in Developing Countries—A Report from the Field." *World Development*, vol. 32, September 2004, pp. 1565–1579. The authors examine regulation of genetically modified crops in six countries. They propose a conceptual framework for biosafety regulation and make recommendations to speed GM safety approvals in developing countries.

Coleman, Gerald D. "Is Genetic Engineering the Answer to Hunger?" *America*, vol. 192, February 21, 2005, p. 14. Coleman reviews Catholic organizations' positions on the use of genetically modified crops in developing countries.

Conko, Gregory. "Modified Crops: Regulate the Product, Not the Process." *Chemical Engineering Progress*, vol. 101, September 30, 2005, pp. 4–5. The safety and usefulness of genetically engineered agricultural products should be judged on a case-by-case basis.

Costa-Font, Joan, Elias Mossialos, and Montserrat Costa-Font. "Erring on the Side of Caution? The Heterogeneity of Public Perceptions of Biotechnology Applications in the European Union." *Journal of Economic Issues*, vol. 40, Sepember 2006, pp. 767–777. Surveys conducted in Europe in 1996 and 1999 regarding perceived risks of biotechnology showed considerable variation among countries and among different biotechnology procedures. Therefore, it may be difficult to establish EU-wide policies on the basis of a single principle, such as the precautionary principle.

"EU's GM Ban Was Illegal, Says WTO." *Agra Europe*, issue 2228, October 6, 2006, pp. EP/4–5. A final report from the World Trade Organisation ruled that the European Union's moratorium on imports of new genetically modified seed products, in place from 1999 to 2004, broke international trade laws.

Fernandez-Cornejo, Jorge. "Use of Genetically Engineered Crops Rising Steadily During First Decade." *Amber Waves*, vol. 3, September 30, 2005, p. 7. Farmers' hopes for increased yields, saved time, and reduced costs have produced a rapid rise in the use of genetically modified crops despite consumer resistance in some countries.

Fiester, Autumn. "Creating Fido's Twin: Can Pet Cloning Be Ethically Justified?" *Hastings Center Report*, vol. 35, July-August 2005, pp. 34–39. Pet cloning may seem frivolous at best, but it may be useful in raising the moral status of companion animals in the public eye because it demonstrates the trouble and expense that some people are willing to endure in an attempt to preserve their pets.

———. "Ethical Issues in Animal Cloning." *Perspectives in Biology and Medicine*, vol. 48, Summer 2005, pp. 328–343. Researchers need to take into account the public's concerns about animal cloning, both those related to consequences (to animals or humans) and those based on principle, if they want their work to be accepted.

Finucane, Melissa L., and J. L. Holup. "Psychosocial and Cultural Factors Affecting the Perceived Risk of Genetically Modified Food: An Overview of the Literature." *Social Science & Medicine*, vol. 60, April 2005, pp. 1603–1612. The authors describe how contrasting socio-psychological and cultural factors in the United States, Europe, and developing countries affect public perceptions of risk from genetically modified foods.

————— and Theresa A. Satterfield. "Risk as Narrative Value: A Theoretical Framework for Facilitating the Biotechnology Debate." *International Journal of Biotechnology*, vol. 7, March 10, 2005, p. 128 ff. The authors present a theoretical framework for understanding risk as value and say that overcoming gaps between opposing values and contradictions between stated values and actual behavior is key to reducing polarization in discussions of risks associated with biotechnology.

Fouassier, Alexandre. "Transgenic Animal Farms." *Chemistry and Industry*, vol. 20, October 18, 2004, pp. 23–24. Transgenic mammals, especially rabbits, engineered to produce medically useful human proteins in their milk could be a more economical source of some protein drugs than bacteria or cells in bioreactors.

"GM Rice, Food Safety, and a Huge Marketing Failure." *Agra Europe*, issue 2224, September 8, 2006, pp. A/1–2. Conflict over genetically modified foods in Europe has been greatly exacerbated by the U.S.-based biotechnology industry's refusal to accept the European public's distrust of these foods and recognize that "the customer is always right" from a commercial point of view, whether backed by scientific evidence or not.

Haribabu, E. "Interests and Meanings: The Socio-Technical Process of Application of Biotechnology to Crop Improvements in India." *International Journal of Biotechnology*, vol. 6, May 26, 2004, pp. 65 ff. Drawing on insights from the sociology of science, Haribabu argues that production of knowledge and its application in industry make up a complex process that involves interplay among conflicting interests and value systems. Biotechnology research in India will continue and farmers will adopt the new technology only if all stakeholders participate in decisionmaking, he says.

Ho, Mae-Wan. "DNA in GM Food and Feed." *Synthesis/Regeneration*, vol. 37, Spring 2005, pp. 7–9. Ho concludes that functional DNA from genetically modified food can be transferred to bacteria and body cells.

Jauhar, Prem P. "Modern Biotechnology as an Integral Supplement to Conventional Plant Breeding: The Prospects and Challenges." *Crop Science*, vol. 46, September-October 2006, pp. 1841–1859. Jauhar shows how gene transfer can effectively supplement plant breeding, provided that resistance to this new technology can be overcome.

Kalaitzandonakes, Nicholas. "Another Look at Biotech Regulation: Are Europe's Labeling Laws for Genetically Modified Foods Cost-Effective, or Even Necessary?" *Regulation*, vol. 27, Spring 2004, pp. 44–50. The author claims that the European Union's stringent regulations affecting genetically modified foods fail all three criteria for justifiable regulation.

Kaplan, Kim. "ARS Leads in Assessing Risk in Transgenics." *Agricultural Research*, vol. 52, September 2004, pp. 4–8. The USDA's Agricultural

Research Service has become a leader in agricultural biotechnology risk assessment research. It is well placed to work with groups on all sides of the GM safety issue.

Kennell, David. "Genetically Engineered Plant Crops: Potential for Disaster." *Synthesis/Regeneration*, vol. 35, September 22, 2004, pp. 10–11. Kennell believes that the spread of genetically engineered crops will encourage monoculture and threaten plant biodiversity.

Krasny, Leslie T. "Second-Generation Biotech: Biotech Foods Are Providing Wellness Benefits in Their Second Stage of Development." *Food Processing*, vol. 66, April 2005, pp. 513–514. Unlike the first generation of agricultural biotechnology products, which chiefly benefited farmers, the generation currently under development promises to provide health benefits to consumers. The author offers examples of such crops, surveys their regulation and risk assessment, and concludes that genetically modified foods are no more risky than products of conventional breeding and, indeed, are perhaps less so.

Krebs, Candace. "Wheat's Tightrope: Growers Balance Technological Advances and Consumer Acceptance." *Agri Marketing*, vol. 42, September 2004, pp. 36-39. Monsanto's abandonment of "Roundup Ready" spring wheat does not mean an end to attempts to genetically modify this crop, but farmers will not accept genetically modified wheat unless the public does. Citizens' acceptance, in turn, is likely to depend on whether the wheat can be segregated from non-GM wheat and on whether modifications add characteristics that benefit consumers.

Lemonick, Michael D. "Woof, Woof! Who's Next?" *Time*, vol. 166, August 15, 2005, p. 54. South Korean researcher Woo Suk Hwang and his team at Seoul National University announce the birth of a cloned Afghan hound puppy, whom they call Snuppy—the first dog to be cloned, an achievement that scientists have long sought.

Lence, Sergio H., and Dermot J. Hayes. "Genetically Modified Crops: Their Market and Welfare Impacts." *American Journal of Agricultural Economics*, vol. 87, November 2005, pp. 931–950. The authors present a framework for examining price and welfare effects of the introduction of genetically modified crops and conclude that introducing this technology increases aggregate welfare over a wide range of scenarios. They also say that both United States and European Union policies on GM foods make sense in light of the views of their respective citizenries.

Lesch, William C., Cheryl J. Wachenheim, and Bard S. Stillerud. "Biotechnology: The Healthy Choice?" *Health Marketing Quarterly*, vol. 22, January 2005, pp. 59–81. This article presents the results of a poll testing consumer reaction to genetically modifed "functional foods" that offer health benefits beyond basic nutrition. The poll showed overall acceptance of such foods.

Lesher, Molly. "Seeds of Change." *Regional Review*, vol. 14, June-September 2004, pp. 12–21. Lesher reviews the progress (or lack of it) in adoption of genetically modified foods and the scientific, economic, and emotional issues these products raise.

Mackey, Maureen, and Jill Montgomery. "Plant Biotechnology Can Enhance Food Security and Nutrition in the Developing World." *Nutrition Today*, vol. 39, March-April 2004, pp. 52–58. Surveying rice, maize, and potatoes, the authors describe experiments to increase resistance of these crops to abiotic stresses such as acidic soil as well as techniques for increasing yields of indigenous staple crops.

Magin, Kimberly. "Improving Foods and Feed Through Biotechnology." *Food Technology*, vol. 60, September 2006, pp. 24–29. Magin shows how second- and third-generation biotechnology crops will benefit farmers, livestock producers, and consumers.

Mayer, Jorge E. "The Golden Rice Controversy: Useless Science or Unfounded Criticism?" *BioScience*, vol. 55, September 2005, pp. 726–727. Mayer supports the use of "golden rice" and other transgenic crops in developing nations. He responds to criticisms of such crops.

"A Meaty Question." *The Economist*, vol. 380, September 23, 2006, p. 12US. Meat grown as muscle cells in vats may eventually replace meat taken from animals, improving nutritional benefits as well as answering the objections of animal rights activists.

"The Men in White Coats Are Winning, Slowly—Non-Food GM." *The Economist*, vol. 373, October 9, 2004, p. 64US. People are more willing to accept nonfood uses of agricultural genetic engineering than GM foods, and such uses are steadily multiplying—but even nonfood GM must deal with public opinion.

Miller, Henry I., and Gregory Conko. "Agricultural Biotechnology: Overregulated and Underappreciated." *Issues in Science and Technology*, vol. 21, Winter 2005, pp. 76–80. Groups opposed to genetic engineering say that genetic modification is not often applied to "orphan crops" because large corporations will not develop products that are not highly profitable. These authors, however, blame overregulation, which makes the cost of research and development unnecessarily high. They say that new crops should be evaluated on the basis of risk rather than according to the process by which they were produced.

Nelkin, Dorothy, and Emily Marden. "The StarLink Controversy: The Competing Frames of Risk Disputes." *International Journal of Biotechnology*, vol. 6, May 26, 2004, pp. 20 ff. The authors examine the controversy in 2000–2001 over the containment of genetically engineered StarLink corn, which had not been approved for human food but managed to enter the food supply anyway. They conclude that, like other disputes in the past and continuing today, the StarLink debate was shaped by conflicting sociopolitical perspectives on the nature of risks, the ability of regulators to predict and control the effects of new technologies, and the best way to contain the products of such technologies.

Palmer, Jay. "Eat up! Why Genetically Modified 'Frankenfood' Is Gaining Ground." *Barron's*, vol. 84, December 6, 2004, pp. 22–24. Genetically engineered crops are pleasing farmers by allowing them to use fewer herbicides and chemical pesticides and to reduce the amount of water needed for irrigation.

Park, Alice. "The Perils of Cloning." *Time*, vol. 168, July 10, 2006, p. 56 ff. Many species of animals have been cloned since the birth of the famous cloned sheep, Dolly, in 1996, but it is becoming increasingly clear that all cloned animals are defective in one way or another. The defects probably arise from abnormal modifications in genes that affect their function. This article includes an interview with Ian Wilmut, Dolly's creator, and a description of the cloning of prime beef cattle.

Pasternak, Shiri. "Stories from Guatemala and North America: Why Indigenous Beliefs Matter in the Debate on Genetically Engineered Food." *Health Law Review*, vol. 15, September 15, 2006, pp. 45–46. Practices relating to food production and consumption are very important to cultures around the world, but they are threatened by the influx of globally homogeneous, processed foods, including genetically modified foods.

Qaim, Matin. "Agricultural Biotechnology Adoption in Developing Countries." *American Journal of Agricultural Economics*, vol. 87, December 2005, pp. 1317–1324. Agricultural biotechnology potentially can help people in developing countries, including poor farmers, but so far it usually has not done so. Bringing these benefits to farmers will require effort from both the private and the public sector.

Quammen, David. "Clone Your Troubles Away: Dreaming at the Frontiers of Animal Husbandry." *Harper's Magazine*, vol. 310, February 2005, pp. 33–43. Quammen investigates the cloning of domestic and wild animals and considers the significance of this technology.

Rayasam, Renuka. "Cloning Around." *U.S. News & World Report*, vol. 142, January 8, 2007, pp. 46–47. Now that the Food and Drug Administration has tentatively approved food products from cloned animals, breeders may soon duplicate their fattest pigs and most productive cows—if they can persuade consumers to accept the results.

Rudenko, Larisa, and John C. Matheson. "The US FDA and Animal Cloning: Risk and Regulatory Approach." *Theriogenology*, vol. 67, January 2007, pp. 198–206. Rudenko and Matheson describe the methods used by the Food and Drug Administration to evaluate the safety of cloned livestock and products made from clones or their progeny. They also discuss how the agency communicated its conclusions to the public.

"Salmon that Grow up Fast." *Business Week Online*, January 11, 2006, n.p. Aqua Bounty may be close to obtaining the Food and Drug Administration's permission to sell its fast-growing, genetically engineered salmon, which environmentalists fear may outcompete wild fish.

Schurman, Rachel. "Fighting 'Frankenfoods': Industry Opportunity Structures and the Efficacy of the Anti-Biotech Movement in Western Europe." *Social Problems*, vol. 51, May 2004, pp. 243–268. Schurman argues that industry structures are a major reason why the anti-biotechnology movement in Western Europe has been so effective. She claims that such structures, which have economic, organizational, and cultural features, enhance or limit social movements' efforts to change industry behavior.

Scott, Dane. "Perspectives on Precaution: The Role of Policymakers in Dealing with the Uncertainties of Agricultural Biotechnology." *International Journal of Global Environmental Issues*, vol. 5, February 12, 2005, p. 10 ff. Differences in regulation of genetically modified products in the United States and Europe reflect differences in approach to the rate of technological transfer. The United States favors a rapid rate of transfer, but Europe prefers a slow and cautious rate, controlled by the so-called precautionary principle. Scott examines the justifications for these conflicting perspectives and the ways policymakers use them to deal with uncertainty.

Scrinis, Gyorgy. "Engineering the Foodchain." *Arena*, vol. 77, June 1, 2005, pp. 37–39. The biotechnology industry hopes that development of genetically modified crops with increased nutritional value, such as a low glycemic-index wheat and rice enhanced with a precursor of vitamin A ("golden rice"), will help to persuade consumers and governments that GM foods are acceptable.

Sheingate, Adam D. "Promotion versus Precaution: The Evolution of Biotechnology Policy in the United States." *British Journal of Political Science*, vol. 36, April 2006, pp. 243–268. Examining congressional committee hearings and the inputs of interest groups, Sheingate traces differences between regulatory policy for agricultural biotechnology (favorable) and that for medical biotechnology (cautious) to policy decisions made in the 1980s.

Stewart, Patrick A., and Andrew J. Knight. "Trends Affecting the Next Generation of U.S. Agricultural Biotechnology." *Technological Forecasting and Social Change*, vol. 72, June 2005, pp. 521–534. Stewart and Knight present the history of policies regulating agricultural biotechnology in the United States. They argue that regulations are currently entering a new phase because of concern about possible health risks from drug-producing plants entering the food supply.

Stokstad, Erik. "Monsanto Pulls the Plug on Genetically Modified Wheat." *Science*, vol. 304, May 21, 2004, pp. 1088–1089. The author explores the reasons for this decision and its likely impact.

Vartan, Starre. "Ah-tchoo! Do Genetically Modified Foods Cause Allergies?" *E*, vol. 17, November 1, 2006, pp. 40–41. The risk that genetically modified foods will trigger allergies is often overlooked.

Weintraub, Arlene. "Crossing the Gene Barrier." *Business Week*, issue 3967, January 16, 2006, p. 72. Goats and mice with human genes are being developed as potential sources of complex biological molecules used as drugs.

Zimmermann, Roukayatou, and Matin Qaim. "Potential Health Benefits of Golden Rice: A Philippine Case Study." *Food Policy*, vol. 29, April 2004, pp. 147–168. The authors conclude that although golden rice, which contains a precursor of vitamin A, will not completely eliminate vitamin A deficiency and should therefore be used with other forms of nutritional supplementation; it could bring significant health benefits to the people of the Philippines.

WEB DOCUMENTS

"Agricultural Biotechnology: Meeting the Needs of the Poor?" Food and Agricultural Organization of the United Nations (FAO). Available online. URL: http://www.fao.org/docrep/006/Y5160E/y5160e00.htm. Posted in May 2004. This report provides background on agricultural biotechnology, comparing the current "gene revolution" to the 1960s Green Revolution; presents facts and statistics about the economic, health, and environmental effects of genetically modified crops and public attitudes toward agricultural biotechnology in different countries; and discusses ways to make agricultural biotechnology work for the poor, stressing capacity-building.

Becker, Geoffrey S. "Agricultural Biotechnology: Background and Recent Issues." Congressional Research Service Reports and Issue Briefs. Available online. URL: http://www.ncseonline.org/NLE/CRSreports/06Sep/RL32809.pdf. Last updated on September 5, 2006. Controversies discussed in this report include risks and benefits of genetically modified crops and products, the impact of GM crops on health and environment, whether foods containing GM material should be labeled, the adequacy of U.S. regulation of GM crops, difficulties in persuading the European Union to accept GM products exported from the United States, and the effects of GM crops on food security in developing countries.

"Biotechnology and Farming." Agriculture and Environment Biotechnology Commission. Available online. URL: http://www.aebc.gov.uk/aebc/about/popular_guide.pdf. Posted on June 30, 2004. This booklet for consumers summarizes the findings that the commission presented to the British government in three major reports.

Birner, Regina, and Gabriela Alcaraz. "Policy Dialogues on Genetically Modified Crops in Europe: Insights for African Policy Dialogues on Biotechnology." International Food Policy Research Institute. Available online. URL: http://www.ifpri.org/africadialogue/pdf/policydialoguespaper. pdf. Posted on September 21, 2004. This background paper examines five European policy dialogues on biotechnology and genetically modified foods and makes recommendations for adopting the techniques used in these dialogues to similar conferences to be held in Africa.

Boa-Amponsem, K., and G. Minozzi. "The State of Development of Biotechnologies as They Relate to the Management of Animal Genetic Resources and Their Potential Application in Developing Countries." Food and Agricultural Organisation of the United Nations (FAO). Available online. URL: ftp://ftp.fao.org/ag/cgrfa/bsp/bsp33e.pdf. Posted in November 2006. This paper reviews literature, particularly that published after 1999, on genetic alteration and reproductive technologies for livestock. It evaluates the potential impact of these technologies on local animal genetic resources and sustainable production.

Chassy, Bruce M., Wayne A. Parrott, and Richard Roush. "Crop Biotechnology and the Future of Food: A Scientific Assessment." Council for Agricultural Science and Technology. Available online. URL: http://www. cast-science.org/websiteUploads/publicationPDFs/QTA2005-2.pdf. Posted in October 2005. This paper answers common questions about genetically engineered crops and food.

"Clonesafety.org FAQs." Clonesafety.org. Available online. URL: http://www.clonesafety.org/cloning/facts/faq. Accessed on March 16, 2007. Frequently asked questions about animal cloning are answered from a pro-cloning point of view.

"Commercial, Safety, and Trade Implications Raised by Importation of Genetically Engineered Ingredients, Grain or Whole Foods for Food, Feed, or Processing." Pew Initiative on Food and Biotechnology. Available online. URL: http://pewagbiotech.org/events/0907/WorkshopReport.pdf. Posted on September 8, 2006. This report comes from a September 2006 workshop held by Pew. Subjects covered include reasons for concern about imported GM crops, testing and identification of genetically modified organisms, regulation, health risks, and import policies.

"Exploring the Moral and Ethical Aspects of Genetically Engineered and Cloned Animals." Pew Initiative on Food and Biotechnology. Available online. URL: http://pewagbiotech.org/events/0124/proceedings.pdf. Posted in January 2005. This report presents the proceedings of a Pew workshop on the topic.

"Exploring the Regulatory and Commercialization Issues Related to Genetically Engineered Animals." Pew Initiative on Food and Biotechnology. Available online. URL: http://pewagbiotech.org/events/0321/proceedings.

pdf. Posted in May 2005. This report offers the proceedings of two workshops on the subject held in early 2005.

Fish, Andrew C., and Larisa Rudenko. "Feeding the World: A Look at Biotechnology and World Hunger." Pew Initiative on Food and Biotechnology. Available online. URL: http://pewagbiotech.org/resources/issuebriefs/feedtheworld.pdf. Posted in March 2004. This paper describes a range of positions on the question of whether agricultural biotechnology can help to meet the developing world's demand for food.

"Food Biotechnology: A Study of U.S. Consumer Attitudinal Trends." International Food Information Council Foundation. Available online. URL: http://ific.org/research/upload/2006%20Biotech%20Consumer%20Research%20Report.pdf. Posted in November 2006. The 11th in this organization's series of studies of U.S. consumer attitudes toward food biotechnology shows that a majority of consumers are confident in the safety of the food supply and are not concerned about agricultural biotechnology, but their knowledge of the subject is superficial and needs to be supplemented with science-based information.

"Genetically Modified Crops: What Do the Scientists Say?" American Society of Plant Biologists. Available online. URL: http://www.aspb.org/publications/plantphys/gmcpub.cfm. Accessed on March 15, 2007. This collection of editor's choice articles published in *Plant Physiology* magazine in 2000 and 2001 is intended to give plant biologists the information they need for educating the public about this issue.

"Genetically Modified Organisms and Biosafety: A Background Paper for Decision-Makers and Others to Assist in Consideration of GMO Issues." World Conservation Union. Available online. URL: http://www.iucn.org/themes/law/pdfdocuments/GMO_English.pdf. Posted in August 2004. This report provides background to enable World Conservation Union members to better investigate the potential effect of genetically modified organisms on biodiversity, societies, and food availability.

"Guidance Document of the Scientific Panel on Genetically Modified Organisms for the Risk Assessment of Genetically Modified Microorganisms and their Derived Products Intended for Food and Feed Use." European Food Safety Authority. Available online. URL: http://www.efsa.europa.eu/EFSA/Scientific_Document/comm_Guidance%20doc_GMM_en0.pdf. Posted in October 2006. This document presents guidelines for evaluating risks from genetically modified organisms and for applying to market such organisms and their products under European Union regulations.

Hallman, William K., et al. "Americans and GM Food: Knowledge, Opinion and Interest in 2004." Food Policy Institute. Available online. URL: http://www.foodpolicyinstitute.org/docs/reports/NationalStudy2004.pdf. Posted in 2004. This report summarizes the Food Policy Institute's third study of the U.S. public's perceptions of genetically modified food. The authors conclude that most citizens are unaware of GM food and the science behind it and are unsure what to think about it. However, they would like to see food labels state whether products contain genetically modified ingredients.

———, and Sarah C. Condry. "Public Opinion and Media Coverage of Animal Cloning and the Food Supply." Food Policy Institute. Available online. URL: http://www.foodpolicyinstitute.org/docs/summary/Animal %20Cloning%20Summary.pdf. Posted on November 21, 2006. After reviewing polls, interviewing opinion leaders, and examining media coverage regarding animal cloning, the authors conclude that few citizens know much about the issue or have made up their minds about it. Their concern is likely to increase now that the FDA has tentatively approved sale of meat and milk from cloned animals and will probably focus more on ethical than on scientific issues.

Hanrahan, Charles E. "Agricultural Biotechnology: The U.S.-EU Dispute." Congressional Research Service Reports and Issue Briefs. Available online. URL: http://www.ncseonline.org/nle/crsreports/06Apr/RS21556. pdf. Last updated on March 10, 2006. This report provides background on the claim, first brought by the United States, Canada, and Argentina in 2003, that the European Union's de facto moratorium on imports of genetically modified crops and agricultural products between 1998 and 2004 violated World Trade Organisation rules for international trade. In 2006, the WTO ruled in favor of the complainants.

"Issues in the Regulation of Genetically Engineered Plants and Animals." Pew Initiative on Food and Biotechnology. Available online. URL: http:// pewagbiotech.org/research/regulation/Regulation.pdf. Posted in April 2004. This report discusses regulation of genetically engineered plants for environmental protection and food safety, as well as regulation of genetically engineered animals. It recommends a "single-door," coordinated regulatory approach.

Jaffe, Gregory. "Creating the Proper Environment for Acceptance of Agricultural Biotechnology." Center for Science in the Public Interest. Available online. URL: http://www.cspinet.org/biotech/proper_environment. pdf. Posted in June 2005. This paper discusses current and future trends in agricultural biotechnology in the United States, especially regarding "biopharming," or producing drugs or other novel chemicals in plants.

———. "Withering on the Vine: Will Agricultural Biotech's Promises Bear Fruit?" Center for Science in the Public Interest. Available online. URL: http://www.cspinet.org/new/pdf/withering_on_the_vine.pdf. Posted on February 2, 2005. Jaffe reports that the number of new genetically engineered crops given pre-commercialization reviews by the Food and Drug Administration and the U.S. Department of Agriculture's Animal and

Plant Health Inspection Service (APHIS) dropped substantially during the preceding decade, while the time needed to receive a regulatory clearance doubled. He finds these trends puzzling and disturbing.

Kolehmainen, Sophia. "Precaution Before Profits: An Overview of Issues in Genetically Engineered Food and Crops." Council for Responsible Genetics. Available online. URL: http://www.gene-watch.org/programs/food/law-sophia.html. Accessed on March 17, 2007. This paper discusses techniques and applications in genetically engineered food, issues in the debate about such foods, and regulation of GM foods in the United States. It makes recommendations for improving the situation surrounding GM food.

"Maize and Biodiversity: The Effects of Transgenic Maize in Mexico, Key Findings and Recommendations." Commission for Environmental Cooperation of North America. Available online. URL: http://www.cec.org/files/PDF//Maize-and-Biodiversity_en.pdf. Posted on November 8, 2004. Scientists found modified genes in corn plants in Oaxaca, even though Mexico had a moratorium on planting of GM crops. This gene transfer was considered "insignificant from a biological point of view" but upset environmentalists.

Mellon, Margaret, and Jane Rissler. "Gone to Seed: Transgenic Contaminants in the Traditional Seed Supply." Union of Concerned Scientists. Available online. URL: http://www.ucsusa.org/assets/documents/food_and_environment/seedreport_fullreport.pdf. Posted in 2004. According to this report, laboratory testing has shown that seeds of traditional crops in the United States are being contaminated by genetic material from genetically modified crops. This finding demonstrates that federal regulation of GM crops is inadequate, the group says.

"NCGA Position on Biotechnology." National Corn Growers Association. Available online. URL: http://www.ncga.com/biotechnology/pdfs/Policy PositionPapers/Position%20in%20English.pdf. Posted in March 2006. The National Corn Growers Association supports agricultural biotechnology because the technology offers potential profits to farmers. The association urges growers to follow careful management practices in order to increase consumer acceptance of genetically modified corn.

Omamo, Steven Were, and Klaus von Grebmer, eds. "Biotechnology, Agriculture, and Food Security in Southern Africa." International Food Policy Research Institute. Available online. URL: http://www.ifpri.org/pubs/books/oc46/oc46toc.pdf. Posted on September 30, 2005. These essays discuss how various types of policy regarding genetically modified crops and foods, including policies on food safety and intellectual property, have been developed.

Onyango, Benjamin, Ramu Govindasamy, and Rodolfo M. Nayga, Jr. "Measuring U.S. Consumer Preferences for Genetically Modified Foods

Using Choice Modeling Experiments: The Role of Price, Product Benefits and Technology." Food Policy Institute. Available online. URL: http://www.foodpolicyinstitute.org/docs/working/AAEAchoicemodeling.pdf. Posted on August 4, 2004. This study models consumers' willingness to make tradeoffs between potential risks and benefits of genetically modified foods. Results suggest that U.S. consumers view such products negatively overall, but they probably will accept them if such foods can be shown to provide direct benefits to health, the environment, or food production.

"Peaceful Coexistence Among Growers of Genetically Engineered, Conventional, and Organic Crops." Pew Initiative on Food and Biotechnology. Available online. URL: http://pewagbiotech.org/events/0301/Workshop Report.pdf. Posted in March 2006. This report offers the proceedings of a workshop attended by speakers representing all three types of crops.

"Plant Biotechnology." Europabio. Available online. URL: http://www.europabio.org/documents/06Benefits%20Brochure.pdf. Accessed on March 16, 2007. This brochure from an industry organization praises the benefits of genetically modified crops for consumers, farmers, the environment, and European competitiveness.

"Position Statement on Genetic Engineering in Agriculture." Ecological Farming Association. Available online. URL: http://www.eco-farm.org/pdfs/Eco-Farm%20GE%20Paper.pdf. Accessed on March 17, 2007. This group is opposed to genetically modified crops and calls for a ban on any new such crops until a variety of conditions covering legal and safety issues are met.

Potrykus, Ingo. "Experience from the Humanitarian Golden Rice Project." AgBioWorld. Available online. URL: http://www.agbioworld.org/biotech-info/articles/biotech-art/potrykus.html. Posted on April 6, 2004. According to Potrykus, governmental overreliance on the "precautionary principle," which rejects any technology not conclusively proven to be safe, has blocked adoption of potentially useful innovations in agricultural biotechnology such as his "golden rice," which can help to prevent blindness by supplementing vitamin A intake.

"Public Sentiment about Genetically Modified Food." Pew Initiative on Food and Biotechnology. Available online. URL: http://pewagbiotech.org/research/2006update/2006summary.pdf. Posted in December 2006. Pew's fifth survey of U.S. consumer attitudes regarding genetically modified food shows that the public is still relatively ignorant about this subject, and opinion is still divided. People are more nervous about cloning and genetic modification of animals than they are about modified plants.

"Raising Risk: Field Testing of Genetic Crops in the United States." U.S. Public Interest Research Group. Available online. URL: http://www.uspirg.org/uploads/SQ/xA/SQxAVr3bFHxrgKQRZAEdyw/Raising

Risk2005.pdf. Posted on April 12, 2005. This organization examined U.S. Department of Agriculture (USDA) data on field tests of genetically engineered crops and concludes that the USDA's regulation is inadequate.

"Recommendations for the Early Food Safety Evaluation of New Non-Pesticidal Proteins Produced by New Plant Varieties Intended for Food Use." Food and Drug Administration. Available online. URL: http://www.cfsan.fda.gov/~dms/bioprgu2.html. Posted in June 2006. These guidelines were issued to the industry to prevent accidental introduction of allergens or toxins, particularly from genetically modified organisms, into the U.S. food and feed supply. The guidelines ask developers to voluntarily provide the FDA with information about the food safety of new proteins at a relatively early stage of crop development.

"Recommendations for Management Practices for Field Trials for Bioengineered Plants." National Agricultural Biotechnology Council. Available online. URL: http://nabc.cals.cornell.edu/pubs/Recomm_final.pdf. Posted on January 25, 2006. The recommendations cover such topics as training of personnel, field site selection, storage and post-harvest disposal of biological materials, monitoring, testing, and reporting.

"Research and Impact: CGIAR and Agricultural Biotechnology." Consultative Group on International Agricultural Research (CGIAR). Available online. URL: http://www.cgiar.org/impact/agribiotech.html. Accessed on March 17, 2007. The authors describe this influential organization's research on agricultural biotechnology.

"Resolution on Food Products from Cloned Animals." Trans Atlantic Consumer Dialogue. Available online. URL: http://www.tacd.org/cgi-bin/db.cgi?page=view&config=admin/docs.cfg&id=308. Posted in February 2007. This organization believes that, if the U.S. and European Union governments fail to adequately regulate the use of cloned animals and their progeny for food, such foods could compromise human health.

"A Risk-Based Approach to Evaluate Animal Clones and Their Progeny." U.S. Food and Drug Administration, Center for Veterinary Medicine. Available online. URL: http://www.fda.gov/cvm/CloneRiskAssessment.htm. Accessed on March 16, 2007. This report describes the technology used to assess the risk of animal clones, their progeny, and products from clones such as meat and milk. It then evaluates risks to animal health and to the health of humans who eat food from clones and concludes that the risks are no greater than with animals created by any other assisted reproductive technology.

"Risk Issues." Eurobarometer. Available online. URL: http://ec.europa.eu/public_opinion/archives/ebs/ebs_238_en.pdf. Posted in February 2006. This report presents the results of a European poll on attitudes toward risks associated with food, including genetically modified food, taken in September and October 2005. Concern about genetically modified organisms in food was found to be in the mid-range of worries about food risks.

Ruane, John, and Andrea Sonnino. "Results from the FAO Biotechnology Forum: Background and Dialogue on Selected Issues." Food and Agricultural Organisation of the United Nations (FAO). Available online. URL: ftp://ftp.fao.org/docrep/fao/009/a0744e/a0744e00.pdf. Posted in 2006. This paper reports on six conferences about agricultural biotechnology. The authors found that people in developing countries are very interested in discussing this subject but have powerful concerns, particularly about genetically modified food and intellectual property issues.

Sankula, Sujatha. "Quantification of the Impacts on U.S. Agriculture of Biotechnology-Derived Crops Planted in 2005." National Center for Food and Agricultural Policy. Available online. URL: http://www.ncfap. org/whatwedo/pdf/2005biotechimpacts-finalversion.pdf. Posted in November 2006. This report analyzes the reasons behind U.S. farmers' widespread adoption of genetically modified crops and the crops' impacts on producers and yield. The author concludes that farmers have received significant benefits from GM crops.

Sosin, Jennifer, and Mark David Richards. "What Will Consumers Do? Understanding Consumer Response When Meat and Milk from Cloned Animals Reaches Supermarkets." Clonesafety.org. Available online. URL: http://clonesafety.org/documents/Analysis_of_Consumers_11_04_05. doc. Posted in November 2005. The authors conclude that some consumers are uneasy about eating foods from cloned animals, but many would at least consider buying such products.

"Twenty Questions on Genetically Modified Foods." World Health Organization. Available online. URL: http://www.who.int/foodsafety/publications/biotech/20questions/en/index.html. Accessed on March 17, 2007. This document provides WHO's answers to the 20 questions most often asked by representatives of member state governments concerning the nature and safety of genetically modified food.

USDA Advisory Committee on Biotechnology and 21st Century Agriculture. "Global Traceability and Labeling Requirements for Agricultural Biotechnology-Derived Products: Impacts and Implications for the United States." U.S. Department of Agriculture. Available online. URL: http://w3.usda.gov/agencies/biotech/ac21/reports/tlpaperv37final.pdf. Posted on May 9, 2005. This paper discusses mandatory requirements for traceability and labeling of genetically modified products established by other countries, how the U.S. food and feed supply chain is addressing those requirements, and relevant marketplace issues and tools.

"What's in It for You? Genetically Modified Food: The New European GM Labeling and Traceability Regulations." Agricultural Biotechnology Council. Available online. URL: http://www.abcinformation.org/incubator/

applications/research_briefing/uploads/Practical%20Guide%20to%20La belling.pdf. Posted in 2004. This guide explains the new European Union regulations for the traceability and labeling of genetically modified food ingredients, which went into effect in April 2004, to consumers and small and medium-sized companies in the European food chain.

PATENTING LIFE

BOOKS

Dutfield, Graham. *Intellectual Property, Biogenetic Resources, and Traditional Knowledge*. London: Earthscan, 2004. Dutfield examines the relationships between intellectual property, biogenetic resources, and knowledge (including traditional knowledge) regarding practical applications of these resources. The book includes case studies from India and Kenya.

Erbisch, F. H., and K. M. Maredia, eds. *Intellectual Property Rights in Agricultural Biotechnology*. Rev. ed. Cambridge, Mass.: CABI Publishing, 2004. This book provides a clear explanation of intellectual property law as it applies to agricultural biotechnology. The revised edition includes updated material and five new chapters.

Merrill, Stephen A., Richard C. Levin, and Mark B. Myers, eds. *A Patent System for the 21st Century*. Washington, D.C.: National Academies Press, 2004. According to these essays, the strain of continual and rapid technological change has revealed weaknesses in the U.S. patent system that have increased the costs of doing research and delayed some scientific work. The authors recommend significant changes in the way the system operates.

Mgbeoji, Ikechi. *Global Biopiracy: Patents, Plants, and Indigenous Knowledge*. Ithaca, N.Y.: Cornell University Press, 2006. Mgbeoji examines the Western assumptions and biases that have shaped international law, the patent system, and related institutions that affect farmers around the world. He shows how these views have led to appropriation of indigenous knowledge and plants without compensation, so-called biopiracy, and have resulted in heavy loss of human cultures and plant diversity. He suggests protective measures to prevent further damage.

National Research Council. *Reaping the Benefits of Genomic and Proteomic Research: Intellectual Property Rights, Innovation, and Public Health*. Washington, D.C.: National Academies Press, 2006. This report concludes that intellectual property restrictions rarely produce significant burdens on biomedical research today, but they might do so in the future if steps are not taken, such as enacting a statutory infringement exception for research.

Office of Technology Assessment, U.S. Congress. *Patenting Life: New Developments in Biotechnology*. Honolulu, Hawaii: University Press of the Pacific, 2006. This special report reviews U.S. patent laws applicable to microorganisms, cells, plants, and animals. It includes a range of options for congressional action.

Perry, Bronwyn. *Trading the Genome: Investigating the Commodification of Bio-Information*. New York: Columbia University Press, 2004. Perry presents her investigation of the operation of the bioprospecting industry and the rise of a global trade in bio-information. Unless this trade is monitored, she says, the consequences for the developing countries that provide much of the world's genetic resources could be disastrous.

Resnik, David B. *Owning the Genome: A Moral Analysis of DNA Patenting*. Albany, N.Y.: State University of New York Press, 2004. After examining the social, ethical, theological, philosophical, and policy issues surrounding DNA patenting, Resnik concludes that only the patenting of a whole human genome would be intrinsically immoral; other DNA patents should be judged by their potential consequences.

Sandor, Judit, ed. *Society and Genetic Information: Codes and Laws in the Genetic Era*. New York: Central European University Press, 2004. The essays in this collection address the legal, social, and ethical implications of collecting, storing, analyzing, and commercializing genetic information, especially in the biomedical field.

ARTICLES

Cayford, Jerry. "Breeding Sanity into the GM Food Debate." *Issues in Science and Technology*, vol. 20, Winter 2004, pp. 49–56. The author claims that the debate over GM foods is so bitter and unproductive because biotechnology advocates misunderstand and oversimplify the arguments of critics. To Cayford, the most important issue is not the safety of these foods but the patenting of plants, which can determine who owns the world's food supply.

Dowie, Mark. "Talking Apes, Flying Pigs, Superhumans with Armadillo Attributes, and Other Strange Considerations of Dr. Stuart Newman's Fight to Patent a Human/Animal Chimera." *Mother Jones*, vol. 29, January-February 2004, pp. 48–55. Anti-biotechnology scientist Stuart Newman's attempt to patent a human-animal chimera (in order to prevent other scientists from creating such beings) raises numerous legal and moral questions.

Drutman, Lee. "It's in the Genes: Patent Barriers to Genetic Research." *Multinational Monitor*, vol. 25, July-August 2004, pp. 17–20. Patents in genetics often have not been enforced, but that may be changing. If it does, cost and concern about being sued for infringement could stifle research.

Hayes, Elizabeth A., and Daphne C. Lainson. "Patenting Life Forms." *Canadian Chemical News*, vol. 58, October 2006, pp. 18–19. This paper provides a review of laws in Canada, the United States, and other countries pertaining to the patenting of living things.

Kim, Mikyung. "An Overview of the Regulation and Patentability of Human Cloning and Embryonic Stem Cell Research in the United States and Anti-Cloning Legislation in Korea." *Santa Clara Computer and High Technology Law Journal*, vol. 21, May 2005, pp. 645–714. After comparing regulation and patent laws related to human cloning and embryonic stem cell research in the two countries, Kim concludes that there is no reason why any invention stemming from these technologies, short of a whole human being, could not be patented.

Kittredge, Clare. "A Question of Chimeras." *The Scientist*, vol. 19, April 11, 2005, pp. 54–55. Two men who oppose the patenting of living things applied for a patent on human-animal chimeras (a blend of two or more species), and the application was denied by the U.S. Patent and Trademark Office. The men hope that this rejection will keep biotechnology companies from obtaining similar patents, but researchers doubt that this will be the case.

Law, Grace S., and Jennifer A. Marles. "*Monsanto v. Schmeiser*: Patent Protection for Genetically Modified Genes and Cells in Canada." *Health Law Review*, vol. 13, Winter 2004, pp. 44–47. The authors analyze this famous Canadian case, which considers whether a farmer saving genetically engineered seeds he did not buy violated Monsanto's patent on the seeds. The analysis covers the majority and minority opinions of the Canadian Supreme Court, which found in favor of Monsanto by a 5-4 vote on May 21, 2004.

Leaf, Clifton, and Doris Burke. "The Law of Unintended Consequences." *Fortune*, September 19, 2005, p. 250 ff. The authors believe that by allowing universities to assume ownership of inventions funded by government money, the Bayh-Dole Act (1980) produced a climate of patent competition and noncommunication that has impeded biomedical research and raised the cost of drug development.

Merz, Jon F., and Neil A. Holtzman, eds. "Genes and Patents." *Community Genetics*, vol. 8, no. 4, 2005, n.p. (whole issue). Papers in this special issue provide a number of viewpoints on controversies surrounding the patenting of human genes.

Morneault, Gregory G. Schlenz, and Amy M. Crout Ziegler. "Patenting Cloning and Stem Cell Technology: Controversy and Comparison in the United States and Europe." *Intellectual Property and Technology Law Journal*, vol. 17, April 2005, pp. 1–6. This article surveys patent-related laws in the United States and Europe that could affect stem cell research. The authors conclude that the laws are unclear, and possible compromises are seldom employed.

Nicol, Dianne. "On the Legality of Gene Patents." *Melbourne University Law Review*, vol. 29, December 2005, pp. 809–842. Nicol examines regulation of patents on genes in Australia, the United States, and Britain. She concludes that, although the legality of gene patents remains unclear, the biotechnology industry can survive by using other mechanisms to deal with intellectual property issues related to genes.

"Patent Law—Utility—Federal Circuit Holds that Expressed Sequence Tags Lack Substantial and Specific Utility Unless Underlying Gene Function Is Identified." *Harvard Law Review*, vol. 119, June 2006, pp. 2604–2611. This article examines the implications of the federal circuit court decision in *In re Fisher*, which states that DNA segments called expressed sequence tags can be patented only if they can be tied to genes with known function. The author says that this decision is good for agricultural biotechnology because it prevents overpatenting.

Reed, Philip A. "Bioprospecting." *Technology Teacher*, vol. 64, December 2004, pp. 14–18. Bioprospecting—scientific research that looks for a useful application, process, or product in nature—is not necessarily either new or abusive. However, some instances of bioprospecting have been called "biopiracy" because they attempted to take resources and knowledge from developing nations without compensation.

Robinson, Douglas, and Nina Medlock. "*Diamond v. Chakrabarty*: A Retrospective on 25 Years of Biotech Patents." *Intellectual Property and Technology Law Journal*, vol. 17, October 2005, pp. 12–15. The authors describe this famous 1980 case, in which the Supreme Court declared that patenting of living things was legal. They also recount some of the important biotechnology patents that followed, including those on the Harvard Oncomouse, the polymerase chain reaction (PCR), genetically modified crops, and primate embryonic stem cells.

Schmieder, Sandra. "Scope of Biotechnology Inventions in the United States and Europe." *Santa Clara Computer and High Technology Law Journal*, vol. 21, November 2004, pp. 163–234. Schmieder analyzes the problem of patent breadth in biotechnology and outlines a form of balanced patent protection that would safeguard inventors' rights without strangling downstream research. Schmieder's system combines broad patent protection for DNA-related subject matter with a compulsory license system, using arbitration rather than regulation as an enforcement tool.

Sulston, John. "Staking Claims in the Biotechnology Klondike." *Bulletin of the World Health Organization*, vol. 84, May 2006, pp. 412–414. A leading British geneticist discusses ways in which patents can inhibit research and suggests possible solutions for these problems.

WEB DOCUMENTS

"Ethics and Gene Patents." Human Genetics Commission. Available online. URL: http://www.hgc.gov.uk/client/Content_wide.asp?ContentId=

372. Accessed on March 18, 2007. This report discusses legal and ethical issues raised by gene patents, including competition, access to information, informed consent, and confidentiality.

"Guidelines for the Licensing of Genetic Inventions." Organisation for Economic Co-operation and Development. Available online. URL: http://www.oecd.org/dataoecd/39/38/36198812.pdf. Posted on August 25, 2006. These guidelines, agreed to by OECD member countries, set out principles and best practices for those in research, business, and health systems who enter into license agreements concerning patented inventions used in health care.

"Human Genetic Materials: Making Canada's Intellectual Property Regime Work for the Health of Canadians." Canadian Biotechnology Advisory Committee. Available online. URL: http://cbac-cccb.ca/epic/site/cbac-cccb.nsf/vwapj/CBAC_FINAL-REPORT_English_Oct14-05.pdf/ $FILE/CBAC_FINAL-REPORT_English_Oct14-05.pdf. Posted in October 2005. The authors of this report conclude that human genetic material should be patentable, but they admit that such patents raise special issues that need to be resolved. They recommend improvements, both within the Canadian patenting system and outside it, to keep patents on human genetic material from limiting research and use of inventions that could aid health care.

Johnston, Josephine, and Angela Wasunna. "Patents, Biomedical Research, and Treatments: Examining Concerns, Canvassing Solutions." Hastings Center. Available online. URL: http://www.thehastingscenter.org/publications/reports.asp. Posted in February 2007. The authors describe practices that reduce barriers to research and health care potentially caused by patents on human genetic material.

King, David. "The Case Against the Patenting of Genes." Human Genetics Alert. Available online. URL: http://www.hgalert.org/topics/lifePatents/Patents2.html. Accessed on March 18, 2007. King claims that patenting genes gives too much control to greedy companies and results in commodification of the human body.

HUMAN GENETICS

BOOKS

Alper, Joseph S., et al., eds. *The Double-Edged Helix: Social Implications of Genetics in a Diverse Society.* Rev. ed. Baltimore, Md.: Johns Hopkins University Press, 2004. This book focuses on the social effects of genetic discoveries, criticizing the attitude of genetic determinism that has risen out of a misunderstanding of these discoveries.

Andrews, Lori B., Maxwell J. Mehlman, and Mark A. Rothstein. *Genetics: Ethics, Law, and Policy.* 2d ed. St. Paul, Minn.: West Group, 2006. This law school casebook covers legal issues raised by genetic aspects of health, employment, criminal justice, insurance, families, and other areas.

Chiu, Lisa Seachrist. *When a Gene Makes You Smell Like a Fish and Other Tales about the Genes in Your Body.* New York: Oxford University Press, 2006. This book uses the catchy titles of popular science writing, but it also provides a detailed and accurate survey of the latest advances in human genetics.

Clark, William R., and Michael Grunstein. *Are We Hardwired? The Role of Genes in Human Behavior.* Rev. ed. New York: Oxford University Press, 2004. The authors attempt to clarify the extent to which human behavior is influenced by genes, environment, and personal choice. They show that some aspects of behavior are probably inherited, but environment and experience influence the way genes are expressed.

Clarke, Angus, and Flo Ticehurst, eds. *Living with the Genome: Ethical and Social Aspects of Human Genetics.* New York: Palgrave Macmillan, 2006. This anthology provides diverse points of view on social, ethical, and legal aspects of recent discoveries about the human genome, including potentials for both benefit and abuse.

Creation and Governance of Human Genetic Research Databases. Paris: Organisation of Economic Co-operation and Development, 2006. This book summarizes the proceedings of a conference on this subject, looking at the way such databases have been established, governed, funded, managed, and commercialized.

de Melo-Martín, Immaculada. *Taking Biology Seriously: What Biology Can and Cannot Tell Us about Moral and Public Policy Issues.* Lanham, Md.: Rowman & Littlefield, 2005. The author points out that people on both sides of controversial issues such as human reproductive cloning often misunderstand the role of genes in human life.

DeSalle, Rob, and Michael Yudell. *Welcome to the Genome: A User's Guide to the Genetic Past, Present, and Future.* Hoboken, N.J.: Wiley, 2004. The authors examine the impacts of genomics on everyday life, including social and ethical issues.

Human Genome Project: Medical Dictionary, Bibliography, and Annotated Research Guide. San Diego, Calif.: Icon Health Publications, 2004. This three-in-one reference book includes Internet resources.

Kerr, Ann. *Genetics and Society: A Sociology of Disease.* New York: Routledge, 2004. This book looks at the wide-ranging impacts of genetics on modern society, discussing such topics as media representation of genetics-related debates and the politics of biomedical regulation.

Krude, Torsten, ed. *DNA: Changing Science and Society.* New York: Cambridge University Press, 2004. These papers explore a variety of subjects related to the social effects of human genetics research, such as DNA identification ("fingerprinting"), the complex role of DNA in cancer research, and how genetic disorders provide information about human communication.

Loeppky, Rodney. *Encoding Capital: The Political Economy of the Human Genome Project*. New York: Routledge, 2005. Loeppky views recent rapid advances in genome sciences, especially the Human Genome Project, through the lens of political economy, showing how the project's emergence was related to political, economic, and social issues.

McKenzie, Wendell H. *Genetics in Human Affairs*. Dubuque, Iowa: Kendall/Hunt Publishing Co., 2006. McKenzie presents the background in genetics that is needed to address the impact of the science on biological and social issues such as prenatal testing and screening, racial variation, and human evolution.

Moss, Lenny. *What Genes Can't Do*. Cambridge, Mass.: MIT Press, 2004. Moss explains that two quite different concepts of the gene exist today: the gene as complete controller of development and the gene as a developmental resource that specifies possibilities rather than certainties. He shows how these concepts apply to current theories about cancer and considers how, in turn, they may be reshaped by future biological discoveries.

Palladino, Michael A. *Understanding the Human Genome Project*. San Francisco, Calif.: Benjamin Cummings, 2005. This brief booklet provides basic information on the project, including its background, findings, and ethical and social implications.

Parfitt, Tudor, and Yulia Egorova. *Genetics, Mass Media, and Identity: A Case Study of Genetic Research on the Lemba and Bene Israel*. New York: Routledge, 2006. This pair of case studies focuses on the influence of genetic tests on the self-identity and narratives of groups, an issue raised by recent international gene-mapping projects such as the Human Genome Diversity Project.

Reardon, Jenny. *Race to the Finish: Identity and Governance in an Age of Genomics*. Princeton, N.J.: Princeton University Press, 2004. Reardon recounts the history of the Human Genome Diversity Project, which attempted to analyze the genes of people from different ethic groups and ran into controversy because the scientists involved could not understand the cultures in which they were attempting to operate. She also explores the effects of genetics and related sciences on racial and ethnic identities.

Redclift, Nanneke, and Sahra Gibbons, eds. *Genetics: Critical Concepts in Social and Cultural Theory*. New York: Routledge, 2006. Articles in this book provide background on molecular biology and modern genetics for researchers in nonbiological disciplines, setting the stage for multidisciplinary studies that explore interactions between biotechnology and culture.

Reilly, Philip R. *The Strongest Boy in the World and Other Adventures in Genetics*. Cold Spring Harbor, N.Y.: Cold Spring Harbor Laboratory Press,

2006. Reilly discusses issues raised by different uses of human genetic information, including forensic DNA databases and gene therapy.

Ridley, Matt. *The Agile Gene: How Nature Turns on Nurture.* New York: HarperCollins, 2004. Ridley shows how genes and environment are involved in an endless feedback loop, with behavior and environment affecting the way genes express themselves.

———. *Genome.* New York: HarperCollins, 2006. This book's 23 chapters, one for each of a human's pairs of chromosomes, provide a "whistle-stop tour" of the genome. The author uses the story of a gene from each chromosome to convey considerable information about genetic science and human development in an entertaining way.

Robinson, Tara Rodden. *Genetics for Dummies.* Hoboken, N.J.: Wiley, 2005. This book explains the science and ethics behind human genetics, including the relationship between genetics and particular diseases.

Sankaran, Neeraja, and Tara Acharya. *The Human Genome Sourcebook.* Westport, Conn.: Greenwood Press, 2005. This book is an encyclopedia with alphabetically arranged entries that cover terms and concepts in genetics, individual chromosomes, genes for normal functions, and genetic diseases.

ARTICLES

Barker, Joanne. "The Human Genome Diversity Project: 'Peoples,' 'Populations,' and the Cultural Politics of Identification." *Cultural Studies*, vol. 18, July 2004, pp. 571–606. The Human Genome Diversity Project, which attempted to discover differences in genes among different ethnic groups, ran into trouble because its scientists failed to understand the relationship between indigenous sovereignty and the self-identification of ethnic and cultural groups.

Berg, Paul. "Origins of the Human Genome Project." *American Journal of Human Genetics*, vol. 79, October 2006, pp. 603–605. Genetics pioneer Berg describes the objections that were raised to the idea of the Human Genome Project in 1986, when the project was first proposed. For example, few scientists saw the point in sequencing the so-called junk DNA that makes up 96 percent of the human genome, yet this information has proved very useful.

Davis, Dena S. "Genetic Research and Communal Narratives." *Hastings Center Report*, vol. 34, July-August 2004, pp. 40–49. Davis uses the Lemba, a southern African people, and the Thomas Jefferson-Sally Hemings family as examples to highlight the effects of genetic research on collective creation stories, an aspect that should be discussed with communities when deciding whether to carry out such research.

Fredericks, Marcel, et al. "Toward an Understanding of 'Genetic Sociology' and Its Relationships to Medical Sociology and Medical Genetics in the

Education Enterprise." *Education*, vol. 125, December 22, 2004, pp. 222–235. Medical sociology urgently needs to establish a new subdiscipline, genetic sociology, which will analyze the effects of genetic discoveries, diagnoses, prognoses, and treatments on the behavior of society.

Lemonick, Michael D. "The Iceland Experiment." *Time*, vol. 167, February 20, 2006, p. 50. In the late 1990s, Iceland native Kari Steffanson started a company, deCODE Genetics, which now has DNA samples from half the island's adult population, many of whom are related to one another. Drawing on this extensive information, researchers at deCODE have identified 15 gene variants affecting 12 diseases and have three drugs under development.

Marks, John. "Biopolitics." *Theory, Culture, and Society*, vol. 23, March-May 2006, pp. 333–334. Biotechnology issues such as stem cell research, the Human Genome Project, and gene therapy all relate to biopolitics, in which biology is drawn into the realm of political power and knowledge.

McElheny, Victor K. "The Human Genome Project + 5." *The Scientist*, vol. 20, February 2006, pp. 42–47. Progress in the five years since publication of the first draft of the human genome sequence has been primarily in the form of consolidation, but cheaper, faster new sequencing technology is likely to speed up advances and lead to new insights in the next five years.

Phimister, Elizabeth G. "Genomic Cartography: Presenting the HapMap." *New England Journal of Medicine*, vol. 353, October 27, 2005, pp. 1766–1768. The recently completed HapMap, a map of haplotypes (groups of genes inherited together), samples the human genome in a way that facilitates the search for genes that affect susceptibility to different diseases.

Wheelwright, Jeff. "Finland's Fascinating Genes." *Discover*, vol. 26, April 2005, pp. 52–59. Because its people are relatively closely related to each other, Finland is proving to be a mother lode for scientists studying inheritance of genes that affect the risk of contracting various diseases.

———. "Study the Clones First." *Discover*, vol. 25, August 2004, pp. 44–50. Studies of identical twins—natural clones—help to reveal what proportion of human characteristics is determined by inheritance and what proportion by environment.

WEB DOCUMENTS

Farabee, M. J. "Human Genetics." Estrella Mountain Community College. Available online. URL: http://www.emc.maricopa.edu/faculty/farabee/biobk/BioBookhumgen.html. Accessed on June 20, 2007. A basic primer on human genetics and the ways in which inherited diseases can be passed from parents to offspring.

"Understanding the Human Genome Project." National Human Genome Research Institute. Available online. URL: http://www.genome. gov/25019879. Accessed on March 18, 2007. This multimedia education kit about the Human Genome Project includes a computer-animated video introducing basic genetic concepts, a glossary, a timeline, two activities on population genetics and human history, information on genome sequencing, a tutorial on bioinformatics, and discussions of the ethical, legal, and social implications of the project.

"Why Should I Be Concerned about Human Genetics?" Human Genetics Alert. Available online. URL: http://www.hgalert.org/briefings/briefing1. PDF. Accessed on March 18, 2007. This pamphlet provides cautionary views on a variety of issues related to new knowledge of human genetics, including genetic determinism, genetic testing, genetic discrimination and invasion of privacy, embryo selection, psychiatric and behavioral genetics, cloning (both reproductive and therapeutic) and stem cell research, gene therapy and human genetic engineering (including germ-line alteration), patents on genes, and gene banks and databases.

DNA "FINGERPRINTING" AND DATABASES

BOOKS

Gerlach, Neil. *The Genetic Imaginary: DNA in the Canadian Criminal Justice System*. Toronto, Canada: University of Toronto Press, 2004. Gerlach explores why Canadians have accepted forensic DNA testing and banking so easily, allowing, in his opinion, an erosion of individual rights in favor of community security.

Junkin, Tim. *Bloodsworth: The True Story of the First Death Row Inmate Exonerated by DNA*. Chapel Hill, N.C.: Algonquin Books, 2004. This book tells the story of Kirk Noble Bloodsworth, falsely accused of the sex-murder of a child in 1984 and exonerated by DNA evidence in 1993.

Kafka, Tina. *DNA on Trial*. San Diego, Calif.: Lucent Books, 2004. This overview of the subject, for young adults, includes material on DNA testing to protect wildlife from exploitation and a chapter on problems with DNA testing.

Lazer, David, ed. *DNA and the Criminal Justice System: The Technology of Justice*. 2d ed. Cambridge, Mass.: MIT Press, 2004. These papers explore the procedural, ethical, and economic challenges posed by the use of DNA evidence; unresolved questions about behavioral genetics; the bioethical issues raised by the collection and maintenance of DNA in databases; and likely tradeoffs between individual (privacy) and societal (control of crime) interests in future use of such databases.

Annotated Bibliography

Marzilli, Alan. *DNA Evidence*. New York: Chelsea House, 2004. This advanced book for young adults, part of the Point Counterpoint series, provides historical background and examples of legal decisions that support different points of view on the topic.

McCartney, Carole. *Forensic Identification and Criminal Justice: Forensic Science, Justice, and Risk*. Uffculme, Cullompton, Devon, England: Willan Publishing, 2006. McCartney recounts the development of forensic identification technologies, especially fingerprinting and forensic DNA typing, and shows how these technologies have affected the legal system. The book includes a detailed discussion of the U.S. national forensic DNA database (NDIS).

ARTICLES

Aronson, Jay D. "DNA Fingerprinting on Trial: The Dramatic Early History of a New Forensic Technique." *Endeavour*, vol. 29, September 2005, pp. 126–131. This article describes the early history of DNA identification in Britain, focusing on two dramatic courtroom trials in the mid-1980s. Aronson maintains that these trials showed more about the effectiveness of threatening to introduce DNA evidence than about the validity of the technology itself, which was still in its infancy.

Bieber, Frederick R. "Turning Base Hits into Earned Runs: Improving the Effectiveness of Forensic Data Bank Programs." *Journal of Law, Medicine, and Ethics*, vol. 34, Summer 2006, pp. 222–233. Discusses criteria on which the effectiveness of DNA databases can be evaluated, reasons why such databases are less effective in solving and preventing crimes than they could be, and ways to make them work better.

"Constitutional Law—Fourth Amendment—Ninth Circuit Upholds Collection of DNA from Parolees." *Harvard Law Review*, vol. 118, December 2004, pp. 818–824. Analyzes the Ninth Circuit Court of Appeals decision in *United States v. Kincade*, which stated that collection of DNA samples from parolees did not violate Fourth Amendment protection against unreasonable searches and seizures. The author opposes the decision because it uses a line of reasoning that encourages expansion of the categories of people who are subject to DNA searches.

Dove, Alan. "Molecular Cops: Forensic Genomics Harnessed for the Law." *Genomics and Proteomics*, vol. 4, June 1, 2004, p. 23. DNA identification is ambiguous in some cases. New technology may reduce such problems, but it will have to prove itself in the courts.

Duffy, Shannon P. "Third Circuit Upholds DNA Testing of Felons, Cites Need for Database." *New Jersey Law Journal*, March 28, 2005, n.p. A recent appeals court ruling in *United States v. Sczubelek* supports the constitutionality of taking a DNA sample from a convict on supervised release (parole).

Fink, Sheri. "Reasonable Doubt: Evidence." *Discover*, vol. 27, July 2006, pp. 54–59. DNA profiles are more reliable than many other forms of forensic evidence, but laboratory errors and other issues can still complicate the identification picture.

Gaensslen, R. E. "Should Biological Evidence or DNA Be Retained by Forensic Science Laboratories after Profiling?" *Journal of Law, Medicine, and Ethics*, vol. 34, Summer 2006, pp. 375–379. The author urges legislatures to clarify what evidence may be retained and what uses may be made of it. Preferably, only DNA profiles, not samples, should be retained. If samples are kept, they should be used only for identification, not extraction of medical or other personal information.

Harlan, Leigh M. "When Privacy Fails: Invoking a Property Paradigm to Mandate the Destruction of DNA Samples." *Duke Law Journal*, vol. 54, October 2004, pp. 179–219. Harlan maintains that the best approach to balancing privacy with law enforcement needs is for police to keep DNA profiles from dragnets but to regard actual DNA samples, which contain sensitive genetic information, as the property of the individuals from whom they were taken.

Herkenham, M. Dawn. "Retention of Offender DNA Samples Necessary to Ensure and Monitor Quality of Forensic DNA Efforts." *Journal of Law, Medicine, and Ethics*, vol. 34, Summer 2006, pp. 380–384. Herkenham feels that retention of DNA samples is necessary because of possible desire for retesting with new technologies and need to maintain laboratories' quality assurance. She says that federal and state laws are adequate to prevent unauthorized disclosure of private information and other forms of misuse.

Jeffreys, Alec J. "Genetic Fingerprinting." *Nature Medicine*, vol. 11, October 2005, pp. 1035–1039. The inventor of DNA profiling provides a personal account of the technology's birth from experiments that had no connection with personal identification or crime solving.

Kruglinski, Susan. "Who's Your Daddy? Don't Count on DNA Testing to Tell You." *Discover*, vol. 27, April 2006, pp. 68–69. Although paternity tests using DNA samples are very reliable, mistakes can occur, and even a low rate of error can seriously undermine confidence in the results.

Lyons, Donna. "Capturing DNA's Crime Fighting Potential." *State Legislatures*, vol. 32, March 2006, pp. 16–17. DNA databases are expanding and solving many cold cases, but laboratories often lack the funding and personnel to deal with backlogs of samples and to provide high-quality testing.

Maclin, Tracey. "Is Obtaining an Arrestee's DNA a Valid Special Needs Search under the Fourth Amendment?" *Journal of Law, Medicine, and Ethics*, vol. 34, Summer 2006, pp. 165–187. This article concludes that laws mandating DNA testing for arrestees are not justified by the special needs exception to the Fourth Amendment and therefore should be declared unconstitutional.

Rothstein, Mark A., and Meghan K. Talbott. "The Expanding Use of DNA in Law Enforcement: What Role for Privacy?" *Journal of Law, Medicine, and Ethics*, vol. 34, Summer 2006, pp. 153–164. Both the scope of forensic DNA databases and the range of uses to which they are put have expanded significantly during the last decade, creating important new legal and policy issues that are surveyed in this article. The authors conclude that privacy considerations require limits to be placed on the uses of DNA in law enforcement.

Simoncelli, Tania. "Dangerous Excursions: The Case Against Expanding Forensic DNA Databases to Innocent Persons." *Journal of Law, Medicine, and Ethics*, vol. 34, Summer 2006, pp. 390–397. Reasons for not expanding DNA databases to cover all citizens include potential violation of the Fourth Amendment and privacy protection laws, diminishing returns to law enforcement, worsening of existing testing backlogs, and cost.

———, and Barry Steinhardt. "California's Proposition 69: A Dangerous Precedent for Criminal DNA Databases." *Journal of Law, Medicine, and Ethics*, vol. 34, Summer 2006, pp. 199–213. The authors believe that the radical expansion of California's forensic DNA database mandated by this proposition, passed in 2004, presents great threats to civil liberties and social justice while doing little to increase the safety of the state's citizens.

"The Sins of the Fathers: DNA Fingerprinting." *The Economist*, vol. 371, April 24, 2004, p. 60US. Some first-offender criminals can now be identified through "family searches" of national forensic DNA databases.

Smith, Michael E. "Let's Make the DNA Identification Database as Inclusive as Possible." *Journal of Law, Medicine, and Ethics*, vol. 34, Summer 2006, pp. 385–389. A national DNA profile database should include all citizens—but only if the tissue samples from which the profile information comes are destroyed to protect privacy.

Wall, Tom. "A Simple Prank by a 13-Year-Old: Now Her Genetic Records are on the National DNA Database For Ever." *New Statesman*, vol. 134, April 25, 2005, pp. 32–33. The author finds the increasing inclusiveness of the British national DNA identity database deeply disturbing.

Weiss, Marcia J. "Beware! Uncle Sam Has Your DNA: Legal Fallout from Its Use and Misuse in the U.S." *Ethics and Information Technology*, vol. 6, March 2004, pp. 55–63. This article explores the ethics and constitutionality of collecting genetic material, balancing the social goods of deterring and solving crime against individuals' right to liberty, privacy, and security.

WEB DOCUMENTS

Asplen, Christopher H., and Smith Alling Lane. "The Non-Forensic Use of Biological Samples Taken for Forensic Purposes: An International Perspective." American Society of Law, Medicine, and Ethics. Available

online. URL: http://www.aslme.org/dna_04/spec_reports/asplen_non_
forensic.pdf. Accessed on March 17, 2007. This article describes regulations and practice concerning retention and nonforensic use of DNA
samples taken for forensic purposes in a variety of countries.

"Police Retention of DNA." GeneWatch UK. Available online. URL: http://
www.genewatch.org/uploads/f03c6d66a9b354535738483c1c3d49e4/
Councillorsbrief07_2.pdf. Posted on February 23, 2007. In order to balance concern for human rights and privacy with the desire of law enforcement agencies to reduce crime, this organization recommends several
changes in the way the British national DNA database is handled, including destruction of samples after profiles have been obtained and restriction of the range of profiles that are kept permanently.

Staley, Kristina. "The Police National DNA Database: Balancing Crime
Detection, Human Rights and Privacy." GeneWatch UK. Available online. URL: http://www.genewatch.org/uploads/f03c6d66a9b3545357384
83c1c3d49e4/NationalDNADatabase.pdf. Posted in January 2005. This
paper supports the existence of Britain's national forensic DNA database
but points out the ethical concerns it raises.

Winickoff, David, and Julie Park. "The Constitutionality of Forensic DNA
Databanks: Fourth Amendment Issues." American Society of Law, Medicine, and Ethics. Available online. URL: http://www.aslme.org/dna_04/
reports/winickoff_update.pdf. Last updated on June 30, 2005. This short
article provides an overview of the important legal issues, particularly
regarding the Fourth Amendment, that are raised by forensic DNA databases.

GENETIC HEALTH TESTING AND DISCRIMINATION

BOOKS

Arnason, Gardar, Salvor Nordal, and Vilhjálmur Arnason, eds. *Blood and
Data: Ethical, Legal, and Social Aspects of Human Genetic Databases*. Oxford,
England: University of Iceland Press (c/o Oxbow Books), 2006. Papers
from a conference held in Iceland in 2004 provide a wide variety of viewpoints on issues concerning nonforensic genetic databases, including
privacy, consent, and genetic discrimination.

Engs, Ruth Clifford. *The Eugenics Movement: An Encyclopedia*. Westport,
Conn.: Greenwood Press, 2005. This book focuses on the eugenics
movement in the United States in the early 20th century, but it also provides information on the movement in Britain and Germany. Its 250 al-
phabetical entries show how the movement developed and spread. In

addition, they cover more modern issues, such as human cloning and genetic alteration.

Featherstone, Katie, et al. *Risky Relations: Family, Kinship, and the New Genetics*. Oxford, England: Berg, 2006. The authors show that genetic testing and related technologies can have profound effects on individuals' feelings of identity and family relationships, for instance when people find out that they or other family members possess a gene that greatly increases risk of developing a certain disease.

Hart, Anne. *Predictive Medicine for Rookies: Consumer Watchdogs, Reviews, and Genetic Testing Firms Online*. New York: ASJA Press, 2005. Hart gives consumers information about at-home genetic testing for predisposition to diseases or ancestry tracing, including the need for proper interpretation.

Jones, Nancy Lee, and Alison M. Smith, eds. *Genetic Information: Legal and Law Enforcement Issues*. New York: Novinka Books, 2004. This book discusses federal laws, state statutes, and legislation related to genetic information, including genetic discrimination.

Miller, Suzanne M., et al., eds. *Individuals, Families, and the New Era of Genetics: Biopsychosocial Perspectives*. New York: Norton, 2006. Mental health practitioners assess the impact of genetic knowledge on family life, covering such topics as genetic testing and inherited diseases.

Parens, Eric, Audrey R. Chapman, and Nancy Press, eds. *Wrestling with Behavioral Genetics: Science, Ethics, and Public Conversation*. Baltimore, Md.: Johns Hopkins University Press, 2005. These essays from an interdisciplinary group of authors discuss the ethics and social implications of behavioral genetics research, including both benefits and harms.

Radetzki, Marcus, Marian Radetzki, and Niklas Juth. *Genes and Insurance: Ethical, Legal, and Economic Issues*. Rev. ed. New York: Cambridge University Press, 2006. The authors examine the relationship between genetic testing and insurance from ethical, legal, and economic standpoints and call for government-funded social insurance to replace private insurance.

Rothstein, Mark A., ed. *Genetics and Life Insurance: Medical Underwriting and Social Policy*. Cambridge, Mass.: MIT Press, 2004. This essay collection, focusing mainly on the United States, discusses issues raised by the conflict between insurers and their customers regarding availability of genetic information.

Schoonmaker, Michele, and Erin D. Williams. *Genetic Testing: Scientific Background and Nondiscrimination Legislation*. New York: Novinka Books, 2006. This pocket-sized text for public policymakers provides a comprehensive overview of the status of genetic testing in the United States. Topics include the definition of genetic information, health information, and evidence of genetic discrimination.

Sharpe, Neil F., and Ronald F. Carter. *Genetic Testing: Care, Consent, and Liability*. Hoboken, N.J.: Wiley, 2006. This book explains the legal framework surrounding genetic testing and counseling, particularly with regard to medical malpractice. It also explores genetic, clinical, and ethical issues raised by such testing, which is now widely available to consumers.

Tutton, Richard, and Oonagh Corrigan, eds. *Genetic Databases: Socio-Ethical Issues in the Collection and Use of DNA*. New York: Routledge, 2004. These essays discuss social and ethical issues raised by nonforensic DNA databases and biobanks in Europe and North America, especially issues related to privacy and consent.

Wailoo, Keith, and Stephen Pemberton. *The Troubled Dream of Genetic Medicine: Ethnicity and Innovation in Tay-Sachs, Cystic Fibrosis, and Sickle-Cell Disease*. Baltimore, Md.: Johns Hopkins University Press, 2006. The authors discuss how these three inherited diseases, each associated with a particular ethnic group, became entangled in racial and ethnic controversies.

Wasserman, David, Jerome Bickenbach, and Robert Wachbroit, eds. *Quality of Life and Human Difference: Genetic Testing, Health Care, and Disability*. New York: Cambridge University Press, 2005. These nine essays discuss prenatal testing for diseases and disabilities from the perspective of disability scholars, who question common assumptions about the quality of life of people with disabilities.

Wertz, Dorothy C., and John C. Fletcher. *Genetics and Ethics in Global Perspective*. New York: Springer, 2004. This international study of social and ethical issues in genetics in 36 countries focuses on vignettes describing situations in medical genetics, including genetic testing, discrimination, and eugenics.

ARTICLES

Crocin, Susan L., et al. "Genetic Tests Are Testing the Law." *Trial*, vol. 42, October 2006, pp. 44–50. This article focuses on lawsuits arising when genetic testing of gametes (eggs and sperm), embryos, or fetuses provides incorrect information that affects decisions about whether to conceive or continue a pregnancy.

Finkelstein, Foel B. "Genetic Privacy." *American Medical News*, vol. 47, April 5, 2004, pp. 5–6. An interview with Francis Collins, head of the National Human Genome Research Institute, discusses genetic discrimination and legislation to prohibit it.

Greely, Henry T. "Banning Genetic Discrimination." *New England Journal of Medicine*, vol. 353, September 1, 2005, pp. 865–867. Greely discusses genetic discrimination in insurance and urges passage of a federal law to ban such discrimination.

Hoel, M., et al. "Genetic Testing in Competitive Insurance Markets with Repulsion from Chance: A Welfare Analysis." *Journal of Health Economics*, September 2006, n.p. After conducting a welfare analysis regarding the extent to which insurance companies should be allowed to use genetic information when making health evaluations for policies, the authors conclude that allowing insurers to use such information increases overall welfare.

Jancin, Bruce. "Survey Dispels Worries of Genetic Discrimination by Insurers." *Ob Gyn News*, vol. 41, May 1, 2006, p. 61. Fears of discrimination based on genetic test results may be overblown.

Javitt, Gail H., and Kathy Hudson. "Federal Neglect: Regulation of Genetic Testing." *Issues in Science and Technology*, vol. 22, Spring 2006, pp. 59–66. Neither the accuracy nor the scientific validity of genetic tests is currently regulated to any great extent, leaving consumers vulnerable to errors and misinformation.

Joly, Yann. "Life Insurers' Access to Genetic Information: A Way Out of the Stalemate?" *Health Law Review*, vol. 14, Spring 2006, pp. 14–21. France, by legislating too soon, and Canada, by ignoring the problem, arrived at similar stalemates in regard to insurers' access to genetic information. The question of genetic discrimination by insurers needs to be considered in the broader context of acccess to health care and the role of the welfare state. Meanwhile, a moratorium on insurers' access to genetic information is recommended.

Lin, Zhen, Art B. Owen, and Russ B. Altman. "Approaches for Protecting Privacy in the Genomic Era." *Genetic Engineering News*, vol. 24, October 1, 2004, pp. 6–8. The authors discuss techniques for protecting genomic data.

Lock, Margaret. "Eclipse of the Gene and the Return of Divination." *Current Anthropology*, vol. 46, December 2005, pp. S47–70. Lock reviews social science literature concerning the social effects of genetic testing for inherited conditions and susceptibilities to disease. She stresses the fact that environment as well as genetics plays a role in the development of most diseases, so estimating risk through genetic tests alone is unrealistic.

Morrison, Patrick J. "Insurance, Unfair Discrimination, and Genetic Testing." *Lancet*, vol. 366, September 10, 2005, pp. 877–880. Morrison claims that prevention of insurers' use of genetic information, as in Great Britain, calms the public's fears about genetic discrimination and does not cause financial hardship for insurers.

Offit, Kenneth, et al. "The 'Duty to Warn' a Patient's Family Members about Hereditary Disease Risks." *Journal of the American Medical Association*, vol. 292, September 22, 2004, pp. 1469–1473. The authors say that medical personnel should encourage but not coerce patients to share information from genetic testing with their families.

"Peering into the Future: Genetic Testing Is Transforming Medicine—and the Way Families Think about Their Health." *Newsweek*, December 11,

2006, p. 52. Genetic testing is beginning to revolutionize the prediction, diagnosis, and treatment of disease. It also raises ethical concerns.

Sherwin, Susan. "BRCA Testing: Ethics Lessons for the New Genetics." *Clinical and Investigative Medicine*, vol. 27, February 2004, pp. 19–22. Genetic testing, as exemplified by testing for the breast cancer gene BRCA, raises complex ethical issues. Society as a whole, not just individuals, needs to make decisions about these matters.

Shute, Nancy. "Unraveling Your DNA's Secrets." *U.S. News & World Report*, vol. 142, January 8, 2007, pp. 50–54, 57–58. Over-the-counter genetic tests, widely available through the Internet, help some people but mislead others. The tests and the laboratories that carry them out are both unregulated.

Spinello, Richard J. "Property Rights in Genetic Information." *Ethics and Information Technology*, vol. 6, March 2004, pp. 29–42. Genetic research can threaten privacy, but classifying genetic information as property is not a good response, Spinello says.

Tavani, Herman T. "Genomic Research and Data-Mining Technology: Implications for Personal Privacy and Informed Consent." *Ethics and Information Technology*, vol. 6, March 2004, pp. 15–28. Data mining in population genetics studies undermines both privacy and the notion of informed consent, according to Tavani.

WEB DOCUMENTS

Baruch, Susannah, et al. "Reproductive Genetic Testing: Issues and Options for Policymakers." Genetics and Public Policy Center. Available online. URL: http://www.dnapolicy.org/images/reportpdfs/ReproGenTestIssuesOptions.pdf. Posted in November 2004. This report presents and evaluates a range of policy options for public and private bodies concerning use, safety and accuracy, access, and delivery of genetic testing.

"Genetic Testing." National Women's Health Resource Center. Available online. URL: http://www.healthywomen.org/healthtopics/genetictesting. Last updated on March 12, 2005. This online pamphlet offers an overview of genetic testing, with links to discussions of specific terms and diseases as well as to pamphlets on facts to know, questions to ask, and commonly asked questions.

"Genetic Testing: Preliminary Policy Guidelines." Council for Responsible Genetics. Available online. URL: http://www.gene-watch.org/programs/privacy/guidelines.html. Posted in June 2006. The guidelines include recommendations for regulation.

Javitt, Gail H., and Kathy Hudson. "Public Health at Risk: Failures in Oversight of Genetic Testing Laboratories." Genetics and Public Policy Center. Available online. URL: http://www.dnapolicy.org/images/reportpdfs/

PublicHealthAtRiskFinalWithCover.pdf. Posted in September 2006. The authors say that the Clinical Laboratories Improvement Amendments of 1988 (CLIA) could ensure the quality of laboratories performing genetic testing, but the Centers for Medicare and Medicaid Services (CMS), charged with implementing this law, have repeatedly failed to do so. They urge creation of a genetic testing specialty under CLIA and frequent proficiency testing for laboratories.

Joint Committee on Medical Genetics. "Consent and Confidentiality in Genetic Practice." British Society for Human Genetics. Available online. URL: http://www.bshg.org.uk/documents/official_docs/Consent_and_confid_corrected_21[1].8.06.pdf. Posted in April 2006. This document provides guidance on genetic testing and sharing genetic information, including who can give consent for testing, sharing the information with other parties, genetic investigations on stored samples, the effect of the (British) Human Tissue Act of 2004 on consent and DNA analysis, and the effect of the Data Protection Act (1998) on the processing of medical information.

Lewis, Celine. "Developing European Quality Standards for Patient Information on Genetic Testing." Genetic Interest Group. Available online. URL: http://www.gig.org.uk/docs/developingeuropeanqualitystandards_genetictestingsept06.pdf. Posted in September 2006. This report describes an assessment of the availability and quality of written information about genetic testing given to patients and their families in Europe.

"Newborn Screening Tests." March of Dimes. Available online. URL: http://www.marchofdimes.com/pnhec/298_834.asp. Accessed on March 18, 2007. This online booklet, aimed at prospective parents, describes standard newborn genetic screening tests.

"Protecting Privacy in the Age of Genetic Information." Canadian Biotechnology Advisory Committee. Available online. URL: http://cbac-cccb.ca/epic/site/cbac-cccb.nsf/vwapj/Privacy_Report_final_e.pdf/$FILE/Privacy_Report_final_e.pdf. Posted in August 2004. This report discusses issues raised by new collections of genetic information such as population biobanking. Topics covered include discrimination and privacy.

"Report on Race and Genetic Determinism." Council for Responsible Genetics. Available online. URL: http://www.gene-watch.org/programs/determinism/RaceReport.html. Posted in May 2006. Biological categories must not be confused with social ones such as race, nor should genes be seen as the only source of differences in susceptibility to disease.

"Understanding Genetics: A New England Guide for Patients and Health Professionals." Genetic Alliance. Available online. URL: http://www.geneticalliance.org/ksc_assets/pdfs/manual_all2.pdf. Posted in 2006. This guide begins with basic genetic concepts and goes on to provide in-depth information about diagnoses of genetic conditions, newborn screening, family-history gathering, genetic counseling, and types of genetic tests.

Van Court, Marian. "The Case for Eugenics in a Nutshell." Future Generations. Available online. URL: http://www.eugenics.net/papers/caseforeugenics.html. Posted in winter 2004. Van Court claims that providing incentives for people with conditions such as mental retardation not to have children would solve a number of social problems by reducing the incidence of these conditions.

MEDICAL BIOTECHNOLOGY, GENE THERAPY, AND HUMAN GENE ALTERATION

BOOKS

Baillie, Harold W., and Timothy K. Casey, eds. *Is Human Nature Obsolete? Genetics, Bioengineering, and the Future of the Human Condition.* Cambridge, Mass.: MIT Press, 2004. Taking a humanistic philosophical perspective, this interdisciplinary group of authors examines the background, goals, and values of technologies that may alter what it means to be human.

Battler, Alexander, and Jonathan Leor, eds. *Stem Cell and Gene-Based Therapy: Frontiers in Regenerative Medicine.* New York: Springer, 2007. These essays cover tissue regeneration, stem cell research, and gene therapy as new approaches in regenerative medicine.

Blazer, Shraga, and Etan Z. Zimmer, eds. *The Embryo: Scientific Discovery and Medical Ethics.* New York: Karger, 2004. The essays in this book, written by religious thinkers, scientists, and ethicists, cover such topics as stem cell research, prenatal and preimplantation genetic testing, surgical intervention to treat fetuses, imaging and monitoring, and "wrongful life" lawsuits.

Castle, David, et al. *Science, Society, and the Supermarket: The Opportunities and Challenges of Nutrigenomics.* Hoboken, N.J.: Wiley, 2006. This book analyzes the potential benefits and risks of nutrigenomics, a new discipline that studies interactions between diet and individual genetic variations. It provides ethics guidelines for regulating such things as nutrigenomic tests sold directly to the public.

Colavito, Mary C. *Gene Therapy.* San Francisco, Calif.: Benjamin Cummings, 2006. This overview presents the history of gene therapy, including both successes and failures, and considers its future.

Espejo, Roman, ed. *Gene Therapy.* San Diego, Calif.: Greenhaven, 2004. This short essay collection for young adults, part of the At Issue series, considers controversial aspects of this branch of biomedical research.

Glover, Jonathan. *Choosing Children: Genes, Disability, and Design*. New York: Oxford University Press, 2006. A moral philosopher examines the issues involved in altering genes to choose the kind of offspring people have, including the question of how *disability* should be defined.

Hughes, James. *Citizen Cyborg: Why Democratic Societies Must Respond to the Redesigned Human of the Future*. Boulder, Colo.: Westview Press, 2004. Unlike groups who unquestioningly embrace the redesign of humanity and those who oppose all such action equally strongly, Hughes recommends "democratic transhumanism," which supports human genetic alteration provided that it is safe, available to all, and done in ways that respects individuals' right to control their own bodies.

Kass, Leon R. *Life, Liberty, and the Defense of Dignity: The Challenge for Bioethics*. Rev. ed. Washington, D.C.: AEI Press, 2004. Prominent conservative philosopher Kass argues that human gene alteration, cloning, stem cell research, and related technologies debase human dignity and should be severely limited.

Lindee, Susan. *Moments of Truth in Genetic Medicine*. Baltimore, Md.: Johns Hopkins University Press, 2005. Lindee surveys the development of genetic medicine between 1950 and 1980, focusing on "moments of truth" that led to concentration on genetic aspects of disease.

McKibben, Bill. *Enough: Staying Human in an Engineered Age*. New York: Owl Books, 2004. Although he does not rule out all fruits of the new genetics, McKibben is in favor of drawing a line that will keep gene alteration from changing the nature of humanity beyond recognition.

Miah, Andy. *Genetically Modified Athletes: Biomedical Ethics, Gene Doping, and Sport*. New York: Routledge, 2004. Miah provides a comprehensive analysis of the technology and ethics of altering genes to provide an advantage in sports.

Naam, Ramez. *More than Human: Embracing the Promise of Biological Enhancement*. New York: Broadway Books, 2005. Naam, who is optimistic about medical technology, surveys an assortment of cutting-edge techniques.

Niewöhner, Jörg, and Christof Tannert, eds. *Gene Therapy: Prospective Technological Assessment in its Societal Context*. New York: Elsevier Science, 2006. This anthology covers scientific, ethical, legal, perception, and communication issues arising from gene therapy.

Panno, Joseph. *Gene Therapy: Treating Disease by Repairing Genes*. New York: Facts On File, 2004. This encyclopedia explores the development and future of gene therapy, with an emphasis on its promising aspects.

Rasko, John E. J., Gabrielle M. O'Sullivan, and Rachel A. Ankeny, eds. *The Ethics of Inheritable Genetic Modification: A Dividing Line?* New York: Cambridge University Press, 2006. Experts in biological and social science, law, and ethics consider how society might deal with ethical concerns raised by germ-line gene alteration.

Rose, Nikolas. *The Politics of Life Itself: Biomedicine, Power, and Subjectivity in the Twenty-First Century.* Princeton, N.J.: Princeton University Press, 2006. Rose analyzes recent developments in genetics and biomedicine that have led to the politicization of medicine, biotechnology, and human life, resulting in important changes in conceptions of personhood and kinship and in the way medicine views disease.

Rothschild, Joan. *The Dream of the Perfect Child.* Bloomington: Indiana University Press, 2005. Rothschild evaluates the use of genetics and prenatal testing to create what parents hope will be the "perfect child," eliminating all who do not seem worthy to be born.

Sasson, Albert. *Medical Biotechnology: Achievements, Prospects, Perceptions.* Tokyo, Japan: United Nations University Press, 2006. This book looks at the drivers of medical and pharmaceutical biotechnology in the United States, Europe, and Japan. It also describes biotechnology tools to fight major global health threats, considers ethical issues and public perceptions of the technology, and shows how developing countries are creating their own bioindustries.

Shannon, Thomas A., ed. *Genetics: Science, Ethics, and Public Policy.* Lanham, Md.: Rowman & Littlefield, 2005. This collection of essays considers ethical issues raised by forms of biotechnology and genetic alteration that affect human beings directly.

Shaw, David. *Genetic Morality.* New York: Peter Lang, 2006. Shaw believes that debates about human genetic modification, cloning, and embryo research can be resolved through careful application of established ethical principles.

Stephenson, Frank. H. *DNA: How the Biotech Revolution Is Changing the Way We Fight Disease.* Amherst, N.Y.: Prometheus Books, 2007. This book, aimed at the general reader, shows how new understandings of genetics and new tools from biotechnology are helping to fight common diseases such as cancer, Alzheimer's disease, and HIV infection.

Tamburrini, Claudio, and Torbjörn Tännsjö, eds. *Genetic Technology and Sport: Ethical Questions.* New York: Routledge, 2005. These papers describe the possibility of athletes manipulating their own genetic code to gain a sports advantage and debate the ethics of such a practice.

ARTICLES

Arnst, Catherine, et al. "Biotech, Finally." *Business Week*, issue 3937, June 13, 2005, p. 30 ff. Medical biotechnology has finally come of age, producing a host of new drugs that are already being used to treat patients. Further advances in cancer treatment, diagnosis, and stem cell research are likely to occur in the near future.

Bayer, Oswald. "Self-Creation? On the Dignity of Human Beings." *Modern Theology* vol. 20, April 2004, pp. 275–290. Bayer examines ethical questions raised by biotechnology within the framework of theological concepts of human dignity.

Baylis, Françoise, and Jason Scott Robert. "The Inevitability of Genetic Enhancement Technologies." *Bioethics*, vol. 18, February 2004, pp. 1–26. The authors present a number of ethical objections to genetic technologies that enhance human capacities and traits, but they see such enhancement as inevitable because of the prevailing worldview of humans as masters of their own evolutionary future.

Behar, Michael. "Will Genetics Destroy Sports?" *Discover*, vol. 25, July 2004, pp. 40–45. A new gene therapy technique could make better bodies for athletes—and telling who is "cheating" by this means might be impossible.

Bostrom, Nick. "In Defense of Posthuman Dignity." *Bioethics*, vol. 19, June 2005, pp. 202–214. Critics say that genetically enhancing humans to produce "posthumans" violates human dignity, but Bostrom argues that the concept of dignity can be broadened to include the posthuman.

Buehler, Lukas K. "Is Genomics Advancing Drug Discovery?" *Pharmaceutical Discovery*, vol. 5, February 2005, pp. 26-28. Genomics and related sciences have contributed much to drug development, but they have by no means solved all its problems.

Burnet, Michael. "Gene to Screen." *Chemistry and Industry*, vol. 1, January 3, 2005, pp. 23–24. Three key sources of promising new drugs are RNA interference (RNAi), which silences harmful genes; optimized natural proteins; and DNA and RNA vaccines.

Cavazzana-Calvo, Marina, Adrian Thrasher, and Fulvio Mavillo. "The Future of Gene Therapy." *Nature*, vol. 427, February 26, 2004, pp. 779–781. This article provides a cautiously optimistic view of gene therapy for immune disorders and inherited diseases that have no conventional treatments.

Cohan, John Alan. "Ethics of Genetic Enhancement of Aptitudes and Personality Traits." *Journal of Philosophical Research*, vol. 30, 2005, pp. 527–535. Cohan considers whether possible negative consequences, such as unequal distribution of treatments across society, are more important than the likely benefits of genetic enhancement.

Cohen, Eric. "Commerce and the Human Body: Selling Bodily Perfection." *Current*, vol. 484, July-August 2006, pp. 3–10. Genetic enhancements, stem cell treatments, and other future forms of altering human beings most likely will result in bodily perfection being sold to anyone with enough disposable income.

DeGrazia, David. "Enhancement Technologies and Human Identity." *Journal of Medicine and Philosophy*, vol. 30, June 2005, pp. 261–283. DeGrazia examines the relationship between genetic enhancement and the concept of human identity and concludes that a "lucid, plausible" understanding of identity is compatible with genetic enhancement.

Fischer, Alain, and Marina Cavazzana-Calvo. "Whither Gene Therapy?" *The Scientist*, vol. 20, February 2006, pp. 36–41. Despite bad publicity, gene therapy has had some successes, and several new techniques promise to make it more efficient and safer.

Goncalves, Manuel A. F. V. "A Concise Peer into the Background, Initial Thoughts, and Practices of Human Gene Therapy." *BioEssays*, vol. 27, May 2005, pp. 506–517. Researchers will need to improve techniques for transferring genes and gain better insight into the way that viral or other vectors (carriers of genes) interact with the body before gene therapy can mature.

Hollingsworth, Leslie Doty. "Ethical Considerations in Prenatal Sex Selection." *Health and Social Work*, vol. 30, May 31, 2005, pp. 126–134. Hollingsworth considers the ethical dilemmas raised by the availability of prenatal sex selection in the areas of social justice, human relationships, and self-determination and individual dignity. Implications for social work practice and policy are also discussed.

Hood, Ernie. "RNAi: What's All the Noise about Gene Silencing?" *Environmental Health Perspectives*, vol. 112, March 15, 2004, pp. A224–A229. RNA interference (RNAi), the technique of using short stretches of RNA to "interfere with" and silence particular genes, is becoming a valuable tool of genomic biology and a potential source of lifesaving drugs.

Judson, Horace Freeland. "The Glimmering Promise of Gene Therapy." *Technology Review*, vol. 109, November-December 2006, pp. 40–47. A history of gene therapy reveals much more hype and failure than success— but hope for this kind of treatment refuses to die.

June, Carl. "Cancer Gene Therapy at the Crossroads: Challenges Abound— Will Blockbusters Ensue?" *The Scientist*, vol. 19, May 9, 2005, pp. 18–19. Despite significant setbacks and continuing challenges, June believes that successful gene therapies for cancer will be developed.

Kalb, Claudia. "Brave New Babies." *Newsweek*, January 26, 2004, p. 44 ff. Several techniques now allow parents to choose the gender of their children—and, despite ethical questions, some U.S. couples are happy to take advantage of them.

Kean, Marcia A. "Biotech and the Pharmaceutical Industry: Back to the Future." *OECD Observer*, issue 243, May 2004, pp. 21–22. The author sees a coming golden age of drug research and development as biotechnology/genomic sciences and the pharmaceutical industry join forces.

Kittredge, Clare. "Gene Therapy . . . Under the Microscope Again." *The Scientist*, vol. 19, May 9, 2005, pp. 14–17. Gene therapy still faces many challenges, but preliminary successes of novel strategies, including induction of the temporary expression of added genes that are not incorporated into the genome, are encouraging investors.

Munro, Neil, and Mark Kukis. "A Brave New World?" *National Journal*, vol. 36, May 22, 2004, pp. 1648–1654. Critics say that the political divide between supporters and opponents of human gene enhancement will never be bridged, but others maintain that the marketplace will allow consumers to buy whatever enhancements they wish and can afford.

Neunder, Mark. "Can Genetic Enhancement Be Obligatory? Four Arguments." *Journal of Philosophical Research*, vol. 30, Annual 2005, pp. 595–604. Neunder argues that, when safely enhancing the genes of offspring becomes possible, parents will be morally obliged to use the technology.

Peng, Zhaohui. "The Genesis of Gendicine." *Biopharm International*, vol. 17, May 2004, p. 42 ff. In this interview, Peng, head of SiBono in China, describes Gendicine, the world's first officially marketed gene therapy. Gendicine treats cancers by introducing functional copies of the p53 tumor suppressor gene.

Rifkin, Jeremy. "Ultimate Therapy: Commercial Eugenics in the 21st Century." *Harvard International Review*, vol. 27, Spring 2005, pp. 44–48. Rifkin is concerned about future misuse of human gene alteration, especially germ-line alteration, in pursuit of supposedly "perfect" humans.

Roberts, Josh P. "Gene Therapy's Fall and Rise (Again)." *The Scientist*, vol. 18, September 27, 2004, pp. 22–24. Gene therapy researchers are improving the viruses and other gene-delivery vectors they use and are cautiously optimistic despite past setbacks and heavy regulation of their field.

Sapolsky, Robert. "Is Any of This a Good Idea?" *Popular Science*, vol. 267, September 1, 2005, p. 96. Future genetic enhancements could present a number of problems, and they cannot offer the social advantages of being "better than average" because the standards for being "average" will simply be raised.

Schu, Bill. "Hype and Hope: The Halting Progress of Gene Delivery." *Genomics and Proteomics*, vol. 5, October 1, 2005, p. 20. Both viral and nonviral methods of gene delivery are improving. Examples include HIV as a gene-delivery vector and "gene pills" that do not use viruses.

Scully, J. L., C. Rippburger, and C. Rehmann-Sutter. "Non-professionals' Evaluations of Gene Therapy Ethics." *Social Science and Medicine*, April 2004, n.p. This paper found that patients differed considerably from medical professionals in moral evaluations of gene therapy; patients, for instance, often saw the therapy as potentially changing an identity that derived in part from their disability or illness.

Sharp, Phillip. "The Prize of RNAi." *Technology Review*, vol. 109, November-December 2006, p. 32. Sharp discusses the importance of RNA interference (RNAi), whose discoverers had just won the Nobel Prize in physiology or medicine, as both a research tool and a possible source of new drugs. RNAi uses short stretches of double-stranded RNA to silence chosen genes.

Verma, Inder M., and Matthew D. Weitzman. "Gene Therapy: Twenty-First Century Medicine." *Annual Review of Biochemistry*, vol. 74, 2005, pp. 711–748. Gene therapy is likely to prosper if researchers can develop delivery systems that are safe and can efficiently transfer genes into the cells of a variety of tissues.

"Who's Afraid of Human Enhancement?" *Reason*, vol. 37, January 2006, pp. 22–32. This article is the text of a debate on the future and ethics of human gene enhancement that occurred on August 25, 2005.

WEB DOCUMENTS

"About the International HapMap Project." International HapMap Project. Available online. URL: http://www.hapmap.org/abouthapmap.html. Accessed on March 18, 2007. The project, an international collaboration, is preparing a map of human genetic variation in the form of haplotypes, or groups of genes that are usually inherited together. The HapMap is expected to help researchers identify variations associated with susceptibility to particular diseases. This article explains the importance of genetic variation and the uses of the HapMap, populations and samples, ethical issues, scientific strategy, data analysis, data access and intellectual property policies, and internal data access policy.

Baruch, Susannah. "Human Germline Genetic Modification: Issues and Options for Policymakers." Genetics and Public Policy Center. Available online. URL: http://www.dnapolicy.org/images/reportpdfs/HumanGermlineGeneticMod.pdf. Posted in May 2005. This report analyzes the scientific, legal, regulatory, ethical, moral, and societal issues raised by genetic modification of the human germ-line, provides data about the U.S. public's views of this procedure, and explores possible policy approaches.

"Biotechnology and the Health of Canadians." Canadian Biotechnology Advisory Committee. Available online. URL: http://cbac-cccb.ca/epic/site/cbac-cccb.nsf/vwapj/BHI-Final_Dec-13-04-E.pdf/$FILE/BHI-Final_Dec-13-04-E.pdf. Posted in December 2004. This report discusses both the promise of biotechnology-based health care innovations and the challenges that must be met. It makes recommendations for research, regulation, commercialization, appraisal of the technologies, and adoption by the Canadian health-care system.

"The Case for Personalized Medicine." Personalized Medicine Coalition. Available online. URL: http://www.personalizedmedicinecoalition.org/communications/TheCaseforPersonalizedMedicine_11_13.pdf. Posted in November 2006. This report provides an optimistic view of personalized medicine, through which physicians will tailor drugs and other treatments to individual patients based on knowledge of their gene variations. It also discusses obstacles that must be overcome before the full impact of this new approach can be realized.

Collins, Francis S., and Anna D. Barker. "Mapping the Cancer Genome." *Scientific American*, March 2007. Also available online. Scientific American. com. URL: http://www.sciam.com/article.cfm?chanID=sa006&colID=1& articleID=CC1E538E-E7F2-99DF-3F44D06D3B292CF3. Posted in March 2007. Francis S. Collins, director of the federal government's National Human Genome Research Institute, and National Cancer Institute Deputy Director for Advanced Technologies and Strategic Partnerships Anna D. Barker explain the Cancer Genome Atlas Project, launched in December 2005, which aims to identify all the genes involved in cancer.

Cramer, Philippe. "Biotechnology for Breast Cancer." Bioimpact.org. Available online. URL: http://www.bioimpact.org/pdf-en/full-report-13. pdf. Posted in March 2007. This descriptive paper shows how Herceptin and other biotechnology drugs have reduced the number of relapses and the rate of death from this common cancer.

———. "Biotechnology for Cardiovascular Diseases." Bioimpact.org. Available online. URL: http://www.bioimpact.org/pdf-en/full-report-1.pdf. Posted in March 2007. This descriptive paper focuses on treatment for blood clots.

"Guidance for Industry: Gene Therapy Clinical Trials—Observing Subjects for Delayed Adverse Events." U.S. Food and Drug Administration. Available online. URL: http://www.fda.gov/cber/gdlns/gtclin.htm. Posted in November 2006. This document provides the FDA's recommendations for long-term follow-up procedures to detect delayed side effects from experimental gene therapy.

Kalfaglou, A., et al. "Reproductive Genetic Testing: What America Thinks." Genetics and Public Policy Center. Available online. URL: http://www. dnapolicy.org/images/reportpdfs/ReproGenTestAmericaThinks.pdf. Posted in 2004. This report summarizes a large series of social science research studies conducted between 2002 and 2004, on U.S. citizens' knowledge and opinions about the use and regulation of reproductive genetic testing.

King, David S. "Preimplantation Genetic Diagnosis and the 'New' Eugenics." Available online. URL: http://www.hgalert.org/topics/genetic Selection/PIDJME.html. Accessed on March 18, 2007. King argues for strict regulation of preimplantation genetic diagnosis, fearing that it will go even further than standard prenatal testing to promoting a new, market-driven form of eugenics because it allows evaluation of more characteristics, including those not related to disease, and does not involve destruction of embryos or fetuses.

"Making Babies: Reproductive Decisions and Genetic Technologies." Human Genetics Commission. Available online. URL: http://www.hgc. gov.uk/UploadDocs/DocPub/Document/Making%20Babies%20Report %20-%20final%20pdf.pdf. Posted in January 2006. This report from a

British government agency discusses prenatal and neonatal screening, preimplantation genetic diagnosis, and other technologies that permit reproductive choice based on genetics.

"Reproduction and Responsibility: The Regulation of New Biotechnologies." President's Council on Bioethics. Available online. URL: http://www.bioethics.gov/reports/reproductionandresponsibility/_pcbe_final_reproduction_and_responsibility.pdf. Posted in March 2004. This report reviews the moral issues raised by new reproductive technologies and the assisted reproduction industry, as well as current methods of regulation, and concludes that the latter are insufficient. It makes recommendations for regulating the most morally questionable practices.

Wallace, Helen. "Your Diet Tailored to Your Genes: Preventing Diseases or Misleading Marketing?" GeneWatch UK. Available online. URL: http://www.genewatch.org/uploads/f03c6d66a9b354535738483c1c3d49e4/Nutrigenomics.pdf. Posted in January 2006. This report concludes that advertisements touting products under the rubric of nutrigenomics, which supposedly can tailor individuals' diets to their genetically determined needs and susceptibilities, are more hype than science.

"What Is Healthcare Biotechnology?" Europabio. Available online. URL: http://www.europabio.org/about_healthcare.htm. Accessed on March 16, 2007. This introduction to health-care biotechnology, from a European biotechnology industry organization, favorably describes stem cells, gene therapy, improvements in diagnosis and drug delivery, new drugs, and other advances.

STEM CELL RESEARCH AND HUMAN CLONING

BOOKS

Allman, Toney. *Great Medical Discoveries: Stem Cells*. San Diego, Calif.: Lucent Books, 2005. This book for young adults presents an overview of stem cell research, the revolutionary medical treatments that may come from it, and the reasons some people think that the research, at least when it involves embryos, is unethical.

Bellomo, Michael. *The Stem Cell Divide: The Facts, the Fiction, and the Fear Driving the Greatest Scientific, Political, and Religious Debate of Our Time*. New York: AMACOM/American Management Association, 2006. In interviews with the scientists at the heart of the stem cell debate, Bellomo encourages them to explain their past and present work and their hopes for the future.

Engdahl, Sylvia Louise, ed. *Cloning*. Farmington Hills, Mich.: Greenhaven Press, 2006. This anthology of articles, part of the Contemporary Issues

Companion series, provides neutral overviews of aspects of cloning as well as opinion pieces that support or decry this technology.

Fox, Cynthia. *The Cell of Cells: The Global Race to Capture and Control the Stem Cell.* New York: Norton, 2006. Stem cells are being studied all over the world, and some treatments involving them are already in use, science journalist Fox explains. She believes that stem cell research will create a revolution in both medicine and politics.

Gralla, Jay D., and Preston Gralla. *The Complete Idiot's Guide to Understanding Cloning.* Indianapolis, Ind.: Alpha, 2004. This book clearly explains the technology of cloning and the controversies that surround it.

Habib, Nagy A., et al., eds. *Stem Cell Repair and Regeneration.* London: Imperial College Press, 2005. In the essays collected in this book, leading scientists describe different applications for which stem cells are being investigated, including organ repair and treatment of such illnesses as Parkinson's disease.

Harris, John. *On Cloning: Thinking in Action.* New York: Routledge, 2004. Harris defends human reproductive cloning, saying that it potentially has the power to improve human life.

Herold, Eve. *Stem Cell Wars: Inside Stories from the Front Lines.* New York: Palgrave Macmillan, 2006. Herold explains clearly what stem cell research is, why it is controversial, and why she thinks the government should support it. She includes information about the rancorous stem cell policy debate in the United States and about the scandal surrounding Korean scientist Woo Suk Hwang's falsified work.

Jones, Gareth, and Mary Byrne, eds. *Stem Cell Research and Cloning: Contemporary Challenges to Our Humanity.* Hindmarsh, South Australia: ATF Press, 2005. These essays view stem cell research from philosophical and theological points of view.

Klotzko, Arlene Judith. *A Clone of Your Own?* New York: Cambridge University Press, 2006. Written in popular-science style, this book presents a positive view of human cloning and stem cell research.

Levick, Stephen E. *Clone Being: Exploring the Psychological and Social Dimensions.* Lanham, Md.: Rowman & Littlefield, 2004. Levick examines psychological and social ramifications of reproductive cloning for the clone, the clone's family, and the community. He concludes that the medical and social risks of cloning are so high that trying to clone a person would be immoral.

Lim, Hwa A. *Multiplicity Yours: Cloning, Stem Cell Research, and Regenerative Medicine.* River Edge, N.J.: World Scientific Publishing Company, 2006. This book discusses reproduction, human cloning, and stem cell research from the points of view of science, society, economics, politics, and ethics.

Macintosh, Kerry Lynn. *Illegal Beings: Human Clones and the Law.* New York: Cambridge University Press, 2005. Macintosh lists points for and against laws banning human reproductive cloning but ultimately argues that such laws are unnecessary and will harm human clones.

Maienschein, Jane. *Whose View of Life? Embryos, Cloning, and Stem Cells*. Rev. ed. Cambridge, Mass.: Harvard University Press, 2005. Maienschein discusses the question of when an embryo or fetus becomes "human," which lies at the heart of the debates surrounding human cloning and embryonic stem cell research.

Marzilli, Alan. *Stem Cell Research and Cloning*. New York: Chelsea House, 2007. This book for young adults, part of the Point/Counterpoint series, describes the controversies surrounding stem cell research and human cloning.

Nardo, Don. *Cloning*. Farmington Hills, Mich.: Lucent Books, 2005. This book for young adults provides an overview of the cloning of plants, animals, and humans, as well as present and possible applications of this technology and the ethical conflicts it has generated.

National Research Council and Institute of Medicine. *Guidelines for Human Embryonic Stem Cell Research*. Washington, D.C.: National Academies Press, 2005. Although they are not legally binding, these guidelines from a prestigious national group help privately funded scientists address the ethical and scientific questions that the research field raises.

Newton, David E. *Stem Cell Research*. New York: Facts On File, 2006. This volume in the Library in a Book series provides an overview of the science behind stem cell research and the ethical controversies the research has aroused, followed by descriptions of laws and court cases, a chronology, capsule biographies of important people in the debate, a glossary, a guide to Internet research, a list of organizations on all sides of the debate, and an extensive annotated bibliography.

Panno, Joseph. *Stem Cell Research: Medical Applications and Ethical Controversy*. New York: Facts On File, 2005. A carefully done, in-depth account of the subject for advanced young adult readers, including a look at legal practices in the United States, Britain, and Europe.

Parson, Anne B. *The Proteus Effect: Stem Cells and Their Promise for Medicine*. Washington, D.C.: Joseph Henry Press, 2006. Parson describes the history of stem cell research, focusing on the process of scientific discovery.

Pence, Gregory E. *Cloning After Dolly: Who's Still Afraid?* Lanham, Md.: Rowman & Littlefield, 2005. Pence explains why he strongly supports human reproductive cloning as well as "therapeutic" cloning for embryonic stem cell research.

Scott, Christopher Thomas. *Stem Cell Now: An Introduction to the Coming Medical Revolution*. New York: Penguin, 2006. Thomas lays out the issues surrounding stem cell research without taking sides.

Smith, Wesley J. *Consumer's Guide to a Brave New World*. New York: Encounter Books, 2004. Focusing on the moral questions that stem cell research, cloning, and related science raise, Smith expresses disapproval of human reproductive cloning and embryonic stem cell research but supports study of adult stem cells and stem cells taken from umbilical cord blood.

Solo, Pam, and Gail Pressberg. *The Promise and Politics of Stem Cell Research*. Westport, Conn.: Praeger, 2006. The authors say that an informed public should have a voice in decisions about how to regulate stem cell research.

Steenblock, David, and Anthony G. Payne. *Umbilical Cord Stem Cell Therapy: The Gift of Healing from Healthy Newborns*. Laguna Beach, Calif.: Basic Health Publications, 2006. Umbilical cord blood stem cells, although technically considered to be "adult," appear to have greater regenerative powers than stem cells from fully adult tissues. They are already used to treat a variety of diseases.

Wilmut, Ian, and Roger Highfield. *After Dolly: The Uses and Misuses of Human Cloning*. New York: Norton, 2006. Wilmut explains how he and his research team in Scotland created the famous cloned sheep, Dolly, in 1996. Now involved with therapeutic cloning and embryonic stem cell research, he argues that these technologies will revolutionize medicine. He is assisted by Roger Highfield, science editor of the British newspaper *Daily Telegraph*.

ARTICLES

Alexander, Brian. "Cloning: A First Amendment Right." New York Times *Upfront*, vol. 136, December 13, 2004, pp. 26–27. This article claims that the First Amendment may protect scientists who want to carry out research on human embryonic stem cells.

Baylis, Françoise. "Canada Bans Human Cloning." *Hastings Center Report*, vol. 34, May-June 2004, p. 5. The author explains why she believes that research on and therapeutic use of embryonic stem cells can proceed without the cloning of embryos.

Bent, Stephen A. "Under the Microscope." *The Scientist*, vol. 19, July 4, 2005, pp. 22–23. This article surveys laws related to stem cell research in the United States and other parts of the world. The author concludes that the U.S. federal government should resume funding for embryonic stem cell research and conduct coherent oversight of the research.

Bowring, Finn. "Therapeutic and Reproductive Cloning: A Critique." *Social Science and Medicine*, vol. 58, January 15, 2004, pp. 40–48. This article is a critical examination of the science and ethics of human cloning. Although Bowring primarily opposes reproductive cloning, he fnds therapeutic cloning unacceptable as well because he feels that the distinction between the two would be impossible to police.

Brownlee, Christen. "Full Stem Ahead." *Science News*, vol. 167, April 2, 2005, pp. 218–220. This article provides an overview of stem cell research.

Bush, George W. "President Bush's Veto Message to Congress." *National Right to Life News*, vol. 33, August 2006, p. 8. This is the complete text of President Bush's message to Congress on July 19, 2006, when he vetoed a bill to mandate federal funding for embryonic stem cell research.

Cameron, C., and R. Williamson. "In the World of Dolly, When Does a Human Embryo Acquire Respect?" *Journal of Medical Ethics*, vol. 31, April 2005, pp. 215–220. The authors argue that the time of implantation is the most useful point for considering an embryo to be due the respect accorded to human beings. This choice point would permit embryonic stem cell research.

Cameron, Nigel M. de S. "Cloning: U.S. and Global Perspectives." *Southern Medical Journal*, vol. 99, December 2006, pp. 1429–1435. This article provides a brief history of government policy and bioethical statements on human embryo research and human cloning, including therapeutic cloning, in the United States, Europe, and the United Nations, ending in the UN Declaration on Human Cloning (adopted on March 23, 2005), which outlawed all human cloning, including therapeutic cloning.

Campbell, Angela. "Ethos and Economics: Examining the Rationale Underlying Stem Cell and Cloning Research Policies in the United States, Germany, and Japan." *American Journal of Law and Medicine*, vol. 31, Spring 2005, pp. 47–86. Campbell examines the ways in which differences in cultural ethos and economic objectives affect the regulation of stem cell research and cloning in these three countries.

Cohen, Jon. "Stem Cell Pioneers." *Smithsonian*, vol. 36, December 2005, pp. 78–87. This article shows how several pioneers of stem cell research are trying to find ways around the ethical debate that work with embryonic stem cells has engendered.

Condic, Maureen L. "What We Know about Embryonic Stem Cells." *First Things*, vol. 169, January 2007, pp. 25–29. Potential treatments with embryonic stem cells face the same serious safety problems, including potential cancer formation, that they did five years ago, Condic points out. She thinks that such treatments probably never will be practical.

Cyranoski, David. "Verdict: Hwang's Human Stem Cells Were All Fakes." *Nature*, vol. 439, January 12, 2006, pp. 122–123. An investigation by a committee from Seoul National University concluded that the research on the basis of which Korean scientist Woo Suk Hwang claimed to have cloned human cells and established embryonic stem cell lines from them was a complete and deliberate fake.

DeBow, Suzanne, Tania Bubela, and Timothy Caulfield. "Stem Cells, Politics, and the Progress Paradigm." *Health Law Review*, vol. 15, September 22, 2006, pp. 50–52. The authors analyze the decade of debates in the

Canadian Parliament that preceded the passage of the Assisted Human Reproduction Act on March 29, 2004. They show how the debate's domination by religious conservatives resulted in a very restrictive policy toward human cloning and embryonic stem cell research.

Dolgin, Janet L. "Embryonic Discourse: Abortion, Stem Cells, and Cloning." *Issues in Law and Medicine*, vol. 19, Spring 2004, pp. 203–261. Debates about abortion and research on human embryonic stem cells both center on the meaning of the human embryo, but they represent different views of the meaning of personhood and the role of the individual, as opposed to those of family and community.

Dresser, Rebecca. "Stem Cell Research: The Bigger Picture." *Perspectives in Biology and Medicine*, vol. 48, Spring 2005, pp. 181–194. Dresser brings out ethical complexities in the controversy about human embryonic stem cells that, she says, both sides of the debate tend to ignore. She takes the position that human embryos have an intermediate moral status, greater than individual body cells but less than complete persons.

Eisner, D. "Just Another Reproductive Technology? The Ethics of Human Reproductive Cloning as an Experimental Medical Procedure." *Journal of Medical Ethics*, vol. 32, October 2006, pp. 596–600. Eisner compares human reproductive cloning with in vitro fertilization and proposes a model to be used in determining when, in a balance between reproductive freedom and safety concerns, reproductive cloning would be acceptable.

Fields, Helen. "What Comes Next?" *U.S. News & World Report*, vol. 140, January 23, 2006, pp. 56–58. Despite the fraud of South Korean researcher Woo Suk Hwang and persistent ethical questions about the technology, research on embryonic stem cells continues to progress.

Friedmann, Theodore. "Lessons for the Stem Cell Discourse from the Gene Therapy Experience." *Perspectives in Biology and Medicine*, vol. 48, Autumn 2005, pp. 585–591. Friedmann says that the ban on federal funding of most embryonic stem cell research means that the field is not being regulated in the same thorough way that gene therapy was, which is unfortunate because it removes a mechanism to control exaggeration and force the careful basic research that will be necessary if this new technology is to thrive.

Gardner, Richard, and Tim Watson. "A Patchwork of Laws." *Scientific American*, vol. 293, July 2005, pp. A16–20. The authors discuss laws in different countries that affect human embryonic stem cell research, reproductive cloning, and therapeutic cloning.

Gibbs, Nancy. "Stem Cells: The Hope and the Hype." *Time*, vol. 168, August 7, 2006, p. 40 ff. Present research on embryonic and adult stem cells is a complex mixture of progress and challenges. Political interference, funding difficulties, and ethical questions complicate the picture further.

Green, Shane K. "Stem Cells: Science, Ethics, and Politics at the Forefront of Biomedical Innovation." *Biotechnology Focus*, vol. 10, January 2007, n.p. This article describes the basics of stem cell research; its ethical, legal, and social implications; and regulation, with a focus on Canada.

Gross, Michael. "New Cells for Old." *Biological Sciences Review*, vol. 18, February 2006, pp. 6–9. This article explains therapeutic cloning and its possible future use as a research tool and a medical treatment. A new technique that would allow treatment with a relatively small number of embryonic stem cell lines, while avoiding immune rejection, is promising.

"Growing Evidence Supports Stem Cell Hypothesis of Cancer." *Oncology News International*, May 1, 2006, p. 24. Stem cells may be at the root of cancer.

Guterman, Lila. "A Silent Scientist Under Fire." *Chronicle of Higher Education*, vol. 52, February 3, 2006, n.p. This article provides background information on the University of Pittsburgh's Gerald Schatten, who was listed as senior author on a landmark paper on human cloning published by South Korean scientist Woo Suk Hwang. Hwang's work was later shown to be a fraud, but how much responsibility Schatten bears for the deception is unclear.

Hall, Stephen S. "Stem Cells: A Status Report." *Hastings Center Report*, vol. 36, January 1, 2006, pp. 16–22. Hall discusses advances such as the creation of a deliberately crippled mouse embryo that lacked the ability to implant in a uterus; equivalent human embryos might be considered acceptable for research on embryonic stem cells.

Hurlbut, William B., Robert P. George, and Markus Grompe. "Seeking Consensus: A Clarification and Defense of Altered Nuclear Transfer." *Hastings Center Report*, vol. 36, September-October 2006, pp. 42–50. The authors provide a precise explanation of altered nuclear transfer, a possible way around the ethical dilemmas presented by standard embryonic stem cell research. They then address some of the major concerns about this approach.

Hyun, Insoo. "Human Research Cloning, Embryos, and Embryo-like Artifacts." *Hastings Center Report*, vol. 36, September 1, 2006, pp. 34–41. Hyun maintains that cloned primate eggs cannot produce true, viable embryos, but rather mere "embryo-like artifacts." If this is true, the ethical objections to therapeutic human cloning are invalid.

Isasi, Rosario M., et al. "Legal and Ethical Approaches to Stem Cell and Cloning Research: A Comparative Analysis of Policies in Latin America, Asia, and Africa." *Journal of Law, Medicine, and Ethics*, vol. 32, Winter 2004, pp. 626–640. This article surveys policies governing human embryonic stem cell research and cloning in 16 countries in Asia, Africa, and Latin America. Some countries have very restrictive policies, while others take a more pragmatic approach and provide little or no governmental regulation.

Annotated Bibliography

Jaenisch, Rudolf. "Human Cloning: The Science and Ethics of Nuclear Transplantation." *The New England Journal of Medicine*, vol. 351, December 30, 2004, pp. 2787–2792. Jaenisch claims that a cloned human embryo would be very unlikely to develop into a healthy human being.

Johnston, Josephine. "Paying Egg Donors: Exploring the Arguments." *Hastings Center Report*, vol. 36, January-February 2006, pp. 28–31. Arguments against paying egg donors for therapeutic stem cell research, including arguments against pressuring for donation and the selling of body parts, may not be valid. For instance, paying for donated eggs does not force women to donate.

Kalb, Claudia, and Debra Rosenberg. "Embryonic War." *Newsweek*, September 4, 2006, p. 42. Robert Lanza of Advanced Cell Technology claims to have derived human embryonic stem cells with a technique that does not harm the embryo, but it remains unclear whether his approach will completely satisfy either fellow scientists or conservative ethicists.

Kimmelman, Jonathan, Françoise Baylis, and Kathleen Cranley Glass. "Stem Cell Trials: Lessons from Gene Transfer Research." *Hastings Center Report*, vol. 36, January-February 2006, pp. 23–26. As the first human trials of therapies using human embryonic stem cells approach, scientists should learn from mistakes during similar trials in another controversial field, gene therapy.

Lanza, Robert, and Nadia Rosenthal. "The Stem Cell Challenge." *Scientific American*, vol. 290, June 2004, pp. 92–99. Scientists are struggling to understand and control embryonic stem cells, which offer hope of replacing or regenerating failing body parts.

Lauritzen, Paul. "Stem Cells, Biotechnology, and Human Rights: Implications for a Posthuman Future." *Hastings Center Report*, vol. 35, March-April 2005, pp. 25–33. Embryonic stem cell research could potentially change perceptions of the meaning of human life and human rights, resulting in a view of the human body, and indeed all of nature, as merely material to be manipulated.

Lemonick, Michael D. "The Rise and Fall of the Cloning King." *Time*, vol. 167, January 9, 2006, p. 40. Scientist Woo Suk Hwang was a national hero in South Korea when he claimed to be the first person to clone human embryos and derive stem cells from them, but he became a pariah when his research was shown to have been faked. Lemonick thinks that Hwang's falsification was more a matter of ambition and pressure—"fudge now, fix later"—than a deliberate attempt at fraud.

Magnus, David. "Stem Cell Research: The California Experience." *Hastings Center Report*, vol. 36, January-February 2006, pp. 26–29. This article surveys ethical issues faced by the California Institute for Regenerative Medicine, the organization established by the state's voters in 2004 to oversee a massive program of human embryonic stem cell research.

Monastersky, Richard. "A Second Life for Cloning." *Chronicle of Higher Education*, vol. 52, February 3, 2006, n.p. Korean scientist Woo Suk Hwang's claim to have cloned human embryos and extracted stem cells from them proved to be a fraud, but other researchers in several countries, including the United States, are continuing therapeutic cloning research and have developed techniques that may allow the research to proceed without destroying human embryos.

Nisbet, Matthew C. "The Polls—Trends: Public Opinion About Stem Cell Research and Human Cloning." *Public Opinion Quarterly*, vol. 68, Spring 2004, pp. 131–154. This article surveys polls conducted between 2000 and 2002 and concludes that the public is easily influenced by the way the media presents stem cell issues.

Paarlberg, Robert L. "The Great Stem Cell Race." *Foreign Policy*, vol. 148, May-June 2005, pp. 44–51. Policies regulating stem cell research are even more restrictive in much of Europe than they are in the United States, and Asia also has limitations. Scientists around the world, however, are finding ways to overcome barriers and continue their studies.

Payne, Anthony G. "New Umbilical Cord Stem Cells Are Helping Life Imitate Myth." *Townsend Letter for Doctors and Patients*, vol. 269, December 2005, pp. 68–71. Treatments with umbilical cord stem cells are already helping some patients with cerebral palsy and other conditions, but challenges remain.

Prainsack, Barbara, and Tim D. Spector. "Twins: A Cloning Experience." *Social Science and Medicine*, vol. 63, August 2006, pp. 2739–2752. Through interviews, the authors compared identical twins—natural clones—with fraternal twins and singletons to explore what it means to be genetically identical to someone else.

Sandel, Michael J. "The Ethical Implications of Human Cloning." *Perspectives in Biology and Medicine*, vol. 48, Spring 2005, pp. 241–247. Sandel examines ethical arguments for and against human reproductive cloning and cloning of embryos for stem cell research. He concludes that embryos should not be regarded as either persons or things, but rather something in between.

Schu, Bill. "Can Science Bridge the Stem Cell Divide?" *Genomics and Proteomics*, vol. 5, April 1, 2005, p. 12 ff. Researchers on embryonic stem cells are looking for techniques to help them bypass the ethical dilemmas and funding restrictions that have limited the field.

Shapiro, Kevin. "Lessons of the Cloning Scandal." *Commentary*, vol. 121, April 1, 2006, pp. 61–64. Shapiro describes and considers the implications of the case of South Korean researcher Woo Suk Hwang, whose claim to have cloned human embryos and derived stem cell lines from them proved to be a fraud.

Shapshay, Sandra. "The Human Being in the Age of Mechanical Reproduction: An Argument Against Human Cloning." *Journal of Philosophical Re-*

search, vol. 30, Annual 2005, pp. S119–133. Shapshay lays out reasons for opposing reproductive cloning. She urges that it be banned because it devalues individual persons.

Singer, Emily. "Stem Cells Reborn." *Technology Review*, vol. 109, May-June 2006, pp. 58–65. Despite the scandal following the revelation that a South Korean scientist's claim to have created stem cells from cloned human embryos was false, researchers are more interested than ever in developing cloned sources of human embryonic stem cells. Political and funding difficulties hamper work in the United States, but new techniques may allow scientists to bypass these hurdles.

Stabile, Bonnie. "National Determinants of Cloning Policy." *Social Science Quarterly*, vol. 87, June 2006, pp. 449–458. Stabile seeks to identify demographic factors that account for the stand taken toward human cloning in different countries. Her findings suggest that nations with higher per-capita incomes and permissive abortion laws are more likely to take a permissive stance toward therapeutic, though not reproductive, cloning than those with lower incomes and less permissive abortion laws.

"Stem Cells: Tapping a Different Source." *Business Week Online*, December 27, 2006, n.p. Umbilical cord blood is a noncontroversial source of life-saving stem cells. The best way to tap this source is for private businesses to offer free cord blood banking and make their money from insurance reimbursements when the blood is used to treat people other than the donors' families.

Stipp, David. "Stem Cells to Fix the Heart." *Fortune*, vol. 150, November 29, 2004, p. 179. In clinical trials, therapy with adult stem cells, usually taken from their own blood, seems to be helping some patients with severe heart disease.

Weed, Matthew. "Discourse on Embryo Science and Human Cloning in the United States and Great Britain: 1984–2002." *Journal of Law, Medicine, and Ethics*, vol. 33, Winter 2005, pp. 802–810. Similar arguments were made for and against therapeutic cloning in the United States and Britain, yet the two countries developed very different policies toward this technology. This article surveys the arguments, especially those made regarding the enactment of two British laws governing research on human embryos.

Weiss, Rick. "The Power to Divide." *National Geographic*, vol. 208, July 2005, pp. 2–27. Stem cells, especially embryonic stem cells, hold great medical promise—but they are also enmired in controversy, especially in the United States.

Wilmut, Ian. "The Case for Cloning Humans." *The Scientist*, vol. 19, April 25, 2005, pp. 16–17. Wilmut, creator of the cloned sheep Dolly, explains why he and other researchers want to clone human embryos to harvest embryonic stem cells. These cells, he says, will be invaluable for research and offer a potential treatment for deadly diseases.

WEB DOCUMENTS

"Alternative Sources of Human Pluripotent Stem Cells." President's Council on Bioethics. Available online. URL: http://www.bioethics.gov/reports/white_paper/alternative_sources_white_paper.pdf. Posted in May 2005. This white paper examines the strengths and weaknesses (particularly from an ethical standpoint) of four new techniques designed to obtain pluripotent stem cells without destroying embryos.

Domestic Policy Council. "Advancing Stem Cell Science Without Destroying Human Life." White House. Available online. URL: http://www.whitehouse.gov/dpc/stemcell/2007/stemcell_04027.pdf. Posted in January 2007. This report stresses work on adult stem cells and other forms of stem cell research that do not destroy embryos.

"Donating Your Eggs for Research." Human Fertilisation and Embryology Authority. Available online. URL: http://www.hfea.gov.uk/docs/Eggs_for_research_factsheet_Feb07.pdf. Posted in February 2007. This British government agency offers its policy statement about egg donation, which is aimed at potential donors.

"Embryo Research." Human Fertilisation and Embryology Authority. Available online. URL: http://www.hfea.gov.uk/docs/Embryo_Research.pdf. Accessed on March 18, 2007. This document explains how the British government regulates research on embryos, what the law permits and prohibits, sources of embryos, present and future research, consent, and more.

"Frequently Asked Questions on Stem Cell Research." International Society for Stem Cell Research. Available online. URL: http://www.isscr.org/public/faq_printversion.html. Accessed on March 17, 2007. This document, from a group that favors stem cell research, covers everything from what stem cells are to sources for further reading.

Hudson, Kathy L., Joan Scott, and Ruth Faden. "Values in Conflict: Public Attitudes on Embryonic Stem Cell Research." Genetics and Public Policy Center. Available online. URL: http://www.dnapolicy.org/images/reportpdfs/2005ValuesInConflict.pdf. Posted in October 2005. A large survey conducted in September 2005 found that two-thirds of the respondents, from a wide variety of political, religious, and socioeconomic backgrounds, supported embryonic stem cell research. Forty-two percent gave early, unimplanted human embryos an intermediate level of moral status.

Javitt, Gail H., Kristen Suthers, and Kathy Hudson. "Cloning: A Policy Analysis." Genetics and Public Policy Center. Available online. URL: http://www.dnapolicy.org/images/reportpdfs/Cloning_A_Policy_Analysis_Revised.pdf. Posted in April 2005. This report discusses government policies and public opinion polls regarding human cloning, including reproductive cloning.

Annotated Bibliography

Johnson, Judith A., and Erin D. Williams. "Human Cloning." Congressional Research Service Reports and Issue Briefs. Available online. URL: http://www.ncseonline.org/NLE/CRSreports/06Aug/RL31358.pdf. Updated on July 20, 2006. This report includes coverage of the Woo Suk Hwang scandal and plans to produce cloned human embryos in order to harvest embryonic stem cells for research and possible therapy.

———. "Stem Cell Research: Federal Research Funding and Oversight." Congressional Research Service Reports and Issue Briefs. Available online. URL: http://www.ncseonline.org/NLE/CRSreports/06Aug/RL33540.pdf. Updated on August 24, 2006. This report reviews the basics of embryonic stem cell research and ethical concerns about it. The authors also cover the history of decisions about federal funding of this research, including President George W. Bush's veto in July 2006 of a funding bill passed by Congress.

———. "Stem Cell Research: State Initiatives." Congressional Research Service Reports and Issue Briefs. Available online. URL: http://www.ncseonline.org/NLE/CRSreports/06Jul/RL33524.pdf. Updated on May 19, 2006. Some states are stepping into the void left by federal decisions not to fund most research on embryonic stem cells by providing their own funding for such research. Some people fear, however, that without the central direction and coordinated research approach that the federal government can provide, the states' actions will result in duplication of effort and a possible lack of ethical oversight.

"Monitoring Stem Cell Research." President's Council on Bioethics. Available online. URL: http://www.bioethics.gov/reports/stemcell/pcbe_final_version_monitoring_stem_cell_research.pdf. Posted in January 2004. This report focuses primarily on ethical issues raised by research on embryonic stem cells and the possibility of cloning embryos for stem cell harvesting.

"Opinion on the Ethical Aspects of Umbilical Cord Blood Banking." European Group on Ethics in Science and New Technologies. Available online. URL: http://ec.europa.eu/european_group_ethics/publications/docs/publop19_en.pdf. Posted on March 16, 2004. The group examined both public (nonprofit) banks, which would make umbilical cord blood cells available to anyone, and private banks, which parents would pay to store the blood for future use by their own families.

"Proposition 71: A Model for State Involvement in Biomedical Research?" FasterCures. Available online. URL: http://www.fastercures.org/pdf/csci_one_year_later.pdf. Posted in fall 2005. This report reviews progress during the first year after voters' passage of a controversial amendment to the California state constitution that earmarked $3 billion for stem cell research, including research on embryonic stem cells and therapeutic cloning, over the following 10 years. The document describes how the proposition came into being, its economic aspects, issues provoking criticism, lawsuits that have held up the program's implementation, and lessons learned.

"Regenerative Medicine 2006." National Institutes of Health. Available online. URL: http://stemcells.nih.gov/staticresources/info/scireport/PDFs/ Regenerative_Medicine_2006.pdf. Posted in August 2006. Written by experts in stem cell research, this report describes advances made since 2001 and outlines expectations for future developments.

"Reproductive Cloning: Ethical and Social Issues." Human Genetics Alert. Available online. URL: http://www.hgalert.org/topics/cloning/cloning. PDF. Posted in January 2004. This document urges an international ban on reproductive cloning.

Smith, Simon. "All the Reasons to Clone Human Beings." Human Cloning Foundation. Available online. URL: http://www.humancloning.org/ allthe.php. Accessed on March 18, 2007. Reproductive cloning is Smith's primary concern; he strongly approves of cloning to make children.

"Stem Cell Basics." National Institutes of Health. Available online. URL: http://stemcells.nih.gov/info/basics. Accessed on March 17, 2007. This report discusses the unique properties of stem cells, similarities and differences between embryonic and adult stem cells, potential uses of stem cells, obstacles that must be overcome before those uses can occur, and sources of further information.

Wolinetz, Carrie D. "Breakthroughs Using Embryonic Stem Cells." Federation of American Societies for Experimental Biology. Available online. URL: http://opa.faseb.org/pdf/BreakthroughsEmbryonicStemCellsJune 2006.pdf. Updated in June 2006. This brochure cites some of the diseases for which treatments with embryonic stem cells are being investigated.

THE FUTURE OF BIOTECHNOLOGY AND NEW FORMS OF BIOTECHNOLOGY

BOOKS

Bainbridge, William Sims, and Mihail C. Roco, eds. *Managing Nano-Bio-Info-Cogno Innovations.* New York: Springer Verlag, 2006. These papers describe the fusion that the authors see occurring among nanotechnology, biotechnology, information technology, and cognitive science and predict that innovations arising from this fusion will have a revolutionary, mostly positive, impact on human society.

Baird, Davis, ed. *Nanotechnology Challenges: Implications for Philosophy, Ethics, and Society.* River Edge, N.J.: World Scientific Publishing, 2006. These essays explore the philosophical underpinnings of nanotechnology and the ethical and sociological issues it brings up.

―――, Alfred Nordmann, and Joachim Schummer, eds. *Discovering the Nanoscale*. Amsterdam, Netherlands: Ios Press, 2005. These essays consider nanotechnology from the perspective of science and technology studies. Topics examined include the history of the field and ethical issues.

Berube, David M. *Nano-Hype: The Truth Behind the Nanotechnology Buzz*. Amherst, N.Y.: Prometheus Books, 2005. This carefully researched overview of nanotechnology and its effects on contemporary culture aims to correct exaggerations made by the media and even some scientists.

Booker, Richard, and Earl Boysen. *Nanotechnology for Dummies*. Hoboken, N. J.: Wiley, 2005. The authors attempt to demystify this technology and its applications for investors, business executives, and the interested public.

Borisenko, Victor E., and Stefano Ossicini. *What Is What in the Nanoworld: A Handbook on Nanoscience and Nanotechnology*. Weinheim, Germany: Wiley-VCH, 2004. This introductory reference handbook summarizes terms and definitions, important phenomena, and regulations relevant to these new fields of science and technology.

Cavalieri, Anthony J. *Biotechnology and Agriculture in 2020*. Washington, D. C.: CSIS Press, 2005. Cavalieri concludes that future populations must be fed by increasing the productivity of existing farmland. He says that many technologies, including biotechnology, are needed to bring about this increase.

Edwards, Steven A. *The Nanotech Pioneers: Where Are They Taking Us?* Weinheim, Germany: Wiley-VCH, 2006. Edwards tries to provide a realistic view of ways that nanotechnology will affect ordinary consumers in the near future.

Gazit, Ehud. *Plenty of Room for Biology at the Bottom: An Introduction to Bionanotechnology*. London: Imperial College Press, 2006. This book takes the reader from the fundamentals of nanobiology (nanotechnology applied to biomedical research) to advanced applications.

Goodsell, David S. *Bionanotechnology: Lessons from Nature*. Wilmington, Del.: Wiley-Liss, 2004. This text introduces the interdisciplinary field of bionanotechnology, which combines the physical and biological sciences, and assesses its future promise.

Hall, J. Storrs. *Nanofuture: What's Next for Biotechnology*. Amherst, N.Y.: Prometheus Books, 2005. Hall foresees nanotechnology passing through five stages of development. He admits that the technology presents risks, but he says that banning it in the United States will simply allow other countries to dominate the field.

Hall, Stephen S. *Merchants of Immortality: Chasing the Dream of Human Life Extension*. Boston: Mariner, 2005. Hall describes scientists and entrepreneurs working to slow the aging process, reduce age-related disease and disability, and extend life.

Biotechnology and Genetic Engineering

Hunt, Geoffrey, and Michael Mehta, eds. *Nanotechnology: Risk, Ethics, and Law*. London: Earthscan Publications, 2006. These essays provide an overview of nanotechnology's interactions with society around the world, including predicted risks and benefits of the technology and the ethical and legal questions it presents to governments, businesses, and the public.

Institute of Medicine and National Research Council. *Globalization, Biosecurity, and the Future of the Life Sciences*. Washington, D.C.: National Academies Press, 2006. This book warns that terrorists or "biohackers" might synthesize and release genetically engineered microorganisms, resulting in disease epidemics.

Jones, Richard A. L. *Soft Machines: Nanotechnology and Life*. New York: Oxford University Press, 2004. Jones explains the science behind nanotechnology for lay readers, focusing on why the nanoworld is so different from the one that humans normally perceive.

Kurzweil, Ray. *The Singularity Is Near: When Humans Transcend Biology*. New York: Penguin, 2006. Kurzweil presents a highly optimistic view of a future in which humans will merge their biology with advances in genetics, nanotechnology, and robotics to create an unimaginably advanced version of the human race. He also considers some "existential risks" and criticisms of such advancement.

Malsch, Neelina H., ed. *Biomedical Nanotechnology*. Washington, D.C.: CRC Press, 2005. These papers focus on applications of nanotechnology in drugs and drug delivery, prostheses and implants, and diagnostic and screening technologies. They consider risks as well as benefits of the technology and examine the social and economic context of biomedical nanotechnology research.

Matsuura, Jeffrey H. *Nanotechnology Regulation and Policy Worldwide*. Boston, Mass.: Artech House, 2006. Matsuura surveys legal and policy initiatives that affect nanotechnology research and looks for a balance between encouraging the research and protecting the public from potentially dangerous consequences. He also discusses ways to protect intellectual property rights in the field.

Miller, G. Wayne. *The Xeno Chronicles: Two Years on the Frontier of Medicine Inside Harvard's Transplant Research Lab*. New York: Public Affairs, 2005. Miller follows the experiments of David H. Sachs and his team, who are trying to develop a genetically engineered pig with organs that can be safely transplanted into other species—baboons at present, but ultimately humans.

National Research Council. *Biotechnology Research in an Age of Terrorism*. Washington, D.C.: National Academies Press, 2004. The council recommends that an international forum on biosecurity be set up to help governments and scientists regulate access to certain types of dangerous information. However, they say that scientific publishing should not be censored.

Annotated Bibliography

Roco, Mihail C., and William S. Bainbridge, eds. *Nanotechnology: Societal Implications, Volume 1: Maximizing Benefits for Humanity*. New York: Springer, 2006. These essays by leading scientists and social scientists survey possible uses of nanotechnology and the legal, ethical, social, economic, and educational issues the technology raises. They consider how to maximize its positive impacts and minimize negative ones.

Sargent, Ted. *The Dance of the Molecules: How Nanotechnology Is Changing Our Lives*. New York: Thunder's Mouth Press, 2006. Sargent presents an extremely optimistic view of nanotechnology's potential contributions to medicine, communications, and environmental protection.

Schmid, Günter, et al. *Nanotechnology: Assessment and Perspectives*. New York: Springer, 2006. This book reports on a two-year study that assessed nanotechnology from various perspectives and prepared integrated recommendations for decisionmakers.

Shelley, Toby. *Nanotechnology: New Promises, New Dangers*. New York: Zed Books, 2006. This book explores the potential benefits and risks of nanotechnology in many areas of life and also considers who controls this new technology.

Silberglitt, Richard. *The Global Technology Revolution 2020*. Santa Monica, Calif.: RAND Corporation, 2006. Areas surveyed include biotechnology and nanotechnology.

Young, Simon. *Designer Evolution: A Transhumanist Manifesto*. Amherst, N. Y.: Prometheus Books, 2005. The emerging philosophical movement of transhumanism holds that humans will soon be able to take control of their own evolution and overcome their collective limitations "through reason, science, and technology"—and that doing so is highly desirable.

ARTICLES

Allhoff, Fritz, and Patrick Lin. "What's So Special about Nanotechnology and Nanoethics?" *International Journal of Applied Philosophy*, vol. 20, Fall 2006, pp. 179–190. This paper defines the concept of nanoethics and explains why it should exist as a separate area of study.

Anderson, M. "Xenotransplantation: A Bioethical Evaluation." *Journal of Medical Ethics*, vol. 32, April 2006, pp. 205–208. Through experimental studies and expert opinion, this review illuminates ethical issues raised by xenotransplantation.

Aston, Adam. "Mega Questions about Nanotech." *Business Week*, issue 3885, May 31, 2004, p. IM4. In this interview, Kristen Kulinowski, executive director for education and public policy at Rice University (Houston)'s Center for Biological and Environmental Nanotechnology, discusses the benefits and the risks of nanotechnology.

"Bioprocessing: Reaping the Benefits of Renewable Resources." *Chemical Week*, vol. 166, February 11, 2004, pp. 15–17. Industrial biotechnology can produce improved chemicals and manufacturing processes that are more energy efficient and environmentally friendly than present ones.

Brown, Stuart F. "Soul of the New Gene Machines." *Fortune*, vol. 151, May 2, 2005, p. 113. Projects using DNA chips are identifying genetic variations that affect disease risk and the way drugs are metabolized.

Brownlee, Christen. "The Sum of the Parts: Synthetic Biologists String Genes into Living Machines." *Science News*, vol. 168, December 10, 2005, pp. 378–380. Synthetic biology aims to make "genetic engineering" into a true engineering discipline by designing biological components that can be combined to produce desired effects. Applications under development showcase the field's potential for both basic science and medicine.

Choffnes, J. Eileen R., Stanley M. Lemon, and David A. Reiman. "A Brave New World in the Life Sciences." *Bulletin of the Atomic Scientists*, vol. 62, September-October 2006, pp. 26–33. The authors warn that advances in biotechnology and similar fields could fall into the hands of terrorists and increase the risk of biological warfare.

Chyba, Christopher F. "Biotechnology and the Challenge to Arms Control." *Arms Control Today*, vol. 36, October 31, 2006, pp. 11–17. Because biotechnology is advancing much faster than international agreements about its regulation, possible misuse of the technology presents serious challenges to arms control. A global web of prevention and response needs to be constructed.

Erickson, Brent. "White Revolution." *Chemistry and Industry*, vol. 7, April 4, 2005, pp. 17–19. The Biotechnology Industry Organization, an industry trade group, says that industrial biotechnology offers remarkable new tools for preventing pollution, reducing the use of energy and resources such as water, cutting production costs, and creating high-quality consumer and intermediate products. However, policy needs to provide incentives for the field's research and development.

Ferber, Dan. "Microbes Made to Order." *Science*, vol. 303, January 9, 2004, pp. 158–161. Synthetic biology aims to let bioengineers custom-design microorganisms "from the ground up," using off-the-shelf components. Researchers already have a good start on being able to do this, but many challenges await them.

Flannery, Maura C. "Synthetic Design." *The American Biology Teacher*, vol. 68, February 2006, pp. 113–117. Standardization, decoupling, and abstraction are the three engineering principles necessary for the development of synthetic biology.

Fletcher, Amy L. "Reinventing the Pig: The Negotiation of Risks and Rights in the USA Xenotransplantation Debate." *International Journal of Risk Assessment and Management*, vol. 7, January 7, 2007, p. 341 ff. Fletcher says that the public, as opposed to scientists, deserves to play a larger role in decision making about xenotransplantation because pig retroviruses passed to humans in transplants could endanger the entire community.

Freedman, Michael. "Nanofear." *Forbes*, vol. 173, September 6, 2004, p. 168. Nanotechnology is becoming a target for activists such as the ETC Group's Pat Mooney, who point to possible health and environmental risks of the new technology and call for precautionary regulation.

Hayes, Richard. "Our Biopolitical Future: Four Scenarios." *World Watch*, vol. 20, March–April 2007, pp. 1–16. Hayes presents four possible scenarios of the human biopolitical future, taking place between 2007 and 2021. They focus on the tension between libertarian and communitarian values.

Hede, Shantesh, and Nagraj Huilgol. "'Nano': The New Nemesis of Cancer." *Journal of Cancer Research and Therapeutics*, vol. 2, October-December 2006, n.p. This review summarizes recent developments that apply nanotechnology to cancer therapy and diagnostics.

Konde, Victor. "Industrial Biotechnology Applications for Food Security in Africa: Opportunities and Challenges." *International Journal of Biotechnology*, vol. 7, March 10, 2005, p. 95. Industrial biotechnology could improve food security in Africa, but a variety of challenges threaten this positive outcome. This article provides an African perspective on the technology and lays out options that the continent could adopt to achieve its benefits.

Kranenburg, Leonieke W., et al. "Reluctant Acceptance of Xenotransplantation in Kidney Patients on the Waiting List for Transplantation." *Social Science and Medicine*, vol. 61, October 15, 2005, pp. 1828–1834. When the theoretical possibility of receiving animal organs was discussed with 61 patients waiting for a kidney transplant in the Netherlands, most said they would accept such organs only in a life-threatening situation where no other treatments were available. Patients' objections centered on possible health risks.

Lemonick, Michael D., Laura A. Locke, and Melissa August. "Mother Nature's DNA." *Time*, vol. 165, June 20, 2005, p. 56. As some scientists struggle to understand the now-sequenced human genome, others search the genomes of microorganisms to find genes for making ethanol, hydrogen, and other useful substances.

Lin, Patrick. "Nanotechnology's Dilemmas: Now Is the Time to Wrestle with the Ethics of Pandora's Box." *The Scientist*, vol. 19, December 5, 2005, p. 10. Lin presents ethical issues that nanotechnology brings up in the areas of regulation, environment and health, society, politics and markets, life extension, and religious and moral values.

Lopez, Jose. "Compiling the Ethical, Legal, and Social Implications of Nanotechnology." *Health Law Review*, vol. 12, September 22, 2004, pp. 24–27. Lopez says that before the ethical, legal, and social implications of nanotechnology can be properly investigated, the narrative used to envision the technology's future must be changed.

Mandel, Michael J. "The Catastrophe that Won't Happen: Why Biotech Gains Will Rein in Medical Costs." *Business Week*, issue 3885, May 31, 2004, p. 57. Drawing on the analogy of information technology, Mandel thinks that gains in productivity will make new drugs and other advances in medical biotechnology drop substantially in price by the end of the 21st century's first decade, reducing overall health care costs per patient.

Maynard, Andrew D. "Nanotechnology: The Next Big Thing, or Much Ado About Nothing?" *Annals of Occupational Hygiene*, vol. 51, January 2007, pp. 1–12. Research suggests that some nanomaterials will present health risks to people working with them in factories because of unique features of their structures and chemistry. Research is needed to further identify these risks and determine the best procedures for deflecting them.

Mehta, Michael D. "The Future of Nanomedicine Looks Promising, But Only if We Learn from the Past." *Health Law Review*, vol. 13, Winter 2004, pp. 16–18. The future of nanomedicine depends on social, regulatory, and clinical issues that must be addressed as soon as possible. Mehta presents an analogy with nuclear medicine to show that public support for a new technology is vital—and fragile. He offers several strategies for developing support for nanomedicine.

Morton, Oliver. "Life, Reinvented." *Wired*, vol. 13, January 2005, pp. 168–175. Engineers at the Massachusetts Institute of Technology are leaders in the new field of synthetic biology, which attempts to treat living organisms as products that can be assembled and modified with standardized component parts.

"Much Ado over 'Lethal Genes.'" *Business Week*, January 12, 2006, n.p. Experimenters in pest control are developing transgenic insects, genetically engineered to, for instance, be unable to reproduce—but some "watchdog" groups fear that the insects themselves could become pests.

Munro, J. Neil. "Creating Life to Order." *National Journal*, vol. 38, June 10, 2006, pp. 54–55. Researchers in synthetic biology are inventing microorganisms that they hope will eventually produce most industrial materials now made from petrochemicals. However, terrorists could also take advantage of this technology if it is not carefully regulated.

Perkel, Jeffrey M. "The Ups and Downs of Nanobiotech." *The Scientist*, vol. 18, August 30, 2004, pp. 14–18. Medical nanotechnology could bring significant improvements in therapeutics, drug delivery, tissue reconstruction, and diagnostics and imaging, but major safety issues must be settled first.

Ritter, Stephen K. "Industrial Biotech Gains Momentum: Growth of Commercial Enzyme-Mediated Processes Points to the Future of the Chemical Industry." *Chemical and Engineering News*, vol. 84, April 3, 2006, pp. 69–71. Industrial biotechnology uses microorganisms, genetically modified and otherwise, to improve manufacturing processes and make new products from renewable materials. A number of products from this new technology are already in wide use.

Schu, Bill. "Gene Expression and the Search for the Fountain of Youth." *Genomics and Proteomics*, vol. 5, June 1, 2005, p. 21 ff. Researchers are discovering the genes and proteins that control aging. They may soon be able to alter these to prevent age-related diseases and extend healthy lifespans in humans.

"Sea of Dreams: Biotechnology." *The Economist*, vol. 371, May 1, 2004, p. 81US. Newly discovered bacteria with unique genes provide the raw material for industrial biotechnology, which potentially can greatly reduce the cost of producing ethanol, plastics, and other industrial chemicals.

Sidawi, Danielle. "Nanotechnology Takes Medical Intervention to a New Level." *R&D*, vol. 46, October 2004, pp. 28–29. Nanotechnology could offer major advances in drug delivery and other medical uses, but biocompatibility and toxicity are potential problems.

Silver, Pamela, and Jeffrey Way. "Cells by Design: The Potential for Synthetic Biology." *The Scientist*, vol. 18, September 27, 2004, pp. 30–31. The authors provide an overview of synthetic biology, which uses fixed pieces of DNA and other biocompounds to build new biological structures and perhaps even new organisms.

Stewart, Patrick A., and Andrew J. Knight. "Trends Affecting the Next Generation of U.S. Agriculture Biotechnology: Politics, Policy, and Plant-Made Pharmaceuticals." *Technological Forecasting and Social Change*, vol. 72, June 2005, pp. 521–534. The introduction of plant-made pharmaceuticals is producing changes in U.S. policies for regulating agricultural biotechnology. The authors discuss the effects of these changes on the industry as a whole.

Strong, C. "Reproductive Cloning Combined with Genetic Modification." *Journal of Medical Ethics*, vol. 31, November 2005, pp. 654–658. Most objections to reproductive human cloning are based on the potential child's lack of unique nuclear DNA, but in the future, genetic engineering may be able to overcome this objection by providing unique DNA, carrying genetic characteristics of both members of an infertile couple, to the cloned child.

"Synthetic Biologists Face Up to Security Issues." *Nature*, vol. 436, August 18, 2005, pp. 894–895. The new field of synthetic biology needs to be carefully regulated to make sure that accidental or deliberate misuse of the technology does not occur.

Tyshenko, Michael G. "Considerations for Using Genetic Material in Medical Nanotechnology." *Health Law Review*, vol. 12, Fall 2004, pp. 19–23. DNA-based nanotechnology is likely to have future medical uses, for instance in drug delivery, but it may also present unique dangers, such as the ability to reach and persist in parts of the body other than those desired. This technology therefore should receive the same sort of early, voluntary self-regulation that occurred with genetic engineering.

Warner, Susan. "Biotech Takes on New Industries." *The Scientist*, vol. 19, February 28, 2005, pp. 45–46. Industrial biotechnology is finally reaching maturity, but it is having trouble persuading businesses to forsake established manufacturing processes in favor of new ones.

Weintraub, Arlene. "What's So Scary about Rice?" *Business Week*, issue 3945, August 1, 2005, p. 58. "Biopharming"—genetically engineering plants to produce drugs—is tempting, but fears that the modified crops might enter the food supply could limit their development.

Williams, Mark. "The Knowledge: Biotechnology's Advance Presents Dark Possibilities." *Technology Review*, vol. 109, March-April 2006, pp. 44–53. The Russian bioweapons development program of the 1980s shows how dangerous weapons can be created even with biotechnology that today would be considered simple or outdated—and is now available widely and cheaply on the Internet. New discoveries in the field are bound to increase the threat of biological weapons.

Williams-Jones, Bryn. "A Spoonful of Sugar Makes the Nanotech Go Down." *Health Law Review*, vol. 12, September 22, 2004, pp. 10–13. Most visions of nanotechnology's future development are either strongly positive or strongly negative. Williams-Jones considers why such extreme visions arise and what problems they create.

Zimmer, Carl. "2006 Scientist of the Year: Jay Keasling." *Discover*, vol. 27, December 2006, pp. 35–37. Zimmer interviews Keasling, who is using synthetic biology to design a microbe that can make a valuable antimalaria drug cheaply.

WEB DOCUMENTS

Andow, David, ed. "A Growing Concern: Protecting the Food Supply in an Era of Pharmaceutical and Industrial Crops." Union of Concerned Scientists. Available online. URL: http://www.ucsusa.org/assets/documents/food_and_environment/Pharma_fullreport.pdf. Posted in December 2004. This report presents the findings of a 2003 expert workshop on protecting the U.S. food supply from contamination by crops engineered to produce pharmaceuticals and industrial chemicals.

"Biopromise? Biotechnology, Sustainable Development, and Canada's Future Economy." Canadian Biotechnology Advisory Committee. Available online. URL: http://cbac-cccb.ca/epic/site/cbac-cccb.nsf/vwapj/BSDE%20Executive%20Report%20-%20ENGLISH.pdf/$file/BSDE%20

Executive%20Report%20-%20ENGLISH.pdf. Posted in September 2006. This report pictures a new relationship between biotechnology, sustainable development, and the Canadian economy in 2020 that will address community, health, economic, and ecological needs. It also makes recommendations for reaching this goal.

"Bugs in the System? Issues in the Science and Regulation of Genetically Modified Insects." Pew Initiative on Food and Biotechnology. Available online. URL: http://pewagbiotech.org/research/bugs/bugs.pdf. Posted on January 22, 2004. This document outlines the development status of genetically modified insects and the benefits and risks associated with them. It also examines the regulatory system and points out gaps in authority and other areas that could be improved.

"Cancer Nanotechnology: Going Small for Big Advances." National Cancer Institute. Available online. URL: http://nano.cancer.gov/resource_center/cancer_nanotechnology_brochure.pdf. Posted in January 2004. The National Cancer Institute is sponsoring research on novel nanodevices for cancer diagnosis, drug delivery, and monitoring. Ultimately, such devices could change the foundations of cancer diagnosis, treatment, and prevention.

"Down on the Farm: The Impact of Nano-scale Technologies on Food and Agriculture." ETC Group. Available online. URL: http://www.etcgroup.org/upload/publication/80/01/etc_dotfarm2004.pdf. Posted in November 2004. This report describes how nano-scale technologies are likely to affect farmers, food, and agriculture before 2025. It expresses concern about the current absence of government regulation and warns about possible social, health, and environmental risks from present and future nanotechnology products associated with agriculture.

"Environmental, Health, and Safety Research Needs for Engineered Nanoscale Materials." National Nanotechnology Initiative. Available online. URL: http://www.nano.gov/NNI_EHS_research_needs.pdf. Posted in September 2006. This document identifies research and information needed for the understanding and management of potential risks of engineered nanoscale materials.

Erickson, Brent, and Christopher J. Hessler. "New Biotech Tools for a Cleaner Environment: Industrial Biotechnology for Pollution Prevention, Resource Conservation, and Cost Reduction." Biotechnology Industry Organization. Available online. URL: http://www.bio.org/ind/pubs/cleaner2004/CleanerReport.pdf. Posted in June 2004. This report describes potential positive effects of expanding biotechnology, considers public policy mechanisms that could provide incentives for developing the technology, and makes recommendations for ways that different groups could encourage such development.

"The Ethical Aspects of Nanomedicine." European Group on Ethics in Science and New Technologies. Available online. URL: http://ec.europa. eu/european_group_ethics/activities/docs/roundt_nano_21march2006_ final_en.pdf. Posted on March 21, 2006. This report, the proceedings of a roundtable debate, includes a scientific overview of the field, technology assessment, a patient's perspective, discussion of nanomedicine and toxicology, and consideration of the relationships among nanotechnology, medicine, and the human condition.

European Commission. "Plants for the Future." European Plant Science Organisation. Available online. URL: http://www.epsoweb.org/catalog/tp/ Plants%20for%20the%20future-Dec04.pdf. Posted in December 2004. This European vision for plant genomics and biotechnology up until 2025 includes discussions of how to achieve the goals of healthy, safe, and sufficient food and feed; sustainable agriculture, forestry, and landscape; "green" products; and competitiveness, consumer choice, and governance.

"Extreme Genetic Engineering: An Introduction to Synthetic Biology." ETC Group. Available online. URL: http://www.etcgroup.org/upload/publication/602/01/synbioreportweb.pdf. Posted in January 2007. This organization concludes that the social, environmental and bioweapons threats of synthetic biology are even greater than those for biotechnology.

"Foresight Nanotechnology Challenges." Foresight Nanotech Institute. Available online. URL: http://www.foresight.org/challenges/index.html. Accessed on March 16, 2007. This web page links to reports on six problems that nanotechnology might solve: providing renewable clean energy, supplying clean water globally, improving health and longevity, healing and preserving the environment, making information technology available to all, and enabling space development.

"Forward Look on Nanomedicine." European Science Foundation. Available online. URL: http://www.esf.org/research-areas/medical-sciences/ publications.html. Posted in December 2005. This report discusses challenges and benefits of nanonmedicine and recommends that Europe develop regulatory guidelines for this new technology quickly.

"The Future of Nanotechnology: We Need to Talk." Nanologue. Available online. URL: http://www.nanologue.net/custom/user/Downloads/Nanologue_we-need-to-talk.pdf. Accessed on March 17, 2007. This report describes the development of nanotechnology to date and presents three possible scenarios for the technology's future advancement and social impact.

Hett, A. "Nanotechnology: Small Size, Large Impact?" Swiss Re. Available online. URL: http://www.swissre.com/resources/of710b80466085b3ba3 bff276a9800c6=Research%20and%20Publications_Risk%20and% 20Expertise_Nanotechnology%20=%20small%20size%20large%20 impact_nanotechnology-report_MBUI=6E7H3N.pdf. Posted in 2004.

This report of a conference on nanotechnology held in December 2004 focuses on risk analysis and risk management, including "phantom" as well as real risks.

"Industrial Biotechnology and Sustainable Chemistry." Royal Belgian Academy Council of Applied Sciences. Available online. URL: http://www. bio-economy.net/bacas_report_en-3.pdf. Posted in January 2004. This report explains how industrial biotechnology can reduce fossil fuel use and greenhouse gas emissions, create sustainable products from renewable resources, and make chemical manufacturing increasingly "green" and nonpolluting as well as more efficient.

Kahan, Dan, et al. "Nanotechnology Risk Perceptions: The Influence of Affect and Values." Available online. URL: http://www.nanotechproject. org/reports. Posted on March 7, 2007. This document reports on a major survey of perceptions of the risks of nanotechnology, conducted by the Cultural Cognition Project in December 2006. The survey found that people knew little about nanotechnology and had not made up their minds about it. They tended to react to the concept in an emotional way that was shaped by cultural factors.

Kreysa, Gerhard, and Rüdiger Marquardt. "Biotechnology 2020: From the Transparent Cell to the Custom-Designed Process." European Commission Directorate-General for Research/DECHEMA. Available online. URL: http://www.6rp.cz/dokums_dokumenty/biotech_2020_418.pdf. Posted in 2005. This report is the product of a team of young experts, who chose important likely advances in a wide range of areas of biotechnology and examined them in detail. Fields covered include industrial biology, regenerative medicine, and foods with added health benefits.

Kuzma, Jennifer, and Peter VerHage. "Nanotechnology in Agriculture and Food Production: Anticipated Applications." Project on Emerging Nanotechnologies. Available online. URL: http://www.nanotechproject.org/ reports. Posted in September 2006. The report estimates possible areas and timeframes for future nanotechnology-based food and agriculture applications, looks at potential benefits and risks, and explores needs for environmental, health, and safety regulation.

Maynard, Andrew D. "Nanotechnology: A Research Strategy for Addressing Risk." Project on Emerging Nanotechnologies. Available online. URL: http://www.nanotechproject.org/reports. Posted in July 2006. Maynard calls for major changes in the U.S. government's handling of research on the possible risks of nanotechnology and proposes a comprehensive framework for systematically exploring such risks.

Miller, Georgia. "Nanomaterials, Sunscreens and Cosmetics: Small Ingredients, Big Risks." Friends of the Earth. Available online. URL: http:// www.foe.org/camps/comm/nanotech/nanocosmetics.pdf. Posted in May 2006. Nanotechnology is being heralded as the basis of the next industrial revolution, yet serious questions exist about the health, environmental, and social impacts of this powerful new technology.

"Nanoscience and Nanotechnologies: Opportunities and Uncertainties." Royal Society. Available online. URL: http://www.nanotec.org.uk/finalreport.htm. Posted on July 29, 2004. Nanotechnologies offer many potential benefits, but research to address uncertainties about their health and environmental effects, public debate about their social implications, and regulation to control exposure and minimize possible damage are all badly needed.

"Nanotech RX: Medical Applications of Nano-scale Technologies: What Impact on Marginalized Communities?" ETC Group. Available online. URL: http://www.etcgroup.org/upload/publication/593/01/etc06nano-techrx.pdf. Posted on September 12, 2006. Nanomedicine and other new technologies are likely to make possible the alteration of human bodies and brains in ways that go beyond therapy into enhancement. This report warns against risks of this technology, especially for already disadvantaged groups such as the disabled and citizens of developing countries, and calls for international monitoring of nanomedical technology.

Peterson, Christine L. "Nanotechnology: From Feynman to the Grand Challenge of Molecular Manufacturing." Institute of Electrical and Electronics Engineers. Available online. URL: http://www.patmedia.net/tbookman/techsoc/Peterson.htm. Posted in winter 2004. This report provides a history of nanotechnology from Richard Feynman's introduction of the concept in 1959 to the possibility of future molecular nanotechnology, or molecular manufacturing.

"Progress Toward Safe Nanotechnology in the Workplace." Centers for Disease Control and Prevention. Available online. URL: http://www.cdc.gov/niosh/docs/2007-123/pdfs/2007-123.pdf. Posted in February 2007. This report from the National Institute for Occupational Safety and Health's Nanotechnology Research Center discusses topics from hazard identification to risk management, lists goals for the research center, and describes progress that has been made toward achieving each goal.

"Sowing Secrecy: The Biotech Industry, USDA, and America's Secret Pharm Belt." Center for Science in the Public Interest. Available online. URL: http://www.cspinet.org/new/pdf/pharmareport.pdf. Posted on June 2, 2004. This report claims that the number of crops genetically engineered to produce drugs, industrial chemicals, or other novel proteins is "secretly" increasing, with the help of U.S. Department of Agriculture permits, and that these crops potentially could produce dangerous contamination in food products.

"Summary Proceedings, Third Annual World Congress on Industrial Biotechnology and Bioprocessing: Linking Biotechnology, Chemistry and Agriculture to Create New Value Chains." Available online. National

Agricultural Biotechnology Council. URL: http://nabc.cals.cornell.edu/ pubs/WCIBB2006_proc.pdf. Posted on July 14, 2006. This report includes discussions of policy, economic issues, biofuels and bioenergy, feedstocks for bioprocessing and biomaterials, biochemicals, manufacturing, bioprocessing and novel applications, and emerging issues in industrial biotechnology.

USDA Advisory Committee on Biotechnology and 21st Century Agriculture. "Opportunities and Challenges in Agricultural Biotechnology: The Decade Ahead." U.S. Department of Agriculture. Available online. URL: http://www.usda.gov/documents/final_main_report-v6.pdf. Posted on July 13, 2006. This report describes types of genetically modified crops that may become available within the specified time period, factors that will determine whether they are successful, risks and their management, changes in public opinion about agricultural biotechnology that have occurred in the past decade, and ways that the agricultural marketplace can adapt to this climate.

———. "Preparing for the Future." U.S. Department of Agriculture. Available online. URL: http://w3.usda.gov/agencies/biotech/ac21/reports/ scenarios-4-5-05final.pdf. Posted on May 9, 2005. This paper analyzes trends and uncertainties that will shape the use of biotechnology in the future and presents three possible scenarios.

"U.S. Environmental Protection Agency Nanotechnology White Paper." Environmental Protection Agency (EPA). Available online. URL: http:// www.epa.gov/osa/pdfs/nanotech/epa-nanotechnology-whitepaper-0207. pdf. Posted in February 2007. This white paper provides a basic description of nanotechnology, reasons for the EPA's interest in the subject, potential environmental benefits, risk assessment issues, needs for further research, the EPA's framework for nanotechnology research, and the agency's possible role in future regulation of nanotechnology.

Wisner, Robert. "The Economics of Pharmaceutical Crops: Potential Benefits and Risks for Farmers and Communities." Union of Concerned Scientists. Available online. URL: http://www.ucsusa.org/assets/documents/food_and_environment/ucs-economics-pharma-crops.pdf. Posted in December 2005. This report evaluates the benefits and risks of pharmaceutical crops, or plants genetically engineered to produce drugs. The Union of Concerned Scientists, which sponsored the research, concludes that claims about the technology's benefits have been exaggerated and that any benefits that do arise will not apply primarily to farmers.

CHAPTER 8

ORGANIZATIONS AND AGENCIES

There are many organizations devoted to biotechnology and genetic engineering, as well as various aspects of genetics. The following entries include professional groups, industry organizations, advocacy groups, and government agencies, both in the United States and abroad. In keeping with the widespread use of the Internet and e-mail, the web site address (URL) and e-mail address are given first when available, followed by the phone number and postal address.

AfricaBio
URL: http://www.africabio.com/index.shtml
E-mail: africabio@mweb.co.za
Phone: (27) 12-667-2689
Promotes safe, ethical, and responsible research, development, and application of biotechnology and its products. Stresses safety and benefits of the technology but also supports labeling of genetically modified foodstuffs.

African Agricultural Technology Foundation (AATF)
URL: http://www.aatf-africa.org
E-mail: aatf@aatf-africa.org
Phone: (650) 833-6660
African Agricultural Technology Foundation c/o ILRI

P.O. Box 30709
Nairobi 00100, Kenya
Africa-based and managed public-private partnership linking large agricultural biotechnology companies such as Monsanto with smallholder farmers in Africa to bring the benefits of biotechnology to that continent and improve plant varieties grown by the country's small farmers.

AgBioWorld Foundation
URL: http://www.agbioworld.org
Phone: (334) 444-7884
P.O. Box 85
Tuskegee Institute, AL 36087-0085
Headed by C. S. Prakash of Tuskegee University. Highly in favor of agricultural biotechnology but does

not accept direct contributions from biotechnology companies or trade associations.

Agricultural Biotechnology Council (abc)
URL: http://www.abcinformation.org
E-mail: Enquiries@abcinformation.org
Phone: (020) 7025 2333
P.O. Box 49710
London WC1V 7WX
United Kingdom
Established in 2002 by the British biotechnology trade industry, including large international companies, to provide information and education that address public concerns about the technology.

Agricultural Groups Concerned About Resources and the Environment (AGCare)
URL: http://www.agcare.org
E-mail: agcare@agcare.org
Phone: (519) 837-1326
Ontario AgriCentre
100 Stone Road West
Suite 106
Guelph, Ontario N1G 5L3
Canada
Coalition of groups representing Ontario's farmers. It provides science and research-based information and policy initiatives on subjects including crop biotechnology, which it supports.

Agricultural Research Service
URL: http://www.ars.usda.gov/main/main.htm
Jamie L. Whitten Building

1400 Independence Avenue, SW
Washington, DC 20250
The Agricultural Research Service is the main in-house research arm of the U.S. Department of Agriculture. It is also one of the world's premiere scientific organizations.

Alliance for Bio-Integrity
URL: http://www.biointegrity.org
E-mail: info@biointegrity.org
Phone: (206) 888-4852
2040 Pearl Lane #2
Fairfield, IA 52556
Believes that genetically engineered foods present "unprecedented dangers to the environment and human health" and should be labeled and carefully regulated.

American Civil Liberties Union (ACLU)
URL: http://www.aclu.org
125 Broad Street
18th Floor
New York, NY 10004-2400
One area of concern for this nonprofit, nonpartisan public interest organization devoted to protecting American civil liberties is genetic privacy. It opposes employment and insurance discrimination based on genetic testing as well as the establishment of national DNA databases. It offers both print and online resources, including books, pamphlets, newsletters, and Web links.

American College of Medical Genetics (ACMG)
URL: http://www.acmg.net//AM/template.cfm?Section=Home3

E-mail: acmg@acmg.net
Phone: (301) 634-7127
9650 Rockville Pike
Bethesda, MD 20814-3998
Professional organization for scientists and health care professionals specializing in medical genetics. Among other things, provides guidelines for genetic health testing and lobbies for effective and fair health policies and legislation.

American Corn Growers Association (ACGA)
URL: http://www.acga.org
E-mail: acga@acga.org
Phone: (202) 835-0330
P.O. Box 18157
Washington, DC 20036
Progressive commodity association, representing thousands of corn producers in 28 states. Questions whether the benefits of genetically modified corn outweigh its economic liability and other risks to farmers.

American Farm Bureau
URL: http://www.fb.org
E-mail: webmaster@fb.org
Phone: (202) 406-3600
600 Maryland Avenue, SW
Suite 1000W
Washington, DC 20024
Implements policies and provides programs that improve the financial well-being and quality of life for farmers and ranchers. Supports agricultural biotechnology and opposes labeling of genetically modified foods.

American Genetic Association
URL: http://www.theaga.org/
overview.html
E-mail: agajoh@oregonstate.edu
Phone: (541) 867-0334
2030 SE Marine Science Drive
Newport, OR 97365
Promotes basic and applied research on the genetics of plants and animals. Publishes *Journal of Heredity.*

American Society for Reproductive Medicine (ASRM)
URL: http://www.asrm.org
E-mail: asrm@asrm.org
Phone: (205) 978-5000
1209 Montgomery Highway
Birmingham, AL 35216-2809
Professional organization devoted to advancing knowledge and expertise in reproductive medicine and biology. Publishes educational and other materials, including some that deal with ethical considerations.

American Society of Gene Therapy (ASGT)
URL: http://www.asgt.org
E-mail: info@asgt.org
Phone: (414) 278-1341
555 East Wells Street
Milwaukee, WI 53202
Fosters education, exchange of information, and research on gene therapy.

American Society of Human Genetics
URL: http://www.ashg.org/
genetics/ashg/ashgmenu.htm

E-mail: society@ashg.org
Phone: (301) 634-7300
9650 Rockville Pike
Bethesda, MD 20814-3998
Professional society of researchers, physicians, genetic counselors, and others interested in human genetics and related social issues. Publishes monthly *American Journal of Human Genetics.*

American Society of Law, Medicine, and Ethics (ASLME)
URL: http://www.aslme.org
E-mail: info@aslme.org
Phone: (617) 262-4990
765 Commonwealth Avenue
Suite 1634
Boston, MA 02215
Members include attorneys, physicians, health care administrators, and others interested in relationship between law, medicine, and ethics. Publishes *American Journal of Law* and *Journal of Law, Medicine, and Ethics,* both quarterly.

American Soybean Association (ASA)
URL: http://www.soygrowers.com
Phone: (800) 688-7692
12125 Woodcrest Executive Drive
Suite 100
St. Louis, MO 63141-5009
Trade association that aims to improve profitability for U.S. soybean farmers and to supply soy products to the world. Supports raising of genetically engineered soybeans.

America's Health Insurance Plans (AHIP)
URL: http://www.ahip.org
E-mail: ahip@ahip.org
Phone: (202) 778-3200
601 Pennsylvania Avenue, NW
South Building, Suite 500
Washington, DC 20004
Advocacy group for the health insurance industry. Opposes legislation that restricts insurers' use of genetic information.

BioImpact.org
URL: http://www.bioimpact.org/en/index.php
E-mail: contact@france-biotech.org
Phone: (33) 0 156-58-1070
Hôpital Saint Louis
1, avenue Claude Vellefaux
75010 Paris
The BioImpact project aims to raise awareness of the current value and importance of biotechnologies for public health. Its site surveys biotechnology treatments for illnesses including breast cancer, cardiovascular diseases, and inflammatory diseases.

BioIndustry Association (BIA)
URL: http://www.bioindustry.org/
E-mail: admin@bioindustry.org
Phone: (44) 0 20-7565-7190
14/15 Belgrave Square
London SW1X 8PS
United Kingdom
British biotechnology trade association that represents the industry and its needs to audiences ranging from patient groups to

regional, national, and pan-European governments.

BIOTECanada
URL: http://www.biotech.ca
E-mail: info@biotech.ca
Phone: (613) 230-5585
130 Albert Street
Suite 420
Ottawa, Ontario K1P 5G4
Canada
National biotechnology industry lobbying group devoted to promoting a better understanding of biotechnology and its benefits to Canadians.

Biotechnology Industry
  Organization (BIO)
URL: http://www.bio.org
E-mail: info@bio.org
Phone: (202) 962-9200
1201 Maryland Avenue, SW
Suite 900
Washington, DC 20024
Chief trade and lobbying organization for the biotechnology industry, including academic institutions as well as commercial companies.

British Society for
  Human Genetics (BSHG)
URL: http://www.bshg.org.uk
E-mail: bshg@bshg.org.uk
Phone: (44) 0 121-627-2634
Clinical Genetics Unit
Birmingham Women's Hospital
Birmingham B15 2TG
United Kingdom
Professional association for British scientists and health professionals involved in clinical and laboratory research on human genetics. Its

Public Policy Committee considers the ethical and social effects of human genetic research and technology. It publishes a quarterly newsletter and issues statements on aspects of human genetics.

The Campaign to Label
  Genetically Engineered Foods
URL: http://www.thecampaign.
  org/index.php
E-mail: label@thecampaign.org
Phone: (425) 771-4049
P.O. Box 55699
Seattle, WA 98155
National grassroots consumer campaign for legislation that will require labeling of genetically engineered food in the United States.

Center for Evolutionary
  Functional Genomics
The Biodesign Institute at
  Arizona State University
URL: http://www.biodesign.asu.
  edu
Phone: (480) 737-8322
P.O. Box 875001
Tempe, AZ 85287-5001

Center for Food Safety (CFS)
URL: http://www.
  centerforfoodsafety.org
E-mail: office@
  centerforfoodsafety.org
Phone: (202) 547-9359
660 Pennsylvania Avenue, SE
Suite 302
Washington, DC 20003
Nonprofit membership organization concerned about impacts of the current industrial food production system on human health, animal

welfare, and the environment. Conducts legal and grassroots campaigns to, among other things, protect consumers from what it sees as the hazards of genetically engineered foods.

Center for Genetics and Society
URL: http://geneticsandsociety.
  org
E-mail: info@geneticsand
  society.org
Phone: (510) 625-0819
436 14th Street
Suite 700
Oakland, CA 94612
Nonprofit information and public affairs organization working to encourage responsible uses and effective governance of new human genetic and reproductive technologies. Supports medical applications of these technologies but opposes applications that objectify and commodify human life and threaten to divide human society or alter the processes of the natural world.

**Center for Science in the
  Public Interest (CSPI)**
URL: http://www.cspinet.org
E-mail: cspi@cspinet.org
Phone: (202) 332-9110
1875 Connecticut Avenue, NW
Suite 300
Washington, DC 20009
Nutrition advocacy organization that publishes *Nutrition Action Healthletter*. Its Biotechnology Project is aimed at accurately identifying the risks and benefits of agricultural biotechnology and strengthening FDA regulation of genetically engineered food crops, for instance by making company submission of safety data mandatory rather than voluntary.

Centre for Applied Bioethics
URL: http://www.nottingham.
  ac.uk/bioethics/
E-mail: kate.millar@nottingham.
  ac.uk
Phone: (44) 0 115-951-6303
School of Biosciences
University of Nottingham
Sutton Bonington Campus
Loughborough
Leicester LE12 5RD
United Kingdom
Concerned with appropriate application of biotechnology to food production, industry, and medical uses of farm animals.

Clone Rights United Front
URL: http://www.clonerights.com
E-mail: rhwicker@optonline.net
Phone: (201) 656-3280
#1 Marine View Plaza
Suite 10E
Hoboken, NJ 07030
Claims that the right to be cloned is part of the right to control reproduction. Opposes bans on human cloning.

CloneSafety.org
URL: http://clonesafety.org
E-mail: natalie@clonesafety.org
Phone: (202) 466-9633
CloneSafety.org provides the public and the press with scientifically accurate information about animal

cloning. It is sponsored by several companies engaged in livestock cloning, in cooperation with the Biotechnology Industry Organization.

Coalition for the Advancement of Medical Research (CAMR)
URL: http://www.
  stemcellfunding.org
E-mail: CAMResearch@yahoo.
  com
Phone: (202) 725-0339
2021 K Street, NW
Suite 305
Washington, DC 20006
The coalition includes patient organizations, universities, scientific societies, foundations, and individuals with life-threatening disorders. It supports advances in regenerative medicine, including stem cell research and somatic cell nuclear transfer, and works to ensure that research cloning remains legal and receives federal funding. It opposes reproductive cloning.

Codex Alimentarius Commission
URL: http://www.
  codexalimentarius.net/web/
  index_en.jsp
E-mail: Codex@fao.org
Phone: (39) 065-7051
Viale delle Terme di Caracalla
00153 Rome
Italy
Established in 1963 by the United Nations Food and Agriculture Organization (FAO) and World Health Organization (WHO) to develop food standards, guidelines,

and codes of practice that protect consumer health and ensure fair trade practices in the food trade. Its work includes safety assessments of genetically modified foods.

Comment on Reproductive Ethics (CORE)
URL: http://www.corethics.org
E-mail: info@corethics.org
Phone: (44) 0 207-581-2623
P.O. Box 4593
London SW3 6XE
United Kingdom
A public interest group focusing on ethical dilemmas surrounding human reproduction, particularly the new technologies of assisted conception. The group focuses on respect for the human embryo and the welfare of children born through assisted reproduction technologies.

Consultative Group on International Agricultural Research (CGIAR)
URL: http://www.cgiar.org
E-mail: cgiar@cgiar.org
Phone: (202) 473-8951
CGIAR Secretariat
The World Bank
MSN G6-601
1818 H Street, NW
Washington, DC 20433
Research organization, with both public and private members, whose mission is to contribute to food security and poverty eradication in developing countries through research, partnerships, capacity building, and policy support, promoting sustainable agricultural develop-

ment based on the environmentally sound management of natural resources. It supports a system of 16 Future Harvest Centers, working in more than 100 countries. Some of its projects apply genetic engineering to crops raised by farmers in the developing world.

Consumer Federation of America (CFA)
URL: http://www.consumerfed.org
Phone: (202) 387-6121
1620 I Street, NW
Suite 200
Washington, DC 20006
Represents more than 285 consumer organizations. Gathers, analyzes, and disseminates information to the public, legislators, and regulators and advocates on behalf of consumers, especially the least affluent. Holds conferences and issues reports, books, brochures, news releases, and a newsletter. Approves of biotechnology but wants it carefully regulated.

Consumer Project on Technology (CPTech)
URL: http://www.cptech.org
Phone: (202) 332-2670
1621 Connecticut Avenue, NW
Suite 500
Washington, DC 20009
Group started by Ralph Nader. Its web site includes a discussion of biotechnology and gene patents related to health care technologies, which it generally opposes or would like to see limited.

Council for Agricultural Science and Technology (CAST)
URL: http://www.cast-science.org
E-mail: info@cast-science.org
4420 West Lincoln Way
Ames, IA 50014-3447
Collects, interprets, and communicates science-based information on food, fiber, agricultural, natural resource, and related societal and environmental issues to the public, the private sector, the media, policy makers, regulators, and legislators. Includes information on agricultural biotechnology.

Council for Biotechnology Information
URL: http://www.whybiotech.com
Phone: (202) 962-9200
1201 Maryland Avenue, SW
Suite 900
Washington, DC 20024
Established by the leading biotechnology companies and trade associations to communicate science-based information about the benefits and safety of agricultural and food biotechnology.

Council for Responsible Genetics (CRG)
URL: http://www.gene-watch.org
E-mail: crg@gene-watch.org
Phone: (617) 868-0870
5 Upland Road, Suite 3
Cambridge, MA 02140
National nonprofit organization of scientists, public health professionals, and others that works to see

that biotechnology develops safely and in the public interest. Its concerns include genetic discrimination, patenting of life-forms, food safety, and environmental quality. Publishes a bimonthly newsletter, *GeneWatch*, and educational materials.

Council of Canadians
URL: http://www.canadians.org
E-mail: inquiries@canadians.org
Phone: (800) 387-7177
700-170 West Laurier Avenue
West Ottawa, Ontario K1P 5V5
Canada
Citizen watchdog organization. Regards claims of biotechnology benefits as "too good to be true."

Cultural Survival
URL: http://www.cs.org
E-mail: culturalsurvival@cs.org
Phone: (617) 441-5400
215 Prospect Street
Cambridge, MA 02139
Defends human rights and cultural autonomy of indigenous peoples and oppressed ethnic minorities, including protection from exploitation by multinational biotechnology companies. Publishes *Cultural Survival Quarterly*, *Cultural Survival Curriculum Resources*, and two other journals.

Department for Environment, Food and Rural Affairs (DEFRA)
URL: http://www.defra.gov.uk
E-mail: helpline@defra.gsi.gov.uk
Phone: (44) 020-7238 6951
Customer Contact Unit

Eastbury House
30-34 Albert Embankment
London SE1 7TL
United Kingdom
British governmental agency that took over the functions of the Ministry of Agriculture, Fisheries, and Food. Among other things, DEFRA is responsible for the control of the deliberate release of genetically modified organisms (GMOs), and for national, EU, and international policy on the environmental safety of GMOs.

Ecological Farming Association (EFA)
URL: http://www.eco-farm.org
E-mail: info@eco-farm.org
Phone: (831) 763-2111
406 Main Street
Suite 313
Watsonville, CA 95076
Nonprofit educational organization that promotes ecologically sound agriculture. Has joined other organic farming and environmental organizations in protesting against genetically modified organisms and the consolidation of farming production into the hands of large multinational corporations.

Environmental Defense
URL: http://www.
  environmentaldefense.org
E-mail: members@
  environmentaldefense.org
Phone: (212) 505-2100
257 Park Avenue South
New York, NY 10010
National nonprofit organization that links science, economics, and

law to create innovative, equitable, and cost-effective solutions to urgent environmental problems. Usually opposes genetically modified crops as threats to the environment and, possibly, human health.

Environmental Protection Agency (EPA)
URL: http://www.epa.gov
Ariel Rios Building
1200 Pennsylvania Avenue, NW
Washington, DC 20460
U.S. government agency that regulates pesticides, including biopesticides, and sets limits for the amounts of such substances that can remain on or in food; also regulates new chemicals, which are considered to include some genetically engineered organisms.

The ETC Group
URL: http://www.etcgroup.org
E-mail: etc@etcgroup.org
Phone: (613) 241-2267
431 Gilmour Street, 2nd Floor
Ottawa, Ontario K2P 0R5
Canada
Formerly Rural Advancement Foundation International (RAFI). Dedicated to conservation and sustainable advancement of cultural and ecological diversity and human rights. Provides research and analysis of technical information, including information about plant genetic resources, biotechnology, and biodiversity, and works to develop strategic options for applying those technologies at the global and regional (inter-

national) levels in ways that will create cooperative and sustainable self-reliance within disadvantaged societies.

European Association for Bioindustries (EuropaBio)
URL: http://www.europabio.org
E-mail : info@europabio.org
Phone: (32 2) 735-0313
Avenue de l'Armée 6
1040 Brussels
Belgium
International trade association representing smaller associations and companies of all sizes working in a variety of biotechnology fields. Works to promote innovative and dynamic yet responsible biotechnology-based industry in Europe and to advocate free and open markets for the technology.

European Community of Consumer Cooperatives (Euro Coop)
URL: http://www.eurocoop.org/home/en/default.asp
E-mail: info@eurocoop.coop
Phone: (32) 2-285-0070
Avenue de Tervueren 12, bte 3
B-1040 Bruxelles
Belgium
Consumer organization that has developed close relations with the European Commission, the European Parliament, and the European Economic and Social Committee; it represents consumer interests in committees that advise these groups. Stresses that most European consumers do not want to

eat genetically modified (GM) foods and expresses concern about keeping organic and conventional farming free of GM contamination.

**European Federation of
Biotechnology (EFB)**
URL: http://www.efb-central.org
E-mail: efb@efb-central.org
Phone: (34) 93-268-7703
Pg. Lluis Companys 23
08010 Barcelona
Spain

Works to increase public and government understanding of biotechnology. Publications include a newsletter and briefing papers on patenting life, biotechnology in foods and drinks, the application of human genetic research, and environmental technology.

**European Food Safety Authority
(EFSA)**
URL: http://www.efsa.europaeu/
EFSA/efsa_locale-
1178620753812_home.htm
E-mail: info@efsa.europa.eu
Phone: (39) 0 521-036111
Largo N. Palli 5/A
I-43100 Parma
Italy

The European Food Safety Authority (EFSA) is a decentralized agency of the European Union. In close collaboration with national authorities, it provides scientific advice on all matters with a direct or indirect impact on food and feed safety, including animal health and welfare and plant protection.

**European Group on Ethics and
Science in New Technologies
(EGE)**
URL: http://ec.europa.eu/
european_group_ethics/index_
en.htm
E-mail: : BEPA-ETHICS-
GROUP@ec.europa.eu
Phone: (32) 2 299-1179
European Commission
Berl 8/143
B-1049 Bruxelles
Belgium

Independent multidisciplinary body that advises the European Commission on ethical aspects of science and new technologies. It has provided opinions on various subjects, including human tissue banking, human embryo research, personal health data in the information society, and human stem cell research.

**European Molecular Biology
Organization (EMBO)**
URL: http://www.embo.org
E-mail: embo@embo.org
Phone: (49) 6221-88910
Postfach 1022.40
D-69012 Heidelberg
Germany

Established in 1964, EMBO promotes molecular biology in Europe and neighboring countries. Publishes the *EMBO Journal* and EMBO Reports. Its Science and Society Committee communicates with the nonscientific community about the effects and benefits of molecular biology. It sponsors research and training through the European

Molecular Biology Laboratory and its outstations.

European Society of Human Genetics (ESHG)
URL: http://www.eshg.org
E-mail: office@eshg.org
Phone: (43) 1-405-138320
c/o Vienna Medical Academy
Alserstrasse 4
1090 Vienna, Austria
Arranges meetings and other scientific gatherings in the field of human genetics and publishes *The European Journal of Human Genetics.*

FasterCures
URL: http://www.fastercures.org/home.php
Phone: (202) 654-7090
509 Seventh Street, NW
Washington, DC 20004
The mission of FasterCures is to identify and implement global solutions to accelerate the process of discovery and clinical development of new therapies for the treatment of deadly and debilitating diseases. Formed under the auspices of the Milken Institute, FasterCures calls itself an "action tank."

Federal Bureau of Investigation, Combined DNA Index System (CODIS) Program
URL: http://www.fbi.gov/hq/lab/codis/index1.htm
Phone: (202) 324-3000
J. Edgar Hoover Building
935 Pennsylvania Avenue, NW
Washington, DC 20535-0001
CODIS, the distributed database and software that coordinates local, state, and national forensic DNA databases, was authorized by the DNA Identification Act of 1994 and went into operation in 1998.

Federation of American Societies for Experimental Biology (FASEB)
URL: http://www.faseb.org
E-mail: webmaster@faseb.org
Phone: (301) 634-7000
9650 Rockville Pike
Bethesda, MD 20814
Coalition of societies that promote advancement in biomedical sciences and provide educational meetings and publications to disseminate biological research results. Advocates for the interests of biomedical and life sciences in matters of public policy.

Food and Agriculture Organization of the United Nations (FAO)
URL: http://www.fao.org
E-mail: FAO-HQ@fao.org
Phone: (39) 065-7051
Viale delle Terme di Caracalla
00153 Rome
Italy
The United Nations's lead agency for agriculture, forestry, fisheries, and rural development, the FAO aims to raise levels of nutrition and standards of living, improve agricultural productivity, and better the condition of rural populations. It

has provided some advice on regulation of genetically modified crops and foods but does not believe that such foods are essential to alleviate world hunger. It has an electronic forum on biotechnology in food and agriculture at http://www.fao.org/biotech/index.asp?lang=en.

Food and Drug Administration (FDA)
URL: http://www.fda.gov
Phone: (888) 463-6332
5600 Fishers Lane
Rockville, MD 20857-0001
U.S. government agency that regulates food, including genetically modified foods, and drugs and medical treatments, including drugs made by biotechnology and gene therapy. Web site has pages on bioengineered food and gene therapy.

Food First/Institute for Food and Development Policy
URL: http://www.foodfirst.org
Phone: (510) 654-4400
398 60th Street
Oakland, CA 94618
This think tank highlights root causes and value-based solutions to world hunger and poverty. Opposes control of agricultural biotechnology and food supply by large multinational corporations.

Food Policy Institute
URL: http://www.foodpolicyinstitute.org
E-mail: info@foodpolicyinstitute.org
Phone: (732) 932-1966

3 Rutgers Plaza
New Brunswick, NJ 08901
The Food Policy Institute is a research unit of Rutgers University that addresses food policy issues. It performs research and provides academic knowledge to public and private decision makers who shape aspects of the food system within which government, agriculture, industry, and the consumer interact.

Foresight Nanotech Institute
URL: http://www.foresight.org
E-mail: foresight@foresight.org
Phone: (650) 289-0860
1455 Adams Drive
Suite 2160
Menlo Park, CA 94025
Founded in 1986, Foresight is a think tank and public interest institute devoted to ensuring the beneficial implementation of nanotechnology. It provides public policy activities, publications, guidelines, networking events, tutorials, conferences, public policy, education, and roadmaps and prizes to guide research.

Foundation on Economic Trends (FOET)
URL: http://www.foet.org
E-mail: office@foet.org
Phone: (301) 656-6272
4520 East West Highway
Suite 600
Bethesda, MD 20814
Headed by Jeremy Rifkin, this non-profit foundation examines emerging trends in science and technology and their impacts on society, culture, the economy, and the environment.

280

Areas of interest include transgenic animals, patents on living things, and human gene alteration. The group urges caution in use of genetic technology.

Friends of the Earth
URL: http://www.foe.org/
E-mail: foe@foe.org
Phone: (877) 843-8687
1717 Massachusetts Avenue, NW
Suite 600
Washington, DC 20036-2002
Part of an international federation of environmental organizations (Friends of the Earth International) that try to work toward sustainable societies, protect the environment (including biological diversity), and promote justice and equal access to resources and opportunities for all the world's people. Among other things, the group fears the effects of multinational corporations' control of genetically modified crops on farmers in developing nations.

Future Generations
URL: http://www.eugenics.net
E-mail: vancourt@comcast.net
Strives to leave a legacy of good health, high intelligence, and noble character to future generations through voluntary "humanitarian eugenics." Web site includes papers explaining and defending the organization's point of view.

Genetic Alliance
URL: http://www.
  geneticalliance.org
E-mail: info@geneticalliance.org

Phone: (202) 966-5557
4301 Connecticut Avenue, NW
Suite 404
Washington, DC 20008-2369
International coalition of consumer, health professional, and patient advocacy organizations working to improve the quality of life of everyone affected by genetics, especially people with genetic conditions and their families.

Genetic Engineering
  Action Network (GEAN)
URL: http://www. geaction.org
E-mail: info@geaction.org
Network of grassroots activists, nongovernmental organizations, farmer groups, academics, and scientists concerned about the risks to the environment, biodiversity, and human health, and the socioeconomic and ethical consequences, of biotechnology and genetic engineering.

Genetic Interest Group (GIG)
URL: http://www.gig.org.uk
E-mail: mail@gig.org.uk
Phone: (44) 0207 704 3141
Unit 4d, Leroy House
436 Essex Road
London N1 3QP
United Kingdom
Alliance of British support groups for people and families with inherited diseases or diseases with a significant genetic component. Among other things, the group works to prevent discrimination based on the misapplication of genetic information, to promote recognition of the health benefits that can result from

genetic research, and to guarantee that such research ultimately benefits individual patients.

Genetic Resources Action International (GRAIN)
URL: http://www.grain.org/front
E-mail: grain@grain.org
Phone: (34) 933-011381
Girona 25, pral.
E-08010 Barcelona
Spain
GRAIN is an international nongovernmental organization that promotes the sustainable management and use of agricultural biodiversity based on people's control over genetic resources and indigenous farmers' knowledge. The group's central concern is genetic erosion and loss of biological diversity, which destroy options for future development and rob people of a key resource base for survival.

Genetics and Public Policy Center
URL: http://www.dnapolicy.org
E-mail: gppcnews@jhu.edu
Phone: (202) 663-5971
1717 Massachusetts Avenue, NW
Suite 530
Washington, DC 20036
The center is funded by the Pew Charitable Trusts and is part of the Berman Bioethics Institute at Johns Hopkins University. It aims to be an independent and objective source of credible information on genetic technologies and policies for the public, media, and policy makers. It focuses on reproductive genetics. It conducts polls, produces reports, sponsors conferences, and performs detailed legal and policy analyses on the subject.

Genetics Society of America (GSA)
URL: http://www.genetics-gsa.org
E-mail: society@genetics-gsa.org
Phone: (301) 634-7300
9650 Rockville Pike
Bethesda, MD 20814-3998
Professional society of scientists and academicians working in the field of genetic studies. Publishes monthly journal, *Genetics*, and educational/career materials.

GeneWatch UK
URL: http://www.genewatch.org
E-mail: mail@genewatch.org
Phone: (44) 01298-24300
60 Lightwood Road
Buxton
Derbyshire SK17 7BB
United Kingdom
Public interest group that aims to ensure genetic technologies are developed and used in ways that promote human health, protect the environment, and respect the rights of humans and interests of animals. It addresses genetically modified (GM) crops and food, GM animals, human genetics, laboratory use, biological weapons, and patenting, focusing on the risks and downsides of genetic technology.

Greenpeace
URL: http://www.
    greenpeace.org/usa
E-mail: info@wdc.greenpeace.
    org
Phone: (800) 326-0959
702 H Street, NW
Washington, DC 20001
Environmental group whose aims include protection of global biodiversity, requirement of careful regulation and labeling of genetically engineered foods, and opposition to patenting of living things. Believes that genetically engineered foods pose environmental and health risks.

Hastings Center
URL: http://www.
    thehastingscenter.org
E-mail: mail@thehastingscenter.
    org
Phone: (845) 424-4040
21 Malcolm Gordon Road
Garrison, NY 10524-4125
The center addresses fundamental ethical issues in health, medicine, and the environment, including issues related to biotechnology and human genetics. Publishes bimonthly journal, *The Hastings Center Report*, a study of ethical issues related to research on human subjects, and other papers.

Human Cloning Foundation
URL: http://www.humancloning.
    org
E-mail: contactus@
    humancloning.org
Stresses the positive aspects of human cloning and promotes education, awareness, and research about human cloning and other biotechnology. Web site offers papers supporting human cloning and lists of books about cloning and genetic engineering.

Human Fertilisation and
    Embryology Authority (HFEA)
URL: http://www.hfea.gov.uk
E-mail: admin@hfea.gov.uk
Phone: (44) 020-7291-8200
21 Bloomsbury Street
London WC1B 3HF
United Kingdom
Governmental body that regulates, licenses, and collects data on fertility treatments and human embryo research in Britain. Provides detailed information and advice to the public on these subjects.

Human Genetics Alert (HGA)
URL: http://www.hgalert.org
E-mail: info@hgalert.org
Phone: (44) 0 207-502-7516
Independent public interest watchdog group funded by a leading British charity. Informs the public and recommends policies about human genetics issues. Does not object to genetic research in general, but does oppose genetic discrimination, human cloning, and germ-line gene modification. Is concerned about "new eugenics," genetic determinism, use of genetic technologies to exacerbate social inequalities, and influence of commercial motives and

sociocultural attitudes on genetic research.

Human Genetics Commission (HGC)
URL: http://www.hgc.gov.uk/
Client/index.asp?ContentId=
E-mail: hgc@dh.gsi.gov.uk
Phone: (44) 020-7972-4351
605, Wellington House
133-155 Waterloo Road
London SE1 8UG
United Kingdom
Advises the British government on ways that new developments in human genetics may affect people and health care. Issues reports and other publications.

Human Genome Organisation (HUGO)
URL: http://www.hugo-
international.org
E-mail: hugo@hugo-
international.org
Phone: (44) 0 20-7249-5167
Unit 16
Hiltongrove N1, 14 Southgate Road
London N1 3LY
United Kingdom
A leading international (chiefly European) professional organization for scientists who study the human genome. Publishes a quarterly newsletter, *Genome Digest.*

The Innocence Project
URL: http://www.
innocenceproject.org
E-mail: info@
innocenceproject.org
Phone: (212) 364-5340
100 Fifth Avenue
3rd Floor
New York, NY 10011
Founded by Barry Scheck and Peter Neufeld at the Cardozo Law School in New York in 1992, the project uses DNA tests and other evidence to show that certain people convicted of crimes are innocent and obtain their release.

Institute for Agriculture and Trade Policy (IATP)
URL: http://www.iatp.org
E-mail: iatp@iatp.org
Phone: (612) 870-0453
2105 First Avenue South
Minneapolis, MN 55404
Nonprofit research and education organization aimed at creating environmentally and economically sustainable communities and regions through sound agriculture and trade policy. Web site contains many essays on farming and trade, including some on genetically modified crops.

Institute of Science in Society
URL: http://www.i-sis.org.uk/
index.php
Phone: (44) 20-8452-2729
P.O. Box 51885
London NW2 9DH
United Kingdom
This organization is devoted to "science, society, and sustainability." Its head is Mae-Wan Ho, a strong opponent of biotechnology and genetically modified crops and foods.

International Center for Technology Assessment (CTA)
URL: http://www.icta.org/ template/index.cfm
E-mail: info@icta.org
Phone: (202) 547-9359
660 Pennsylvania Avenue, SE
Suite 302
Washington, DC 20003
Analyzes impacts of technology on society. Concerns include limiting genetic engineering, halting the patenting of life, and defending the integrity of food. Sponsors the Campaign for Food Safety and Biotechnology Watch Human Applications.

International Centre for Genetic Engineering and Biotechnology (ICGEB)
URL: http://www.icgeb.trieste.it
E-mail: kerbav@icgeb.org
Phone: (39) 040-37571
AREA Science Park
Padriciano 99
34012 Trieste
Italy
Established by the UN Industrial Development Organization in 1995 to provide research, training, and scientific services in the safe use of molecular biology and biotechnology, especially as it applies to the needs of developing nations.

International Cloning Society
URL: http://www.angelfire.com/ la/jfled/ics.html
E-mail: adlafferty@msn.com
Acts as agency to preserve cell/DNA specimens of people who want to be cloned in the future, particularly for purposes of space travel.

International Embryo Transfer Society
URL: http://www.iets.org
E-mail: iets@assochq.org
Phone: (217) 398-4697
1111 North Dunlap Avenue
Savoy, IL 61874
This group focuses on cloning and assisted reproduction for livestock and other animals.

International Food Information Council Foundation (IFIC)
URL: http://ific.org
E-mail: foodinfo@ific.org
Phone: (202) 296-6540
1100 Connecticut Avenue, NW
Suite 430
Washington, DC 20036
Communicates science-based information on food safety and nutrition to health professionals, media, and others providing information to consumers. Supported by the food, beverage, and agricultural industries. Has information on food biotechnology.

International Food Policy Research Institute
URL: http://www.ifpri.org
E-mail: ifpri@cgiar.org
Phone: (202) 862-5600
2033 K Street, NW
Washington, DC 20006-1002
IFPRI's mission is to provide policy solutions that reduce malnutrition and hunger in developing countries, especially for vulnerable groups as

influenced by caste, class, religion, ethnicity, and gender. It carries out research, disseminates the result, and guides decision makers in forming sound and appropriate local, national, and international food policies.

International Service for the Acquisition of Agri-Biotech Applications (ISAAA)
URL: http://www.isaaa.org
E-mail: americenter@isaaa.org
Phone: (607) 255-1724
ISAAA AmeriCenter
417 Bradfield Hall
Cornell University
Ithaca, NY 14853
Nonprofit organization that delivers the benefits of new agricultural biotechnologies to the poor in developing countries. It aims to share these powerful technologies with those who stand to benefit from them and at the same time establish an enabling environment for their safe use. The group has centers in the United States, Kenya, and the Philippines.

International Society for Environmental Biotechnology (ISEB)
URL: http://www.iseb-web.org
E-mail: iseb@cape.uwaterloo.ca
Fax: (519) 746-4979
Department of Chemical Engineering
University of Waterloo
Waterloo, Ontario N2L 3G1
Canada
Interdisciplinary communication network of scientists and others interested in environmental biotechnology, or the development, use, and regulation of biological systems for remediation of contaminated environments and for environment-friendly processes.

International Society for Stem Cell Research
URL: http://www.isscr.org
E-mail: isscr@isscr.org
Phone: (847) 509-1944
60 Revere Drive
Suite 500
Northbrook, IL 60062
The International Society for Stem Cell Research is an independent, nonprofit organization established to promote the dissemination of information relating to stem cells, to encourage this field of research, and to aid professional and public education in all areas of stem cell research and application.

J. Craig Venter Institute
URL: http://www.tigr.org
Phone: (301) 795-7000
9712 Medical Center Drive
Rockville, MD 20850
TIGR is a nonprofit research institute that analyzes the genomes and gene products of a wide variety of organisms, from viruses to humans.

Kennedy Institute of Ethics, Georgetown University
URL: http://kennedyinstitute. georgetown.edu/site/index.htm
Phone: (202) 687-8099
4th Floor, Healy Hall

Georgetown University
Washington, DC 20057
This teaching and research center sponsors research on medical ethics and related policy issues, including issues related to human genetics and gene alteration. Publishes a journal, newsletter, a bibliography, and other materials.

March of Dimes Birth Defects
Federation
URL: http://www.marchofdimes.
com
Phone: (914) 997-4488
1275 Mamaroneck Avenue
White Plains, NY 10605
Works to prevent and treat birth defects, including those caused by genetic mutations. Offers educational material on genetic and other birth defects and reports on such subjects as newborn screening programs.

Nanoethics Group
URL: http://www.nanoethics.org
E-mail: hello@nanoethics.org
A nonpartisan, independent organization that studies the ethical and societal implications of nanotechnology and encourages the industry to talk openly about possible misuses and unintended consequences of nanotechnology so that the critical war for public opinion will not be lost.

National Agricultural
Biotechnology Council
(NABC)
URL: http://nabc.cals.cornell.
edu
E-mail: NABC@cornell.edu
Phone: (607) 254-4856

419 Boyce Thompson Institute
Tower Road
Ithaca, NY 14853
Provides an open forum to discuss issues related to agricultural biotechnology and encourage the field's safe, ethical, efficacious, and equitable development. Composed of major nonprofit agricultural biotechnology research and/or teaching institutions in Canada and the United States. Offers reports of annual meetings, papers, and a newsletter, *NABC News.*

National Center for
Biotechnology Information
(NCBI)
URL: http://www.ncbi.nlm.nih.
gov
E-mail: info@ncbi.nlm.nih.gov
Phone: (301) 496-2475
National Library of Medicine
Building 38A
Bethesda, MD 20894
Offers information about biotechnology, including a newsletter, *NCBI News,* and genetic databases and analysis software.

National Center for Food and
Agricultural Policy
(NCFAP)
URL: http://www.ncfap.org
E-mail: ncfap@ncfap.org
Phone: (202) 328-5048
1616 P Street, NW
1st Floor
Washington, DC 20036
Private nonprofit research organization that conducts studies in

biotechnology, pesticides, international trade and development, and farm and food policy. Although officially a nonadvocacy group, the organization favors biotechnology, and several news stories claim that it is associated with the biotechnology and grocery industries.

National Corn Growers Association (NCGA)
URL: http://www.ncga.com
E-mail: corninfo@ncga.com
Phone: (636) 733-9004
632 Cepi Drive
Chesterfield, MO 63005
Corn growers' trade association. Believes that genetically altered corn holds great promise for corn farmers but would like to see the use of such crops spread more slowly. Warns that acceptance of genetically modified corn as food depends on better methods of informing consumers and on better management and regulation of crops.

National Environmental Trust (NET)
URL: http://www.net.org
E-mail: cdelany@net.org
Phone: (202) 887-8800
1200 18th Street, NW
5th Floor
Washington, DC 20036
Nonprofit, nonpartisan group that aims to educate citizens about environmental problems and how they affect human health and quality of life. It opposes genetically modified foods.

National Family Farm Coalition (NFFC)
URL: http://www.nffc.net
E-mail: nffc@nffc.net
Phone: (202) 543-5675
110 Maryland Avenue, NE
Suite 307
Washington, DC 20002
Links grassroots organizations working on family farm issues. Strongly opposes genetically modified crops. Conducts Farmer to Farmer Campaign on Genetic Engineering, calling for sustainable production and consumers' right to GMO-free food.

National Farmers Union (NFU)
URL: http://www.nfu.org
Phone: (303) 337-5500
5619 DTC Parkway
Suite 300
Greenwood Village, CO 80111-3136
Protects and enhances the economic interests and quality of life of family farmers and ranchers. Believes that farmers must make individual decisions about whether to plant genetically modified crops.

National Human Genome Research Institute (NHGRI)
URL: http://www.genome.gov
Phone: (301) 402-0911
Communications and Public Liaison Branch
National Human Genome Research Institute
National Institutes of Health
Building 31, Room 4B09
31 Center Drive, MSC 2152
9000 Rockville Pike

Bethesda, MD 20892-2152
Part of the National Institutes of Health, the NHGRI heads the Human Genome Project. It offers reports and databases of genetic sequencing and other research, including links to other groups doing genome research. Its Ethical, Legal, and Social Implications (ELSI) Working Group has studied such issues as privacy and fairness in use of genetic information. It offers reports and fact sheets on its work.

National Institutes of Health
  Office of Biotechnology
  Activities (NIH OBA)
URL: http://www4.od.nih.gov/
  oba
E-mail: oba@od.nih.gov
Phone: (301) 496-9838
6705 Rockledge Drive
Suite 750, MSC 7985
Bethesda, MD 20892-7985
U.S. governmental agency that monitors scientific progress in human genetics research in order to anticipate future developments, including ethical, legal, and social concerns; manages the operation of, and provides analytical support to, the NIH Recombinant DNA Advisory Committee (RAC) and two committees that advise the Secretary of Health and Human Services on genetic issues; coordinates and provides liaison between the government and federal and nongovernmental national and international organizations concerned with recombinant DNA, human gene transfer, genetic technologies, and xenotransplantation; provides advice to the NIH director, other federal agencies, and state regulatory organizations concerning genetic technologies; and develops and implements NIH policies and procedures for the safe conduct of recombinant DNA activities and human gene transfer.

National Nanotechnology
  Initiative
URL: http://www.nano.gov
E-mail: info@nnco.nano.gov
A federal research and development program that coordinates the efforts of many agencies in investigating nanoscale science, engineering, and technology. Aims to create products for economic growth, jobs, and other public benefit and to develop educational resources, a skilled workforce, and the supporting infrastructure and tools to advance nanotechnology in a responsible way.

National Newborn Screening
  and Genetics Resource
  Center (NNSGRC)
URL: http://genes-r-us.uthscsa.
  edu
E-mail: therrell@uthscsa.edu
Phone: (512) 454-6419
1912 West Anderson Lane
Suite 210
Austin, TX 78757
Formed by a cooperative agreement between the federal government's Maternal and Child Health Bureau and the University of Texas. Provides a forum for interaction among consumers, health care professionals,

researchers, organizations, and policy makers in refining and developing public health newborn screening and genetics programs. Serves as a national resource center for information and education in the areas of newborn screening and genetics. Stresses states' role in ensuring effective and nondiscriminatory genetic testing of newborns.

National Organization for Rare Disorders (NORD)
URL: http://www.rarediseases. org
E-mail: orphan@rarediseases.org
Phone: (800) 999-6673
55 Kenosia Avenue
P.O. Box 1968
Danbury, CT 06813-1968
A federation of voluntary health organizations dedicated to helping people with rare ("orphan") diseases and to assisting the organizations that serve them. Offers several databases online.

National Society of Genetic Counselors (NSGC)
URL: http://www.nsgc.org
E-mail: nsgc@nsgc.org
Phone: (312) 321-6834
401 North Michigan Avenue
Chicago, IL 60611
Professional organization of genetic counselors. Web site includes consumer information and press releases.

Navdanya (Research Foundation for Science, Technology, and Ecology)
URL: http://www.navdanya.org
E-mail: vshiva@vsnl.com
Phone: (91) 11-26968077
A-60, Hauz Khas
New Delhi 110016
India
Headed by activist Vandana Shiva, this group works to conserve biodiversity, protect the rights of indigenous peoples, and oppose centralized systems of monoculture in agriculture, forestry, and fisheries. Shiva opposes patents on living things and "biopiracy," or exploitation of native plants, animals, and cultural knowledge by international biotechnology corporations.

Nuffield Council on Bioethics
URL: http://www. nuffieldbioethics.org
E-mail: bioethics@ nuffieldbioethics.org
Phone: (44) 020-7681-9619
28 Bedford Square
London WC1B 3JS
United Kingdom
Established by the Trustees of the Nuffield Foundation in 1991 to identify, examine, and report on the ethical questions raised by recent advances in biological and medical research. Publishes papers on such subjects as genetically modified foods, stem cell research, and patents on living things.

Organic Consumers Association (OCA)
URL: http:// www.organicconsumers.org

Phone: (218) 226-4164
6771 South Silver Hill Drive
Finland, MN 55603
Grassroots nonprofit organization that represents the interests of organic consumers in dealing with crucial issues of food safety, industrial agriculture, genetic engineering, corporate accountability, and environmental sustainability. Opposes genetically engineered food, human cloning, and patenting of living things.

Organic Trade Association (OTA)
URL: http://www.ota.com/index.
    html
E-mail: info@ota.com
Phone: (413) 774-7511
P.O. Box 547
Greenfield, MA 01302
Industry association of organic farmers. Calls for a moratorium on growing GMOs and mandatory labeling of existing ones. Opposes Bt crops because they may lead to insect resistance to one of the few pesticides that organic farmers can use.

Organisation for Economic Co-operation and Development (OECD)
URL: http://www.oecd.org/
    home/0,2987,en_2649_
    201185_1_1_1_1_1,00.html
Phone: (33) 1-4524-8200
2, rue André Pascal
F-75775 Paris Cedex 16
France

Includes 30 member countries that share a commitment to democratic government and the market economy. The group has relationships with some 70 other countries, nongovernmental organizations, and civil society. It is well known for its publications and statistics on a variety of economic and social issues, and governments often rely on its advice. Its Internal Co-ordination Group for Biotechnology has considerable information on this subject at http://www.oecd.org/topic/
0,2686,en_2649_37437_1_1_1_1_
37437,00.html.

Personalized Medicine Coalition
URL: http://www.personalized
    medicinecoalition.org
1225 New York Avenue, NW
Suite 450
Washington, DC 20005
An independent, nonprofit group that seeks to advance the understanding and adoption of personalized medicine concepts and products by educating opinion leaders and the public.

Pesticide Action Network of North America (PANNA)
URL: http://www.panna.org
E-mail: panna@panna.org
Phone: (415) 981-1771
49 Powell Street
Suite 500
San Francisco, CA 94102
Works to replace pesticide use with ecologically sound and socially just alternatives. Opposes genetically modified foods.

Pew Initiative on Food and Biotechnology
URL: http://pewagbiotech.org
E-mail: crusten@pewtrusts.org
Aims to be an independent and objective source of credible information about agricultural biotechnology for the public, media, and policy makers. Sponsors workshops and conferences and produces reports in order to encourage debate on the subject. The initiative concluded its work in March 2007, but its publications are archived on its Web site.

President's Council on Bioethics
URL: http://www.bioethics.gov
E-mail: info@bioethics.gov
Phone: (202) 296-4669
1425 New York Avenue, NW
Suite C100
Washington, DC 20005
Seventeen-member council of renowned bioethicists and other scientists that advises the president of the United States on bioethical issues arising from advances in biomedical science and technology. In July 2002 it produced a highly publicized report that evaluated the ethics of human reproductive and research cloning and made recommendations for policy.

Project on Emerging Nanotechnologies
URL: http://www.nanotechproject.org
E-mail: nano@wilsoncenter.org
Phone: (202) 691-4282

Woodrow Wilson International Center for Scholars
Ronald Reagan Building and International Trade Center
One Woodrow Wilson Plaza
1300 Pennsylvania Avenue, NW
Washington, DC 20004-3027
Established in April 2005 as a partnership between the Woodrow Wilson International Center for Scholars and the Pew Charitable Trusts. The project provides independent, objective knowledge and analysis to inform critical decisions affecting the development and commercialization of nanotechnologies. Its goal is to ensure that possible risks of the technologies are minimized and potential benefits are realized.

Soil Association
URL: http://www.soilassociation.org
E-mail: info@soilassociation.org
Phone: (44) 0117-314-5000
South Plaza, Marlborough Street
Bristol BS1 3NX
United Kingdom
Supports organic food and farming and sustainable forestry. Opposes genetically modified crops and claims that farmers lose money on them.

Stem Cell Network
URL: http://www.stemcellnetwork.ca
E-mail: info@stemcellnetwork.ca
Phone: (613) 562-5855

451 Smyth Road
Room 3105
Ottawa, Ontario K1H 8M5
Canada
Brings together more than 70 leading Canadian scientists, clinicians, engineers, and ethicists to investigate the therapeutic potential of stem cells for the treatment of presently incurable diseases.

Third World Network
URL: http://www.twnside.org.sg
E-mail: twnet@po.jaring.my
Phone: (60) 4-2266728
131 Jalan Macalister
10400 Penang
Malaysia
An independent, nonprofit international network of organizations and individuals involved in issues relating to developing countries. Objectives are to conduct research on economic, social, and environmental issues pertaining to developing countries in the Southern Hemisphere; to publish books and magazines; to organize and participate in seminars; and to provide a platform representing broadly southern interests and perspectives at international fora such as UN conferences. Opposes the introduction of genetically modified crops into developing countries.

Trans Atlantic Consumer Dialogue (TACD)
URL: http://www.tacd.org
E-mail: tacd@consint.org
Phone: (44) 207-226-66-63
TACD Secretariat

Consumers International, Office for Developed Economies
24 Highbury Crescent
London N5 1RX
United Kingdom
Forum of U.S. and European Union (EU) consumer organizations that develops joint consumer policy recommendations to the U.S. government and EU to promote consumers' interests in EU and U.S. policy making. In late 2002 it reproached the United States for failing to establish mandatory labeling and safety reviews for GM crops and praised the EU for its new traceability and labeling requirements and for promoting the "precautionary principle" in trade.

Union of Concerned Scientists (UCS)
URL: http://www.ucsusa.org
Phone: (617) 547-5552
Two Brattle Square
Cambridge, MA 02238-9105
Works to improve the environment and protect human health, safety, and quality of life. Urges strong government regulation of agricultural biotechnology.

U.K. Forum for Genetics and Insurance (UKFGI)
URL: http://www.ukfgi.org.uk/corporates.htm
E-mail: ukfgi@actuaries.org.uk
Phone: (44) 020-7632-2177
Staple Inn Hall, High Holborn
London WC1V 7QJ
United Kingdom
Sponsored chiefly by the insurance industry but also includes other

businesses, caring professions, charities, scientific organizations, national and local governments, and consumer groups. Aims to bring together medical and statistical research on the extra risks to people with conditions to which there is a significant genetic predisposition, to consider the value of the results of genetic tests in the assessment of people's insurability, and to encourage further research and discussion in these areas.

United States Department of Agriculture (USDA)
Animal and Plant Health Inspection Service (APHIS)
URL: http://www.aphis.usda.gov/
E-mail: biotechquery@aphis.usda.gov
Phone: (301) 734-5301
USDA, APHIS, BRS
Unit 147
4700 River Road
Riverdale, MD 20737-1236
U.S. government agency that regulates any organisms, including genetically altered organisms, that are or might be plant pests. Offers publications on genetically engineered plants and related subjects.

U.S. Public Interest Research Group (U.S. PIRG)
URL: http://www.uspirg.org

Phone: (617) 747-4370
44 Winter Street
4th Floor
Boston, MA 02108
The state PIRGs created the U.S. PIRG to act as a watchdog for the national public interest, much as the state PIRGs have worked to safeguard the public interest in state capitals. Aims to influence national policy by means of investigative research, media exposés, grassroots organizing, advocacy, and litigation. Opposes agricultural biotechnology.

World Intellectual Property Organization (WIPO)
URL: http://www.wipo.int/portal/index.html.en
Phone: (41) 22-338-91-11
P.O. Box 18
CH-1211 Geneva 20
Switzerland
WIPO is developing global services, units, and initiatives to respond efficiently and rapidly to the existing and emerging needs of its member states, in particular those flowing from the burst of knowledge-based economy. It seeks to provide an integrated approach to protecting intellectual property and promoting creativity, an alternative that will be more favored by developing countries than the WTO/GATT/TRIPS approach.

PART III

APPENDICES

APPENDIX A

BUCK V. BELL,
274 U.S. 200 (1927)

1. The Virginia statute providing for the sexual sterilization of inmates of institutions supported by the State who shall be found to be afflicted with an hereditary form of insanity or imbecility, is within the power of the State under the Fourteenth Amendment. P. 207.

2. Failure to extend the provision to persons outside the institutions named does not render it obnoxious to the Equal Protection Clause. P. 208.

143 Va. 310, Affirmed.

Opinions

Error to a judgment of the Supreme Court of Appeals of the State of Virginia which affirmed a judgment ordering the Superintendent of the State Colony of Epileptics and Feeble Minded to perform the operation of salpingectomy on Carrie Buck, the plaintiff in error.

Holmes, J., Opinion of the Court

Mr. JUSTICE HOLMES delivered the opinion of the Court.

This is a writ of error to review a judgment of the Supreme Court of Appeals of the State of Virginia affirming a judgment of the Circuit Court of Amherst County by which the defendant in error, the superintendent of the State Colony for Epileptics and Feeble Minded, was ordered to perform the operation of salpingectomy upon Carrie Buck, the plaintiff in error, for the purpose of making her sterile. [143 Va. 310.] The case comes here upon the contention that the statute authorizing the judgment is void under the Fourteenth Amendment as denying to the plaintiff in error due process of law and the equal protection of the laws.

Biotechnology and Genetic Engineering

Carrie Buck is a feeble minded white woman who was committed to the State Colony above mentioned in due form. She is the daughter of a feeble minded mother in the same institution, and the mother of an illegitimate feeble minded child. She was eighteen years old at the time of the trial of her case in the Circuit Court, in the latter part of 1924. An Act of Virginia, approved March 20, 1924, recites that the health of the patient and the welfare of society may be promoted in certain cases by the sterilization of mental defectives, under careful safeguard, &c.; that the sterilization may be effected in males by vasectomy and in females by salpingectomy, without serious pain or substantial danger to life; that the Commonwealth is supporting in various institutions many defective persons who, if now discharged, would become a menace, but, if incapable of procreating, might be discharged with safety and become self-supporting with benefit to themselves and to society, and that experience has shown that heredity plays an important part in the transmission of insanity, imbecility, &c. The statute then enacts that, whenever the superintendent of certain institutions, including the above-named State Colony, shall be of opinion that it is for the best interests of the patients and of society that an inmate under his care should be sexually sterilized, he may have the operation performed upon any patient afflicted with hereditary forms of insanity, imbecility, &c., on complying with the very careful provisions by which the act protects the patients from possible abuse.

The superintendent first presents a petition to the special board of directors of his hospital or colony, stating the facts and the grounds for his opinion, verified by affidavit. Notice of the petition and of the time and place of the hearing in the institution is to be served upon the inmate, and also upon his guardian, and if there is no guardian, the superintendent is to apply to the Circuit Court of the County to appoint one. If the inmate is a minor, notice also is to be given to his parents, if any, with a copy of the petition. The board is to see to it that the inmate may attend the hearings if desired by him or his guardian. The evidence is all to be reduced to writing, and, after the board has made its order for or against the operation, the superintendent, or the inmate, or his guardian, may appeal to the Circuit Court of the County. The Circuit Court may consider the record of the board and the evidence before it and such other admissible evidence as may be offered, and may affirm, revise, or reverse the order of the board and enter such order as it deems just. Finally any party may apply to the Supreme Court of Appeals, which, if it grants the appeal, is to hear the case upon the record of the trial in the Circuit Court, and may enter such order as it thinks the Circuit Court should have entered. There can be no doubt that, so far as procedure is concerned, the rights of the patient are most carefully considered, and, as every step in this case was taken in scrupulous compliance with the statute and after months of observation, there is no doubt that, in that respect, the plaintiff in error has had due process of law.

The attack is not upon the procedure, but upon the substantive law. It seems to be contended that in no circumstances could such an order be justified. It certainly is contended that the order cannot be justified upon the existing grounds. The judgment finds the facts that have been recited, and that Carrie Buck is the probable potential parent of socially inadequate offspring, likewise afflicted, that she may be sexually sterilized without detriment to her general health, and that her welfare and that of society will be promoted by her sterilization, and thereupon makes the order. In view of the general declarations of the legislature and the specific findings of the Court, obviously we cannot say as matter of law that the grounds do not exist, and, if they exist, they justify the result. We have seen more than once that the public welfare may call upon the best citizens for their lives. It would be strange if it could not call upon those who already sap the strength of the State for these lesser sacrifices, often not felt to be such by those concerned, in order to prevent our being swamped with incompetence. It is better for all the world if, instead of waiting to execute degenerate offspring for crime or to let them starve for their imbecility, society can prevent those who are manifestly unfit from continuing their kind. The principle that sustains compulsory vaccination is broad enough to cover cutting the Fallopian tubes. *Jacobson v. Massachusetts*, 197 U.S. 11. Three generations of imbeciles are enough.

But, it is said, however it might be if this reasoning were applied generally, it fails when it is confined to the small number who are in the institutions named and is not applied to the multitudes outside. It is the usual last resort of constitutional arguments to point out shortcomings of this sort. But the answer is that the law does all that is needed when it does all that it can, indicates a policy, applies it to all within the lines, and seeks to bring within the lines all similarly situated so far and so fast as its means allow. Of course, so far as the operations enable those who otherwise must be kept confined to be returned to the world, and thus open the asylum to others, the equality aimed at will be more nearly reached.

APPENDIX B

DIAMOND V. CHAKRABARTY, 447 U.S. 303 (1980)

DIAMOND, COMMISSIONER OF PATENTS AND TRADEMARKS V. CHAKRABARTY CERTIORARI TO THE UNITED STATES COURT OF CUSTOMS AND PATENT APPEALS. NO. 79-136

Argued March 17, 1980
Decided June 16, 1980

Title 35 U.S.C. 101 provides for the issuance of a patent to a person who invents or discovers "any" new and useful "manufacture" or "composition of matter." Respondent filed a patent application relating to his invention of a human-made, genetically engineered bacterium capable of breaking down crude oil, a property which is possessed by no naturally occurring bacteria. A patent examiner's rejection of the patent application's claims for the new bacteria was affirmed by the Patent Office Board of Appeals on the ground that living things are not patentable subject matter under 101. The Court of Customs and Patent Appeals reversed, concluding that the fact that micro-organisms are alive is without legal significance for purposes of the patent law.

Held:

1. A live, human-made micro-organism is patentable subject matter under 101. Respondent's micro-organism constitutes a "manufacture" or "composition of matter" within that statute. Pp. 308–318.

(a) In choosing such expansive terms as "manufacture" and "composition of matter," modified by the comprehensive "any," Congress contemplated that the patent laws should be given wide scope, and the relevant legislative history also supports a broad construction. While laws of nature, physical phenomena, and abstract ideas are not patentable, respondent's claim is not to a hitherto unknown natural phenomenon, but to a nonnaturally occurring manufacture or composition of matter—a product of human ingenuity "having a distinctive name, character [and] use." *Hartranft v. Wiegmann*, 121 U.S. 609, 615. *Funk Brothers Seed Co. v. Kalo Inoculant Co.*, 333 U.S. 127, distinguished. Pp. 308–310.

(b) The passage of the 1930 Plant Patent Act, which afforded patent protection to certain asexually reproduced plants, and the 1970 Plant Variety Protection Act, which authorized protection for certain sexually reproduced plants but excluded bacteria from its protection, does not evidence congressional understanding that the terms "manufacture" or "composition of matter" in 101 do not include living things. Pp. 310–314. [447 U.S. 303, 304].

(c) Nor does the fact that genetic technology was unforeseen when Congress enacted 101 require the conclusion that micro-organisms cannot qualify as patentable subject matter until Congress expressly authorizes such protection. The unambiguous language of 101 fairly embraces respondent's invention. Arguments against patentability under 101, based on potential hazards that may be generated by genetic research, should be addressed to the Congress and the Executive, not to the Judiciary. Pp. 314–318.

596 F.2d 952, affirmed.

BURGER, C. J., delivered the opinion of the Court, in which STEWART, BLACKMUN, REHNQUIST, and STEVENS, JJ., joined. BRENNAN, J., filed a dissenting opinion, in which WHITE, MARSHALL, and POWELL, JJ., joined, post, p. 318.

Deputy Solicitor General Wallace argued the cause for petitioner. With him on the brief were Solicitor General McCree, Assistant Attorney General Shenefield, Harriet S. Shapiro, Robert B. Nicholson, Frederic Freilicher, and Joseph F. Nakamura.

Edward F. McKie, Jr., argued the cause for respondent. With him on the brief were Leo I. MaLossi, William E. Schuyler, Jr., and Dale H. Hoscheit.

MR. CHIEF JUSTICE BURGER delivered the opinion of the Court.

We granted certiorari to determine whether a live, human-made microorganism is patentable subject matter under 35 U.S.C. 101.

I

In 1972, respondent Chakrabarty, a microbiologist, filed a patent application, assigned to the General Electric Co. The application asserted 36 claims related to Chakrabarty's invention of "a bacterium from the genus *Pseudomonas* containing therein at least two stable energy-generating plasmids, each of said plasmids providing a separate hydrocarbon degradative pathway." This human-made, genetically engineered bacterium is capable of breaking down multiple components of crude oil. Because of this property, which is possessed by no naturally occurring bacteria, Chakrabarty's invention is believed to have significant value for the treatment of oil spills.

Chakrabarty's patent claims were of three types: first, process claims for the method of producing the bacteria; [447 U.S. 303, 306] second, claims for an inoculum comprised of a carrier material floating on water, such as straw, and the new bacteria; and third, claims to the bacteria themselves. The patent examiner allowed the claims falling into the first two categories, but rejected claims for the bacteria. His decision rested on two grounds: (1) that micro-organisms are "products of nature," and (2) that as living things they are not patentable subject matter under 35 U.S.C. 101.

Chakrabarty appealed the rejection of these claims to the Patent Office Board of Appeals, and the Board affirmed the examiner on the second ground. Relying on the legislative history of the 1930 Plant Patent Act, in which Congress extended patent protection to certain asexually reproduced plants, the Board concluded that 101 was not intended to cover living things such as these laboratory created micro-organisms.

The Court of Customs and Patent Appeals, by a divided vote, reversed on the authority of its prior decision in *In re Bergy*, 563 F.2d 1031, 1038 (1977), which held that "the fact that microorganisms . . . are alive . . . [is] without legal significance" for purposes of the patent law. Subsequently, we granted the Acting Commissioner of Patents and Trademarks' petition for certiorari in *Bergy*, vacated the judgment, and remanded the case "for further consideration in light of *Parker v. Flook*, 437 U.S. 584 (1978)" 438 U.S. 902 (1978). The Court of Customs and Patent Appeals then vacated its judgment in *Chakrabarty* and consolidated the case with *Bergy* for reconsideration. After re-examining both cases in the light of our holding in *Flook*, that court, with one dissent, reaffirmed its earlier judgments 596 F.2d 952 (1979). [447 U.S. 303, 307].

The Commissioner of Patents and Trademarks again sought certiorari, and we granted the writ as to both *Bergy* and *Chakrabarty*. 444 U.S. 924 (1979). Since then, *Bergy* has been dismissed as moot, 444 U.S. 1028 (1980), leaving only *Chakrabarty* for decision.

II

The Constitution grants Congress broad power to legislate to "promote the Progress of Science and useful Arts, by securing for limited Times to Authors and Inventors the exclusive Right to their respective Writings and Discoveries." Art. I, 8, cl. 8. The patent laws promote this progress by offering inventors exclusive rights for a limited period as an incentive for their inventiveness and research efforts. *Kewanee Oil Co. v. Bicron Corp.*, 416 U.S. 470, 480–481 (1974); *Universal Oil Co. v. Globe Co.*, 322 U.S. 471, 484 (1944). The authority of Congress is exercised in the hope that "[t]he productive effort thereby fostered will have a positive effect on society through the introduction of new products and processes of manufacture into the economy, and the emanations by way of increased employment and better lives for our citizens." *Kewanee*, supra, at 480.

The question before us in this case is a narrow one of statutory interpretation requiring us to construe 35 U.S.C. 101, which provides:

"Whoever invents or discovers any new and useful process, machine, manufacture, or composition of matter, or any new and useful improvement thereof, may obtain a patent therefor, subject to the conditions and requirements of this title."

Specifically, we must determine whether respondent's micro-organism constitutes a "manufacture" or "composition of matter" within the meaning of the statute [447 U.S. 303, 308].

III

In cases of statutory construction we begin, of course, with the language of the statute. *Southeastern Community College v. Davis*, 442 U.S. 397, 405 (1979). And "unless otherwise defined, words will be interpreted as taking their ordinary, contemporary, common meaning." *Perrin v. United States*, 444 U.S. 37, 42 (1979). We have also cautioned that courts "should not read into the patent laws limitations and conditions which the legislature has not expressed." *United States v. Dubilier Condenser Corp.*, 289 U.S. 178, 199 (1933).

Guided by these canons of construction, this Court has read the term "manufacture" in 101 in accordance with its dictionary definition to mean "the production of articles for use from raw or prepared materials by giving to these materials new forms, qualities, properties, or combinations, whether by hand-labor or by machinery." *American Fruit Growers, Inc. v. Brogdex Co.*, 283 U.S. 1, 11 (1931). Similarly, "composition of matter" has been construed consistent with its common usage to include "all compositions of two or more substances and . . . all composite articles, whether they be the results of chemical union, or of mechanical mixture, or whether they be gases, fluids, powders or solids." *Shell Development Co. v. Watson*, 149 F. Supp. 279, 280 (DC 1957) (citing 1 A. Deller, *Walker on Patents* 14, p. 55 (1st ed. 1937)). In choosing such expansive terms as "manufacture" and "composition of matter," modified by the comprehensive "any," Congress plainly contemplated that the patent laws would be given wide scope.

The relevant legislative history also supports a broad construction. The Patent Act of 1793, authored by Thomas Jefferson, defined statutory subject matter as "any new and useful art, machine, manufacture, or composition of matter, or any new or useful improvement [thereof]." Act of Feb. 21, 1793, 1, 1 Stat. 319. The Act embodied Jefferson's philosophy that "ingenuity should receive a liberal encouragement." [447 U.S. 303, 309] (*Writings of Thomas Jefferson* 75–76) (Washington ed. 1871). See *Graham v. John Deere Co.*, 383 U.S. 1, 7–10 (1966). Subsequent patent statutes in 1836, 1870, and 1874 employed this same broad language. In 1952, when the patent laws were recodified, Congress replaced the word "art" with "process," but otherwise left Jefferson's language intact. The Committee Reports accompanying the 1952 Act inform us that Congress intended statutory subject matter to "include anything under the sun that is made by man." S. Rep. No. 1979, 82d Cong., 2d Sess., 5 (1952); H. R. Rep. No. 1923, 82d Cong., 2d Sess., 6 (1952).

This is not to suggest that 101 has no limits or that it embraces every discovery. The laws of nature, physical phenomena, and abstract ideas have been held not patentable. See *Parker v. Flook*, 437 U.S. 584 (1978); *Gottschalk v. Benson*, 409 U.S. 63, 67 (1972); *Funk Brothers Seed Co. v. Kalo Inoculant Co.*, 333 U.S. 127, 130 (1948); *O'Reilly v. Morse*, 15 How. 62, 112–121 (1854); *Le Roy v. Tatham*, 14 How. 156, 175 (1853). Thus, a new mineral discovered in the earth or a new plant found in the wild is not patentable subject matter. Likewise, Einstein could not patent his celebrated law that $E=mc^2$.; nor could Newton have patented the law of gravity. Such discoveries are "manifestations of . . . nature, free to all men and reserved exclusively to none." *Funk*, supra, at 130.

Judged in this light, respondent's micro-organism plainly qualifies as patentable subject matter. His claim is not to a hitherto unknown natural phenomenon, but to a nonnaturally occurring manufacture or composition of matter—a product of human ingenuity "having a distinctive name, character [and] [447 U.S. 303, 310] use" *Hartranft v. Wiegmann*, 121 U.S. 609, 615 (1887). The point is underscored dramatically by comparison of the invention here with that in *Funk*. There, the patentee had discovered that there existed in nature certain species of root-nodule bacteria which did not exert a mutually inhibitive effect on each other. He used that discovery to produce a mixed culture capable of inoculating the seeds of leguminous plants. Concluding that the patentee had discovered "only some of the handiwork of nature," the Court ruled the product nonpatentable:

"Each of the species of root-nodule bacteria contained in the package infects the same group of leguminous plants which it always infected. No species acquires a different use. The combination of species produces no new bacteria, no change in the six species of bacteria, and no enlargement of the range of their utility. Each species has the same effect it always had. The bacteria perform in their natural way. Their use in combination does not improve in any way their natural functioning. They serve the ends nature originally provided and act quite independently of any effort of the patentee" [333 U.S., at 131].

Here, by contrast, the patentee has produced a new bacterium with markedly different characteristics from any found in nature and one having the potential for significant utility. His discovery is not nature's handiwork, but his own; accordingly it is patentable subject matter under 101.

IV

Two contrary arguments are advanced, neither of which we find persuasive.

(A)

The petitioner's first argument rests on the enactment of the 1930 Plant Patent Act, which afforded patent protection to certain asexually reproduced plants, and the 1970 Plant [447 U.S. 303, 311] Variety Protection Act, which authorized protection for certain sexually reproduced plants but excluded bacteria from its protection. In the petitioner's view, the passage of these Acts evidences congressional understanding that the terms "manufacture" or "composition of matter" do not include living things; if they did, the petitioner argues, neither Act would have been necessary.

We reject this argument. Prior to 1930, two factors were thought to remove plants from patent protection. The first was the belief that plants, even those artificially bred, were products of nature for purposes of the patent law. This position appears to have derived from the decision of the Patent Office in *Ex parte* Latimer, 1889 Dec. Com. Pat. 123, in which a patent claim for fiber found in the needle of the *Pinus australis* was rejected. The Commissioner reasoned that a contrary result would permit "patents [to] be obtained upon the trees of the forest and the plants of the earth, which of course would be unreasonable and impossible" Id., at 126. The Latimer case, it seems, came to "se[t] forth the general stand taken in these matters" that plants were natural products not subject to patent protection. Thorne, *Relation of Patent Law to Natural Products*, 6 J. Pat. Off. Soc. 23, 24 [447 U.S. 303, 312] (1923). The second obstacle to patent protection for plants was the fact that plants were thought not amenable to the "written description" requirement of the patent law. See 35 U.S.C. 112. Because new plants may differ from old only in color or perfume, differentiation by written description was often impossible. See Hearings on H. R. 11372 before the House Committee on Patents, 71st Cong., 2d Sess., 7 (1930) (memorandum of Patent Commissioner Robertson).

In enacting the Plant Patent Act, Congress addressed both of these concerns. It explained at length its belief that the work of the plant breeder "in aid of nature" was patentable invention. S. Rep. No. 315, 71st Cong., 2d Sess., 6–8 (1930); H. R. Rep. No. 1129, 71st Cong., 2d Sess., 7–9 (1930). And it relaxed the written description requirement in favor of "a description

. . . as complete as is reasonably possible" 35 U.S.C. 162. No Committee or Member of Congress, however, expressed the broader view, now urged by the petitioner, that the terms "manufacture" or "composition of matter" exclude living things. The sole support for that position in the legislative history of the 1930 Act is found in the conclusory statement of Secretary of Agriculture Hyde, in a letter to the Chairmen of the House and Senate Committees considering the 1930 Act, that "the patent laws . . . at the present time are understood to cover only inventions or discoveries in the field of inanimate nature." See S. Rep. No. 315, supra, at Appendix A; H. R. Rep. No. 1129, supra, at Appendix A. Secretary Hyde's opinion, however, is not entitled to controlling weight. His views were solicited on the administration of the new law and not on the scope of patentable [447 U.S. 303, 313] subject matter—an area beyond his competence. Moreover, there is language in the House and Senate Committee Reports suggesting that to the extent Congress considered the matter it found the Secretary's dichotomy unpersuasive. The Reports observe:

"There is a clear and logical distinction between the discovery of a new variety of plant and of certain inanimate things, such, for example, as a new and useful natural mineral. The mineral is created wholly by nature unassisted by man. . . . On the other hand, a plant discovery resulting from cultivation is unique, isolated, and is not repeated by nature, nor can it be reproduced by nature unaided by man. . . ." S. Rep. No. 315, supra, at 6; H. R. Rep. No. 1129, supra, at 7.

Congress thus recognized that the relevant distinction was not between living and inanimate things, but between products of nature, whether living or not, and human-made inventions. Here, respondent's micro-organism is the result of human ingenuity and research. Hence, the passage of the Plant Patent Act affords the Government no support.

Nor does the passage of the 1970 Plant Variety Protection Act support the Government's position. As the Government acknowledges, sexually reproduced plants were not included under the 1930 Act because new varieties could not be reproduced true-to-type through seedlings. Brief for Petitioner 27, n. 31. By 1970, however, it was generally recognized that true-to-type reproduction was possible and that plant patent protection was therefore appropriate. The 1970 Act extended that protection. There is nothing in its language or history to suggest that it was enacted because 101 did not include living things.

In particular, we find nothing in the exclusion of bacteria from plant variety protection to support the petitioner's position. See n. 7, supra. The legislative history gives no reason for this exclusion. As the Court of Customs and [447 U.S. 303, 314] Patent Appeals suggested, it may simply reflect congressional agreement with the result reached by that court in deciding *In re Arzberger*, 27 C. C. P. A. (Pat.) 1315, 112 F.2d 834 (1940), which held that bacteria were not plants for the purposes of the 1930 Act. Or it may reflect the fact that prior to 1970 the Patent Office had issued patents for bacteria under 101. In any event, absent some clear indication that Congress "focused on [the] issues . . . directly related to the one presently before the Court," *SEC v. Sloan,* 436 U.S. 103, 120–121 (1978), there is no basis for reading into its actions an intent to modify the plain meaning of the words found in 101. See *TVA v. Hill,* 437 U.S. 153, 189–193 (1978); *United States v. Price,* 361 U.S. 304, 313 (1960).

(B)

The petitioner's second argument is that micro-organisms cannot qualify as patentable subject matter until Congress expressly authorizes such protection. His position rests on the fact that genetic technology was unforeseen when Congress enacted 101. From this it is argued that resolution of the patentability of inventions such as respondent's should be left to Congress. The legislative process, the petitioner argues, is best equipped to weigh the competing economic, social, and scientific considerations involved, and to determine whether living organisms produced by genetic engineering should receive patent protection. In support of this position, the petitioner relies on our recent holding in *Parker v. Flook,* 437 U.S. 584 (1978), and the statement that the judiciary "must proceed cautiously when . . . asked to extend [447 U.S. 303, 315] patent rights into areas wholly unforeseen by Congress." [Id., at 596].

It is, of course, correct that Congress, not the courts, must define the limits of patentability; but it is equally true that once Congress has spoken it is "the province and duty of the judicial department to say what the law is." *Marbury v. Madison,* 1 Cranch 137, 177 (1803). Congress has performed its constitutional role in defining patentable subject matter in 101; we perform ours in construing the language Congress has employed. In so doing, our obligation is to take statutes as we find them, guided, if ambiguity appears, by the legislative history and statutory purpose. Here, we perceive no ambiguity. The subject-matter provisions of the patent law have been cast in broad terms to fulfill the constitutional and statutory goal of promoting "the Progress of Science and the useful Arts" with all that means for the social and economic benefits envisioned by Jefferson. Broad general language is not necessarily ambiguous when congressional objectives require broad terms.

Nothing in *Flook* is to the contrary. That case applied our prior precedents to determine that a "claim for an improved method of calculation, even when tied to a specific end use, is unpatentable subject matter under 101." [437 U.S., at 595, n. 18]. The Court carefully scrutinized the claim at issue to determine whether it was precluded from patent protection under "the principles underlying the prohibition against patents for 'ideas' or phenomena of nature." [Id., at 593]. We have done that here. *Flook* did not announce a new principle that inventions in areas not contemplated by Congress when the patent laws were enacted are unpatentable *per se*.

To read that concept into *Flook* would frustrate the purposes of the patent law. This Court frequently has observed that a statute is not to be confined to the "particular application[s] . . . contemplated by the legislators." *Barr v. United States*, 324 U.S. 83, 90 (1945). Accord, *Browder v. United States*, 312 U.S. 335, 339 (1941); *Puerto Rico v. Shell Co.*, [447 U.S. 303, 316] 302 U.S. 253, 257 (1937). This is especially true in the field of patent law. A rule that unanticipated inventions are without protection would conflict with the core concept of the patent law that anticipation undermines patentability. See *Graham v. John Deere Co.*, 383 U.S., at 12–17. Mr. Justice Douglas reminded that the inventions most benefiting mankind are those that "push back the frontiers of chemistry, physics, and the like." *Great A. & P. Tea Co. v. Supermarket Corp.*, 340 U.S. 147, 154 (1950) (concurring opinion). Congress employed broad general language in drafting 101 precisely because such inventions are often unforeseeable.

To buttress his argument, the petitioner, with the support of amicus, points to grave risks that may be generated by research endeavors such as respondent's. The briefs present a gruesome parade of horribles. Scientists, among them Nobel laureates, are quoted suggesting that genetic research may pose a serious threat to the human race, or, at the very least, that the dangers are far too substantial to permit such research to proceed apace at this time. We are told that genetic research and related technological developments may spread pollution and disease, that it may result in a loss of genetic diversity, and that its practice may tend to depreciate the value of human life. These arguments are forcefully, even passionately, presented; they remind us that, at times, human ingenuity seems unable to control fully the forces it creates—that, with Hamlet, it is sometimes better "to bear those ills we have than fly to others that we know not of."

It is argued that this Court should weigh these potential hazards in considering whether respondent's invention is [447 U.S. 303, 317] patentable subject matter under 101. We disagree. The grant or denial of patents on micro-organisms is not likely to put an end to genetic research or to its attendant risks. The large amount of research that has already occurred when no researcher had sure knowledge that patent protection would be available suggests that legislative or judicial fiat as to patentability will not deter the scientific mind from probing into the unknown any more than Canute could command the tides. Whether respondent's claims are patentable may determine whether research efforts are accelerated by the hope of reward or slowed by want of incentives, but that is all.

What is more important is that we are without competence to entertain these arguments—either to brush them aside as fantasies generated by fear of the unknown, or to act on them. The choice we are urged to make is a matter of high policy for resolution within the legislative process after the kind of investigation, examination, and study that legislative bodies can provide and courts cannot. That process involves the balancing of competing values and interests, which in our democratic system is the business of elected representatives. Whatever their validity, the contentions now pressed on us should be addressed to the political branches of the Government, the Congress and the Executive, and not to the courts. [447 U.S. 303, 318]

We have emphasized in the recent past that "[o]ur individual appraisal of the wisdom or unwisdom of a particular [legislative] course . . . is to be put aside in the process of interpreting a statute." *TVA v. Hill*, 437 U.S., at 194. Our task, rather, is the narrow one of determining what Congress meant by the words it used in the statute; once that is done our powers are exhausted. Congress is free to amend 101 so as to exclude from patent protection organisms produced by genetic engineering. Cf. 42 U.S.C. 2181 (a), exempting from patent protection inventions "useful solely in the utilization of special nuclear material or atomic energy in an atomic weapon." Or it may choose to craft a statute specifically designed for such living things. But, until Congress takes such action, this Court must construe the language of 101 as it is. The language of that section fairly embraces respondent's invention.

Accordingly, the judgment of the Court of Customs and Patent Appeals is

Affirmed. . . .

MR. JUSTICE BRENNAN, with whom MR. JUSTICE WHITE, MR. JUSTICE MARSHALL, and MR. JUSTICE POWELL join, dissenting.

I agree with the Court that the question before us is a narrow one. Neither the future of scientific research, nor even the ability of respondent Chakrabarty to reap some monopoly profits from his pioneering work, is at stake. Patents on the processes by which he has produced and employed the new living organism are not contested. The only question we need decide is whether Congress, exercising its authority under Art. I, 8, of the Constitution, intended that he be able to secure a monopoly on the living organism itself, no matter how produced or how used. Because I believe the Court has misread the applicable legislation, I dissent. [447 U.S. 303, 319]

The patent laws attempt to reconcile this Nation's deep-seated antipathy to monopolies with the need to encourage progress. *Deepsouth Packing Co. v. Laitram Corp.*, 406 U.S. 518, 530–531 (1972); *Graham v. John Deere Co.*, 383 U.S. 1, 7–10 (1966). Given the complexity and legislative nature of this delicate task, we must be careful to extend patent protection no further than Congress has provided. In particular, were there an absence of legislative direction, the courts should leave to Congress the decisions whether and how far to extend the patent privilege into areas where the common understanding has been that patents are not available. Cf. *Deepsouth Packing Co. v. Laitram Corp.*, supra.

In this case, however, we do not confront a complete legislative vacuum. The sweeping language of the Patent Act of 1793, as re-enacted in 1952, is not the last pronouncement Congress has made in this area. In 1930 Congress enacted the Plant Patent Act affording patent protection to developers of certain asexually reproduced plants. In 1970 Congress enacted the Plant Variety Protection Act to extend protection to certain new plant varieties capable of sexual reproduction. Thus, we are not dealing—as the Court would have it—with the routine problem of "unanticipated inventions." Ante, at 316. In these two Acts Congress has addressed the general problem of patenting animate inventions and has chosen carefully limited language granting protection to some kinds of discoveries, but specifically excluding others. These Acts strongly evidence a congressional limitation that excludes bacteria from patentability. [447 U.S. 303, 320]

First, the Acts evidence Congress' understanding, at least since 1930, that 101 does not include living organisms. If newly developed living organisms not naturally occurring had been patentable under 101, the plants included in the scope of the 1930 and 1970 Acts could have been patented without new legislation. Those plants, like the bacteria involved in this case, were new varieties not naturally occurring. Although the Court, ante, at 311, rejects this line of argument, it does not explain why the Acts were necessary unless to correct a pre-existing situation. I cannot share the Court's implicit assumption that Congress was engaged in either idle exercises or mere correction of the public record when it enacted the 1930 and 1970 Acts. And Congress certainly thought it was doing something significant. The Committee Reports contain expansive prose about the previously unavailable benefits to be derived from extending patent protection to plants. H. R. [447 U.S. 303, 321] Rep. No. 91-1605, pp. 1–3 (1970); S. Rep. No. 315, 71st Cong., 2d Sess., 1–3 (1930). Because Congress thought it had to legislate in order to make agricultural "human-made inventions" patentable and because the legislation Congress enacted is limited, it follows that Congress never meant to make items outside the scope of the legislation patentable.

Second, the 1970 Act clearly indicates that Congress has included bacteria within the focus of its legislative concern, but not within the scope of patent protection. Congress specifically excluded bacteria from the coverage of the 1970 Act. 7 U.S.C. 2402 (a). The Court's attempts to supply explanations for this explicit exclusion ring hollow. It is true that there is not mention in the legislative history of the exclusion, but that does not give us license to invent reasons. The fact is that Congress, assuming that animate objects as to which it had not specifically legislated could not be patented, excluded bacteria from the set of patentable organisms.

The Court protests that its holding today is dictated by the broad language of 101, which cannot "be confined to the 'particular application[s] . . . contemplated by the legislators.'" Ante, at 315, quoting *Barr v. United States*, 324 U.S. 83, 90 (1945). But as I have shown, the Court's decision does not follow the unavoidable implications of the statute. Rather, it extends the patent system to cover living material [447 U.S. 303, 322] even though Congress plainly has legislated in the belief that 101 does not encompass living organisms. It is the role of Congress, not this Court, to broaden or narrow the reach of the patent laws. This is especially true where, as here, the composition sought to be patented uniquely implicates matters of public concern.

[footnotes omitted]

APPENDIX C

NORMAN-BLOODSAW V. LAWRENCE BERKELEY LABORATORY, 135 F.3D 1260 (1998)

Note: This case was in the United States Court of Appeals for the Ninth Circuit.

MARYA S. NORMAN-BLOODSAW; EULALIO R. FUENTES; VERTIS B. ELLIS; MARK E. COVINGTON; JOHN D. RANDOLPH; ADRIENNE L. GARCIA; and BRENDOLYN B. SMITH, Plaintiffs-Appellants, v. No. 96-16526 LAWRENCE BERKELEY LABORATORY; CHARLES V. SHANK, Director of D.C. No. Lawrence Berkeley Laboratory; CV-95-03220-VRW HENRY H. STAUFFER, M.D.; LISA SNOW, M.D.; T. F. BUDINGER, M.D.; WILLIAM G. DONALD, JR., M.D.; FEDERICO PENA, Secretary of the Department of Energy;* and THE REGENTS OF THE UNIVERSITY OF CALIFORNIA, a non-profit public corporation, Defendants-Appellees. Appeal from the United States District Court for the Northern District of California Vaughn R. Walker, District Judge, Presiding Argued and Submitted June 10, 1997—San Francisco, California Filed February 3, 1998

Before: Stephen Reinhardt, Thomas G. Nelson, and Michael Daly Hawkins, Circuit Judges. Opinion by Judge Reinhardt.

This appeal involves the question whether a clerical or administrative worker who undergoes a general employee health examination may, with-

---

*Federico Pena has been substituted for his predecessor in office, Hazel O'Leary, pursuant to Fed. R. App. P. 43(c)(1). 1149

out his knowledge, be tested for highly private and sensitive medical and genetic information such as syphilis, sickle-cell trait, and pregnancy.

Lawrence Berkeley Laboratory is a research institution jointly operated by state and federal agencies. Plaintiffs-appellants, present and former employees of Lawrence, allege that in the course of their mandatory employment entrance examinations and on subsequent occasions, Lawrence, without their knowledge or consent, tested their blood and urine for intimate medical conditions—namely, syphilis, sickle-cell trait, and pregnancy. Their complaint asserts that this testing violated Title VII of the Civil Rights Act of 1964, the Americans with Disabilities Act (ADA), and their right to privacy as guaranteed by both the United States and State of California Constitutions. The district court granted the defendants-appellees' motions for dismissal, judgment on the pleadings, and summary judgment on all of plaintiffs-appellants' claims. We affirm as to the ADA claims, but reverse as to the Title VII and state and federal privacy claims.

BACKGROUND

Plaintiffs Marya S. Norman-Bloodsaw, Eulalio R. Fuentes, Vertis B. Ellis, Mark E. Covington, John D. Randolph, Adrienne L. Garcia, and Brendolyn B. Smith are current and former administrative and clerical employees of defendant Lawrence Berkeley Laboratory ("Lawrence"), a research facility operated by the appellee Regents of the University of California pursuant to a contract with the United States Department of Energy (the Department). Defendant Charles V. Shank is the director of Lawrence, and defendants Henry H. Stauffer, Lisa Snow, T. F. Budinger, and William G. Donald, Jr., are all current or former physicians in its medical department. The named defendants are sued in both their official and individual capacities.

The Department requires federal contractors such as Lawrence to establish an occupational medical program. Since 1981, it has required its contractors to perform "preplacement examinations" of employees as part of this program, and until 1995, it also required its contractors to offer their employees the option of subsequent "periodic health examinations." The mandatory preplacement examination occurs after the offer of employment but prior to the assumption of job duties. The Department actively oversees Lawrence's occupational health program, and, prior to 1992, specifically required syphilis testing as part of the preplacement examination.

With the exception of Ellis, who was hired in 1968 and underwent an examination after beginning employment, each of the plaintiffs received written offers of employment expressly conditioned upon a "medical examination," "medical approval," or "health evaluation." All accepted these offers and underwent preplacement examinations, and Randolph and Smith underwent subsequent examinations as well.

In the course of these examinations, plaintiffs completed medical history questionnaires and provided blood and urine samples. The questionnaires asked, inter alia, whether the patient had ever had any of sixty-one medical conditions, including "[s]ickle cell anemia," "[v]enereal disease," and, in the case of women, "[m]enstrual disorders."

The blood and urine samples given by all employees during their pre-placement examinations were tested for syphilis; in addition, certain samples were tested for sickle-cell trait; and certain samples were tested for pregnancy. Lawrence discontinued syphilis testing in April 1993, pregnancy testing in December 1994, and sickle-cell trait testing in June 1995. Defendants assert that they discontinued syphilis testing because of its limited usefulness in screening healthy populations, and that they discontinued sickle-cell trait testing because, by that time, most African-American adults had already been tested at birth. Lawrence continues to perform pregnancy testing, but only on an optional basis. Defendants further contend that "for many years" signs posted in the health examination rooms and "more recently" in the reception area stated that the tests at issue would be administered.

Following receipt of a right-to-sue letter from the EEOC, plaintiffs filed suit in September 1995 on behalf of all past and present Lawrence employees who have ever been subjected to the medical tests at issue. Plaintiffs allege that the testing of their blood and urine samples for syphilis, sickle-cell trait, and pregnancy occurred without their knowledge or consent, and without any subsequent notification that the tests had been conducted. They also allege that only black employees were tested for sickle-cell trait and assert the obvious fact that only female employees were tested for pregnancy.

Finally, they allege that Lawrence failed to provide safeguards to prevent the dissemination of the test results. They contend that they did not discover that the disputed tests had been conducted until approximately January 1995, and specifically deny that they observed any signs indicating that such tests would be performed. Plaintiffs do not allege that the defendants took any subsequent employment-related action on the basis of their test results, or that their test results have been disclosed to third parties.

On the basis of these factual allegations, plaintiffs contend that the defendants violated the ADA by requiring, encouraging, or assisting in medical testing that was neither job-related nor consistent with business necessity. Second, they contend that the defendants violated the federal constitutional right to privacy by conducting the testing at issue, collecting and maintaining the results of the testing, and failing to provide adequate safeguards against disclosure of the results. Third, they contend that the testing violated their right to privacy under Article I, [section] 1 of the California Constitution. Finally, plaintiffs contend that Lawrence and the Regents violated Title VII

by singling out black employees for sickle-cell trait testing and by performing pregnancy testing on female employees generally.

The state defendants moved for judgment on the pleadings or, in the
alternative, for summary judgment. The sole federal defendant (the "Secretary"), then-Secretary of Energy Hazel O'Leary, moved to dismiss the
various claims against her for lack of subject matter jurisdiction and for
failure to state a claim. Turning first to the ADA claims, the district court
reasoned that because the medical questionnaires inquired into information
such as venereal disease and reproductive status, plaintiffs were on notice at
the time of their examinations that Lawrence was engaging in medical inquiries that were neither job-related nor consistent with business necessity.
Thus, given that the most recent examination occurred over two years before the filing of the complaint, the district court held that all of the ADA
claims were time-barred. It also rejected the argument that storage of the
test results constitutes a "continuing violation" of the ADA that tolls the
limitations period.

The district court next concluded that the federal privacy claims were
also time-barred and, in the alternative, failed on the merits. On the
grounds that the tests were "part of a comprehensive medical examination
to which plaintiffs had consented," and that plaintiffs had completed a
medical history form of "highly personal questions" that included inquiries
concerning "venereal disease," "sickle-cell anemia," and "menstrual problems," it concluded that plaintiffs were aware at the time of their examinations "of sufficient facts to put them on notice" that their blood and urine
would be tested for syphilis, sickle-cell trait, and pregnancy, and that their
claims were thus time-barred. The district court then held, in the alternative, that the testing had not violated plaintiffs' due process right to privacy. Relying again on the fact that the tests were performed as part of a
general medical examination "that covered the same areas as the tests
themselves," it concluded that any "additional incremental intrusion" from
the tests was so minimal that no constitutional violation could have occurred despite defendants' failure to identify "an undisputed legitimate
governmental purpose" for the tests.

Finally, the district court held that the Title VII claims, even if viable,
were time-barred for the same reasons as were the privacy and ADA claims.
It also concluded that plaintiffs had failed to state a cognizable Title VII
claim, reasoning that plaintiffs had "neither alleged nor shown any connection between these discontinued confidential tests and [their] employment
terms or conditions, either in the past or in the future"; and finding that
"[p]laintiffs' charge of stigmatic harm, stripped of hyperbole, speculation,
and conjecture . . . evaporates."

This appeal followed.

DISCUSSION

I. STATUTE OF LIMITATIONS

[1] The district court dismissed all of the claims on statute of limitations grounds because it found that the limitations period began to run at the time the tests were taken, in which case each cause of action would be time-barred. Federal law determines when the limitations period begins to run, and the general federal rule is that "a limitations period begins to run when the plaintiff knows or has reason to know of the injury which is the basis of the action." *Trotter v. International Longshoremen's & Warehousemen's Union*, 704 F.2d 1141, 1143 (9th Cir. 1983). Because the district court resolved the statute of limitations question on summary judgment, we must determine, viewing all facts in the light most favorable to plaintiffs and resolving all factual ambiguities in their favor, whether the district court erred in determining that plaintiffs knew or should have known of the particular testing at issue when they underwent the examinations.

[2] We find that whether plaintiffs knew or had reason to know of the specific testing turns on material issues of fact that can only be resolved at trial. Plaintiffs' declarations clearly state that at the time of the examination they did not know that the testing in question would be performed, and they neither saw signs nor received any other indications to that effect. The district court had three possible reasons for concluding that plaintiffs knew or should have expected the tests at issue: (1) they submitted to an occupational preplacement examination; (2) they answered written questions as to whether they had had "venereal disease," "menstrual problems," or "sickle-cell anemia"; and (3) they voluntarily gave blood and urine samples. Given the present state of the record, these facts are hardly sufficient to establish that plaintiffs either knew or should have known that the particular testing would take place.

The question of what tests plaintiffs should have expected or foreseen depends in large part upon what preplacement medical examinations usually entail, and what, if anything, plaintiffs were told to expect. The record strongly suggests that plaintiffs' submission to the exam did not serve to afford them notice of the particular testing involved. The letters that plaintiffs received informed them merely that a "medical examination," "medical approval," or "health evaluation" was an express condition of employment. These letters did not inform plaintiffs that they would be subjected to comprehensive diagnostic medical examinations that would inquire into intimate health matters bearing no relation to their responsibilities as administrative or clerical employees.

The record, indeed, contains considerable evidence that the manner in which the tests were performed was inconsistent with sound medical practice. Plaintiffs introduced before the district court numerous expert declarations by medical scholars roundly condemning Lawrence's alleged practices and explaining, inter alia, that testing for syphilis, sickle-cell trait, and pregnancy is not an appropriate part of an occupational medical examination and is rarely if ever done by employers as a matter of routine; that Lawrence lacked any reasonable medical or public health basis for performing these tests on clerical and administrative employees such as plaintiffs; and that the performance of such tests without explicit notice and informed consent violates prevailing medical standards.

The district court also appears to have reasoned that plaintiffs knew or had reason to know of the tests because they were asked questions on a medical form concerning "venereal disease," "sickle-cell anemia," and "menstrual disorders," and because they gave blood and urine samples. The fact that plaintiffs acquiesced in the minor intrusion of checking or not checking three boxes on a questionnaire does not mean that they had reason to expect further intrusions in the form of having their blood and urine tested for specific conditions that corresponded tangentially if at all to the written questions. First, the entries on the questionnaire were neither identical to nor, in some cases, even suggestive of the characteristics for which plaintiffs were tested. For example, sickle-cell trait is a genetic condition distinct from actually having sickle-cell anemia, and pregnancy is not considered a "menstrual disorder" or a "venereal disease." Second, and more important, it is not reasonable to infer that a person who answers a questionnaire upon personal knowledge is put on notice that his employer will take intrusive means to verify the accuracy of his answers. There is a significant difference between answering on the basis of what you know about your health and consenting to let someone else investigate the most intimate aspects of your life. Indeed, a reasonable person could conclude that by completing a written questionnaire, he has reduced or eliminated the need for seemingly redundant and even more intrusive laboratory testing in search of highly sensitive and non-job-related information.

Furthermore, if plaintiffs' evidence concerning reasonable medical practice is to be credited, they had no reason to think that tests would be performed without their consent simply because they had answered some questions on a form and had then, in addition, provided bodily fluid samples: Plaintiffs could reasonably have expected Lawrence to seek their consent before running any tests not usually performed in an occupational health exam—particularly tests for intimate medical conditions bearing no relationship to their responsibilities or working conditions as clerical employees. The mere fact that an employee has given a blood or urine sample does not provide notice that an employer will perform any and all tests on that specimen that it desires,—no matter how invasive—particularly where, as here, the employer has yet to offer a valid reason for the testing.

[3] In sum, the district court erred in holding as a matter of law that the plaintiffs knew or had reason to know of the nature of the tests as a result of their submission to the preemployment medical examinations. Because the question of what testing, if any, plaintiffs had reason to expect turns on material factual issues that can only be resolved at trial, summary judgment on statute of limitations grounds was inappropriate with respect to the causes of action based on an invasion of privacy in violation of the Federal and California Constitutions, and also on the Title VII claims.

II. FEDERAL CONSTITUTIONAL DUE PROCESS RIGHT OF PRIVACY

The district court also ruled, in the alternative, on the merits of all of plaintiffs' claims except the ADA claims. We first examine its ruling with respect to the claim for violation of the federal constitutional right to privacy. While acknowledging that the government had failed to identify any "undisputed legitimate governmental purpose" for the three tests, the district court concluded that no violation of plaintiffs' right to privacy could have occurred because any intrusions arising from the testing were *de minimis* in light of (1) the "large overlap" between the subjects covered by the medical questionnaire and the three tests and (2) the "overall intrusiveness" of "a full-scale physical examination." We hold that the district court erred.

Because the ADA claims fail on the merits, as discussed below, we do not determine whether the district court erred in dismissing those claims on statute of limitations grounds.

[4] The constitutionally protected privacy interest in avoiding disclosure of personal matters clearly encompasses medical information and its confidentiality. *Doe v. Attorney General of the United States*, 941 F.2d 780, 795 (9th Cir. 1991) (citing *United States v. Westinghouse Elec. Corp.*, 638 F.2d 570, 577 (3d Cir. 1980)); *Roe v. Sherry*, 91 F.3d 1270, 1274 (9th Cir. 1996); see also *Doe v. City of New York*, 15 F.3d 264, 267–69 (2d Cir. 1994). Although cases defining the privacy interest in medical information have typically involved its disclosure to "third" parties, rather than the collection of information by illicit means, it goes without saying that the most basic violation possible involves the performance of unauthorized tests—that is, the non-consensual retrieval of previously unrevealed medical information that may be unknown even to plaintiffs. These tests may also be viewed as searches in violation of Fourth Amendment rights that require Fourth Amendment scrutiny. The tests at issue in this case thus implicate rights protected under both the Fourth Amendment and the Due Process Clause of the Fifth or Fourteenth Amendments. *Yin v. California*, 95 F.3d 864, 870 (9th Cir. 1996), cert. denied, 117 S. Ct. 955 (1997).

[5] Because it would not make sense to examine the collection of medical information under two different approaches, we generally "analyze [medical tests and examinations] under the rubric of [the Fourth] Amendment. " Id. at 871 & n.12. Accordingly, we must balance the government's interest in conducting these particular tests against the plaintiffs' expectations of privacy. Id. at 873. Furthermore, "application of the balancing test requires not only considering the degree of intrusiveness and the state's interests in requiring that intrusion, but also 'the efficacy of this [the state's] means for meeting' its needs." Id. (quoting *Vernonia Sch. Dist. 47J v. Acton*, 515 U.S. 646, 660 (1995)).

[6] The district court erred in dismissing the claims on the ground that any violation was *de minimis*, incremental, or overlapping. The latter two grounds are actually just the court's explanations for its adoption of its "*de minimis*" conclusion. They are not in themselves reasons for dismissal. Nor if the violation is otherwise significant does it become insignificant simply because it is overlapping or incremental. We cannot, therefore, escape a scrupulous examination of the nature of the violation, although we can, of course, consider whether the plaintiffs have in fact consented to any part of the alleged intrusion.

[7] One can think of few subject areas more personal and more likely to implicate privacy interests than that of one's health or genetic make-up. Doe, 15 F.3d at 267 ("Extension of the right to confidentiality to personal medical information recognizes there are few matters that are quite so personal as the status of one's health"); see *Vernonia Sch. Dist. 47J*, 515 U.S. at 658 (noting under Fourth Amendment analysis that "it is significant that the tests at issue here look only for drugs, and not for whether the student is, for example, epileptic, pregnant, or diabetic"). Furthermore, the facts revealed by the tests are highly sensitive, even relative to other medical information. With respect to the testing of plaintiffs for syphilis and pregnancy, it is well established in this circuit "that the Constitution prohibits unregulated, unrestrained employer inquiries into personal sexual matters that have no bearing on job performance." *Schowengerdt v. General Dynamics Corp.*, 823 F.2d 1328, 1336 (9th Cir. 1987) (citing *Thorne v. City of El Segundo*, 726 F.2d 459, 470 (9th Cir. 1983)). The fact that one has syphilis is an intimate matter that pertains to one's sexual history and may invite tremendous amounts of social stigma. Pregnancy is likewise, for many, an intensely private matter, which also may pertain to one's sexual history and often carries far-reaching societal implications. See *Thorne*, 726 F.2d at 468–70; Doe, 15 F.3d at 267 (noting discrimination and intolerance to which HIV-positive persons are exposed). Finally, the carrying of sickle-cell trait can pertain to sensitive information about family history and reproductive decisionmaking. Thus, the conditions tested for were aspects of one's health in which one enjoys the highest expectations of privacy.

[8] As discussed above, with respect to the question of the statute of limitations, there was little, if any, "overlap" between what plaintiffs consented to and the testing at issue here. Nor was the additional invasion only incremental. In some instances, the tests related to entirely different conditions. In all, the information obtained as the result of the testing was qualitatively different from the information that plaintiffs provided in their answers to the questions, and was highly invasive. That one has consented to a general medical examination does not abolish one's privacy right not to be tested for intimate, personal matters involving one's health—nor does consenting to giving blood or urine samples, or filling out a questionnaire. As we have made clear, revealing one's personal knowledge as to whether one has a particular medical condition has nothing to do with one's expectations about actually being tested for that condition. Thus, the intrusion was by no means de minimis. Rather, if unauthorized, the testing constituted a significant invasion of a right that is of great importance, and labelling it minimal cannot and does not make it so.

[9] Lawrence further contends that the tests in question, even if their intrusiveness is not de minimis, would be justified by an employer's interest in performing a general physical examination. This argument fails because issues of fact exist with respect to whether the testing at issue is normally part of a general physical examination. There would of course be no violation if the testing were authorized, or if the plaintiffs reasonably should have known that the blood and urine samples they provided would be used for the disputed testing and failed to object. However, as we concluded in Section I, material issues of fact exist as to those questions.

Summary judgment in the alternative on the merits of the federal constitutional privacy claim was therefore incorrect.

III. RIGHT TO PRIVACY UNDER ARTICLE I, [SECTION] 1 OF THE CALIFORNIA CONSTITUTION

With respect to the state privacy claims, defendants argue, as they did with respect to the federal privacy claims, that the intrusions occasioned by the testing were so minimal that the government need not demonstrate a legitimate interest in performing the tests. In the alternative, they argue that the intrusions were so minimal that plaintiffs' privacy interests were necessarily overcome by the government's interest in performing the pre-placement examinations. We understand this argument to be essentially the same as the argument that these tests are a part of an ordinary general medical examination. Defendants urge no additional governmental interest but appear to rely entirely on the interest that any employer might assert in requiring potential employees to undergo general medical testing. The district court did not adopt either of the defendants' positions expressly but simply ruled that plaintiffs "could not proceed" because the "undisputed facts"—namely, completion of the medical questionnaire, consent to the preplacement examination, and the voluntary giving of blood and urine samples—showed that the tests had inflicted "only a *de minimis* privacy invasion."

[10] To assert a cause of action under Article I, § 1 of the California Constitution, one must establish three elements: (1) a legally protected privacy interest; (2) a reasonable expectation of privacy under the circumstances; and (3) conduct by the defendant that amounts to a "serious invasion" of the protected privacy interest. *Loder v. City of Glendale*, 927 P.2d 1200, 1228 (Cal. 1997) (quoting *Hill v. National Collegiate Athletic Ass'n*, 865 P.2d 633, 657 (Cal. 1994), cert. denied, 118 S.Ct. 44 (1997)). These elements must be "viewed simply as 'threshold elements,'" after which the court must conduct a balancing test between the "countervailing interests" for the conduct in question and the intrusion on privacy resulting from the conduct. A showing of "countervailing interests" may, in turn, be rebutted by a showing that there were "feasible and effective alternatives" with a "lesser impact on privacy interests." *Hill*, 865 P.2d at 657.

[11] For much the same reasons as we have discussed above with respect to the statute of limitations and federal privacy claims, the district court erred in dismissing the state constitutional privacy claim. The only possible difference between the state claim and the federal claim is the threshold requirement that the invasion be serious, and for purposes of summary judgment, that requirement has been more than met.

For the reasons discussed above, we find that material issues of fact exist with respect to whether the defendants had any interest at all in obtaining the information and whether plaintiffs had a reasonable expectation of privacy under the circumstances. Both these questions involve a factual dispute regarding the ordinary or accepted medical practice regarding general or pre-employment medical exams.

Accordingly, the district court also erred in dismissing the state constitutional privacy claims.

IV. TITLE VII CLAIMS

The district court also dismissed the Title VII counts on the merits on the grounds that plaintiffs had failed to state a claim because the "alleged classifications, standing alone, do not suffice to provide a cognizable basis for relief under Title VII" and because plaintiffs had neither alleged nor demonstrated how these classifications had adversely affected them.

[12] Section 703(a) of Title VII of the Civil Rights Act of 1964 provides that it is unlawful for any employer:

(1) to fail or refuse to hire or to discharge any individual, or otherwise to discriminate against any individual with respect to his compensation, terms, conditions, or privileges of employment, because of such individual's race, color, religion, sex, or national origin; or

(2) to limit, segregate, or classify his employees or applicants for employment in any way which would deprive or tend to deprive any individual of employment opportunities or otherwise adversely affect his status as an employee, because of such individual's race, color, religion, sex, or national origin.

42 U.S.C. § 2000e-2(a)

The Pregnancy Discrimination Act further provides that discrimination on the basis of "sex" includes discrimination "on the basis of pregnancy, childbirth, or related medical conditions." 42 U.S.C. § 2000e(k). "In accordance with Congressional intent, the above language is to be read in the broadest possible terms. The intent of Congress was not to list specific discriminatory practices, nor to definitively set out the scope of the activities covered." EEOC Compliance Manual (CCH) § 613.1, at P 2901 (citing *Rogers v. EEOC*, 454 F.2d 234 (5th Cir. 1971)).

Despite defendants' assertions to the contrary, plaintiffs' Title VII claims fall neatly into a Title VII framework: Plaintiffs allege that black and female employees were singled out for additional nonconsensual testing and that defendants thus selectively invaded the privacy of certain employees on the basis of race, sex, and pregnancy. The district court held that

(1) the tests did not constitute discrimination in the "terms" or "conditions" of plaintiffs' employment; and that (2) plaintiffs have failed to show any "adverse effect" as a result of the tests. It also granted the plaintiffs leave to amend their complaint to show adverse effect.

[13] Under [section] 2000e-2(a)(1), supra, an employer who "otherwise . . . discriminate[s]" with respect to the "terms" or "conditions" of employment on account of an illicit classification is subject to Title VII liability. It is well established that Title VII bars discrimination not only in the "terms" and "conditions" of ongoing employment, but also in the "terms" and "conditions" under which individuals may obtain employment. See, e.g., *Griggs v. Duke Power Co.*, 401 U.S. 424, 432–37 (1971) (facially neutral educational and testing requirements that are not reasonable measures of job performance and have disparate impact on hiring of minorities violate Title VII). Thus, for example, a requirement of preemployment health examinations imposed only on female employees, or a requirement of preemployment background security checks imposed only on black employees, would surely violate Title VII.

[14] In this case, the term or condition for black employees was undergoing a test for sickle-cell trait; for women it was undergoing a test for pregnancy. It is not disputed that the preplacement exams were, literally, a

condition of employment: the offers of employment stated this explicitly. Thus, the employment of women and blacks at Lawrence was conditioned in part on allegedly unconstitutional invasions of privacy to which white and/or male employees were not subjected. An additional "term or condition" requiring an unconstitutional invasion of privacy is, without doubt, actionable under Title VII. Furthermore, even if the intrusions did not rise to the level of unconstitutionality, they would still be a "term" or "condition" based on an illicit category as described by the statute and thus a proper basis for a Title VII action. Thus, the district court erred in ruling on the leadings that the plaintiffs had failed to assert a proper Title VII claim under [section] 2000e-2(a)(1).

[15] The district court also erred in finding as a matter of law that there was no "adverse effect" with respect to the tests as required under [section] 2000e-2(a)(2). The unauthorized obtaining of sensitive medical information on the basis of race or sex would in itself constitute an "adverse effect," or injury, under Title VII.

Thus, it was error to rule that as a matter of law no "adverse effect" could arise from a classification that singled out particular groups for unconstitutionally invasive, non-consensual medical testing, and the district court erred in dismissing the Title VII claims on this ground as well.

V. THE ADA CLAIMS

Plaintiffs may challenge only the medical examinations that occurred "on or after January 26, 1992," which is the effective date of the ADA for public entities. The only plaintiffs who underwent any examinations or testing on or after that date are Fuentes and Garcia, who were tested in April 1992 and August 1993, respectively. The complaint alleges that defendants violated the ADA by requiring medical examinations and making medical inquiries that were "neither job-related nor consistent with business necessity." (Compl.P 64 (citing 42 U.S.C. [section] 12112(c)(4)). On appeal, plaintiffs also argue that "the ADA limits medical record keeping by an employer to the results of job-related examinations consistent with business necessity." Appellant Br. at 49 (citing 42 U.S.C.[section] 12112(d)). Plaintiffs do not allege that defendants made use of information gathered in the examinations to discriminate against them on the basis of disability; indeed, neither Garcia nor Fuentes received any positive test results.

[16] The ADA creates three categories of medical inquiries and examinations by employers: (1) those conducted prior to an offer of employment ("preemployment" inquiries and examinations); (2) those conducted "after an offer of employment has been made" but "prior to the commencement of . . . employment duties" ("employment entrance examinations"); and (3) those conducted at any point thereafter. It is undisputed that the second

category, employment entrance examinations, as governed by [section] 12112(d)(3), are the examinations and inquiries to which Fuentes and Garcia were subjected. Unlike examinations conducted at any other time, an employment entrance examination need not be concerned solely with the individual's "ability to perform job-related functions," [section] 12112(d)(2); nor must it be "job-related or consistent with business necessity," [section] 12112(d)(4). Thus, the ADA imposes no restriction on the scope of entrance examinations; it only guarantees the confidentiality of the information gathered, [section] 12112(d)(3)(B), and restricts the use to which an employer may put the information. [section] 12112(d)(3)(C); see 42 U.S.C.[section] 12112(d)(1) (medical examinations and inquiries must be consistent with the general prohibition in [section] 12112(a) against discrimination on the basis of disability); 29 C.F.R. [section] 1630.14(b)(3) (if the results of the examination exclude an individual on the basis of disability, the exclusionary criteria themselves must be job-related and consistent with business necessity). Because the ADA does not limit the scope of such examinations to matters that are "job-related and consistent with business necessity," dismissal of the ADA claims was proper.

Plaintiffs' new argument on appeal that the ADA limits medical record-keeping to "the results of job-related examinations consistent with business necessity" also lacks merit. Section 12112(d)(3)(B) sets forth the conditions under which information obtained during the entrance examination must be kept but clearly does not purport to restrict the records that may be kept to matters that are "job-related and consistent with business necessity." Thus plaintiffs' ADA claims also fail in this respect.

The only possible ADA claim is directed at the defendants' alleged failure to maintain plaintiffs' medical records in the manner required by [section] 12112(d)(3)(B). The allegations in plaintiffs' complaint do not explicitly set forth such a violation but incorporate by reference the factual allegation that the defendants "[f]ail[ed] to provide safeguards to prevent the dissemination to third parties of sensitive medical information regarding the plaintiffs." On appeal the plaintiffs argue only that the defendants have "failed to describe the procedures by which a third party might gain access to the records, and the enforcement of any rules, policies, regulations or procedures to prevent third parties from gaining access to the records." To the extent that one can construe the complaint to allege that the defendants are in violation of [section] 12112(d)(3)(B), the bare allegation that defendants have not provided, or adequately described, safeguards fails to state a violation of the ADA requirements as set forth in [section] 12112(d)(3)(B) or as implemented in Department orders. See DOE Order 440.1 (Sep. 30, 1995); DOE Order 5480.8A (June 6, 1992); DOE Order 5480.8 (May 22, 1981). Accordingly, dismissal of the ADA claims was proper.

Appendix C

VI. PLAINTIFFS' CLAIMS ARE NOT MOOT

The Secretary contends that the claims against him in his official capacity for injunctive and declaratory relief are moot because (1) the only testing that the Department ever required was syphilis testing, and (2) the DOE order that required syphilis testing was cancelled on June 22, 1992, and replaced by a different order that requires "[u]rinalysis and serology" only "when indicated." Compare DOE Order 5480.8 (May 22, 1981), with DOE Order 5480.8A (June 26, 1992).

Although the state defendants do not raise the issue, a similar argument can be made on their behalf: Lawrence discontinued syphilis testing in April 1993, pregnancy testing in December 1994, and sickle-cell trait testing in June 1995.

[17] "[A] case is moot when the issues presented are no longer 'live' or the parties lack a legally cognizable interest in the outcome." *County of Los Angeles v. Davis*, 440 U.S.625, 631 (1979) [quoting *Powell v. McCormack*, 395 U.S. 486, 496 (1969)]. "Mere voluntary cessation of allegedly illegal conduct does not moot a case; if it did, the courts would be compelled to leave [t]he defendant . . . free to return to his old ways." *United States v. Concentrated Phosphate Export Ass'n*, 393 U.S. 199, 203 (1968) (quoting *United States v. W. T. Grant Co.*, 345 U.S. 629, 632 (1953)). Nevertheless, part or all of a case may become moot if (1) "subsequent events [have] made it absolutely clear that the allegedly wrongful behavior [cannot] reasonably be expected to recur," *Concentrated Phosphate*, 393 U.S. at 203, and (2) "interim relief or events have completely and irrevocably eradicated the effects of the alleged violation." *Lindquist v. Idaho State Bd. of Corrections*, 776 F.2d 851, 854 (9th Cir. 1985) (quoting *Davis*, 440 U.S. at 631). "The burden of demonstrating mootness 'is a heavy one.'" *Davis*, 440 U.S. at 631 (quoting *W. T. Grant*, 345 U.S. at 632–33).

[18] Defendants have not carried their heavy burden of establishing either that their alleged behavior cannot be reasonably expected to recur, or that interim events have eradicated the effects of the alleged violation. First, they do not contend that the Department will never again require or permit, or that Lawrence will never again conduct, the tests at issue. They assert only that syphilis testing was discontinued because of its limited usefulness in screening healthy populations, and that sickle-cell trait testing was discontinued as redundant of testing that most African Americans now receive at birth. Moreover, in the case of pregnancy testing, they do not even argue that such testing is no longer medically useful; rather, they have simply made it optional. Defendants have neither asserted nor demonstrated that they will never resume mandatory testing for intimate medical conditions; nor have they offered any reason why they might not return in the future to their original views on the utility of mandatory testing. In contrast, plaintiffs

325

have introduced evidence, in the form of correspondence between Lawrence and the department, that the syphilis tests were discontinued merely for reasons of "cost-effectiveness." See *Concentrated Phosphate*, 393 U.S. at 203 (holding that mere statement that it would be "uneconomical" for defendants to continue their allegedly wrongful conduct "cannot suffice to satisfy the heavy burden" of establishing mootness).

[19] Second, defendants also have not asserted that any "interim relief or events have completely and irrevocably eradicated the effects of the alleged violation." *Lindquist*, 776 F.2d at 854. Indeed, it is undisputed that the Department requires Lawrence to retain plaintiffs' test results and that Lawrence does in fact do so. See DOE Order 440.1, dated September 30, 1995 ("Employee medical records shall be adequately protected and stored permanently.") Even if the continued storage, against plaintiffs' wishes, of intimate medical information that was allegedly taken from them by unconstitutional means does not itself constitute a violation of law, it is clearly an ongoing "effect" of the allegedly unconstitutional and discriminatory testing, and expungement of the test results would be an appropriate remedy for the alleged violation. Cf. *Fendler v. United States Parole Comm'n*, 774 F.2d 975, 979 (9th Cir. 1985) ("Federal courts have the equitable power 'to order the expungement of Government records where necessary to vindicate rights secured by the Constitution or by statute.") [quoting *Chastain v. Kelley*, 510 F.2d 1232, 1235 (D.C. Cir. 1975)]; *Maurer v. Pitchess*, 691 F.2d 434, 437 (9th Cir. 1982). Accordingly, plaintiffs' claims for injunctive and declaratory relief are not moot.

VII. IRREPARABLE INJURY

[20] Finally, the Secretary contends that plaintiffs cannot seek injunctive relief because they have not alleged irreparable injury. To obtain injunctive relief, " '[a] reasonable showing' of a 'sufficient likelihood' that plaintiff will be injured again is necessary." *Kruse v. State of Hawaii*, 68 F.3d 331, 335 (9th Cir. 1995) (internal quotation marks omitted); see *City of Los Angeles v. Lyons*, 461 U.S. 95, 111 (1983). "The likelihood of the injury recurring must be calculable and if there is no basis for predicting that any future repetition would affect the present plaintiffs, there is no case or controversy." *Sample v. Johnson*, 771 F.2d 1335, 1340 (9th Cir. 1985). In this case, plaintiffs seek not only to enjoin future illegal testing, but also to require defendants, inter alia, to notify all employees who may have been tested illegally; to destroy the results of such illegal testing upon employee request; to describe any use to which the information was put, and any disclosures of the information that were made; and to submit Lawrence's medical department to "independent oversight and monitoring."

[21] At the very least, the retention of undisputedly intimate medical information obtained in an unconstitutional and discriminatory manner

would constitute a continuing "irreparable injury" for purposes of equitable relief. Moreover, the Department orders still require Lawrence to conduct preplacement examinations. DOE Order 440.1 (Sep. 30, 1995). Thus, there seems to be at least a reasonable possibility that Lawrence would again conduct undisclosed medical testing of its employees for intimate medical conditions. For these reasons, a request for injunctive relief is proper.

CONCLUSION

Because material and disputed issues of fact exist with respect to whether reasonable persons in plaintiffs' position would have had reason to know that the tests were being performed, and because the tests were a separate and more invasive intrusion into their privacy than the aspects of the examination to which they did consent, the district court erred in granting summary judgment on statute of limitations grounds with respect to the Title VII claims and the federal and state constitutional privacy claims. The district court also erred in dismissing the federal and state constitutional privacy claims and the Title VII claims on the merits. The district court's dismissal of the ADA claims was proper. None of the Secretary's arguments with respect to the claims brought against him in his official capacity has merit.

AFFIRMED IN PART, REVERSED IN PART, AND REMANDED.

[footnotes omitted]

APPENDIX D

BRAGDON V. ABBOTT, 97 U.S. 156 (1998)

[portions are omitted]

ON WRIT OF CERTIORARI TO THE UNITED STATES COURT OF APPEALS FOR THE FIRST CIRCUIT

Justice Kennedy delivered the opinion of the Court.

We address in this case the application of the Americans with Disabilities Act of 1990 (ADA), 104 Stat. 327, 42 U.S.C. § 12101 et seq., to persons infected with the human immunodeficiency virus (HIV). We granted certiorari to review, first, whether HIV infection is a disability under the ADA when the infection has not yet progressed to the so-called symptomatic phase; and, second, whether the Court of Appeals, in affirming a grant of summary judgment, cited sufficient material in the record to determine, as a matter of law, that respondent's infection with HIV posed no direct threat to the health and safety of her treating dentist.

I

Respondent Sidney Abbott has been infected with HIV since 1986. When the incidents we recite occurred, her infection had not manifested its most serious symptoms. On September 16, 1994, she went to the office of petitioner Randon Bragdon in Bangor, Maine, for a dental appointment. She disclosed her HIV infection on the patient registration form. Petitioner completed a dental examination, discovered a cavity, and informed respondent of his policy against filling cavities of HIV-infected patients. He offered to perform the work at a hospital with no added fee for his services, though respondent would be responsible for the cost of using the hospital's facilities. Respondent declined.

Respondent sued petitioner under state law and §302 of the ADA, 104 Stat. 355, 42 U.S.C. § 12182 alleging discrimination on the basis of her disability. The state law claims are not before us. Section 302 of the ADA provides:

Appendix D

"No individual shall be discriminated against on the basis of disability in the full and equal enjoyment of the goods, services, facilities, privileges, advantages, or accommodations of any place of public accommodation by any person who . . . operates a place of public accommodation." §12182(a).

The term "public accommodation" is defined to include the "professional office of a health care provider." §12181(7)(F).

A later subsection qualifies the mandate not to discriminate. It provides:

"Nothing in this subchapter shall require an entity to permit an individual to participate in or benefit from the goods, services, facilities, privileges, advantages and accommodations of such entity where such individual poses a direct threat to the health or safety of others." §12182(b)(3).

The United States and the Maine Human Rights Commission intervened as plaintiffs. After discovery, the parties filed cross-motions for summary judgment. The District Court ruled in favor of the plaintiffs, holding that respondent's HIV infection satisfied the ADA's definition of disability. 912 F. Supp. 580, 585–587 (Me. 1995). The court held further that petitioner raised no genuine issue of material fact as to whether respondent's HIV infection would have posed a direct threat to the health or safety of others during the course of a dental treatment. Id., at 587–591. The court relied on affidavits submitted by Dr. Donald Wayne Marianos, Director of the Division of Oral Health of the Centers for Disease Control and Prevention (CDC). The Marianos affidavits asserted it is safe for dentists to treat patients infected with HIV in dental offices if the dentist follows the so-called universal precautions described in the *Recommended Infection-Control Practices for Dentistry* issued by CDC in 1993 (1993 CDC *Dentistry Guidelines*). 912 F. Supp., at 589.

The Court of Appeals affirmed. It held respondent's HIV infection was a disability under the ADA, even though her infection had not yet progressed to the symptomatic stage. 107 F.3d 934, 939–943 (CA1 1997). The Court of Appeals also agreed that treating the respondent in petitioner's office would not have posed a direct threat to the health and safety of others. Id., at 943–948. Unlike the District Court, however, the Court of Appeals declined to rely on the Marianos affidavits. Id., at 946, n. 7. Instead the court relied on the 1993 CDC *Dentistry Guidelines*, as well as the *Policy on AIDS, HIV Infection and the Practice of Dentistry*, promulgated by the American Dental Association in 1991 (1991 American Dental Association *Policy on HIV*). 107 F.3d, at 945–946.

II

We first review the ruling that respondent's HIV infection constituted a disability under the ADA. The statute defines disability as:

"(A) a physical or mental impairment that substantially limits one or more of the major life activities of such individual;

"(B) a record of such an impairment; or

"(C) being regarded as having such impairment." §12102(2).

We hold respondent's HIV infection was a disability under subsection (A) of the definitional section of the statute. In light of this conclusion, we need not consider the applicability of subsections (B) or (C).

Our consideration of subsection (A) of the definition proceeds in three steps. First, we consider whether respondent's HIV infection was a physical impairment. Second, we identify the life activity upon which respondent relies (reproduction and child bearing) and determine whether it constitutes a major life activity under the ADA. Third, tying the two statutory phrases together, we ask whether the impairment substantially limited the major life activity. In construing the statute, we are informed by interpretations of parallel definitions in previous statutes and the views of various administrative agencies which have faced this interpretive question.

A

The ADA's definition of disability is drawn almost verbatim from the definition of "handicapped individual" included in the Rehabilitation Act of 1973, 29 U.S.C. § 706(8)(B) (1988 ed.), and the definition of "handicap" contained in the Fair Housing Amendments Act of 1988, 42 U.S.C. § 3602(h)(1) (1988 ed.). Congress' repetition of a well-established term carries the implication that Congress intended the term to be construed in accordance with pre-existing regulatory interpretations. See *FDIC v. Philadelphia Gear Corp.*, 476 U.S. 426, 437–438 (1986); *Commissioner v. Estate of Noel*, 380 U.S. 678, 681–682 (1965); *ICC v. Parker*, 326 U.S. 60, 65 (1945). In this case, Congress did more than suggest this construction; it adopted a specific statutory provision in the ADA directing as follows:

"Except as otherwise provided in this chapter, nothing in this chapter shall be construed to apply a lesser standard than the standards applied under title V of the Rehabilitation Act of 1973 (29 U.S.C. 790 *et seq.*) or the regulations issued by Federal agencies pursuant to such title." 42 U.S.C. § 12201(a).

The directive requires us to construe the ADA to grant at least as much protection as provided by the regulations implementing the Rehabilitation Act.

1. The first step in the inquiry under subsection (A) requires us to determine whether respondent's condition constituted a physical impairment. The Department of Health, Education and Welfare (HEW) issued the first regulations interpreting the Rehabilitation Act in 1977. The regulations are of particular significance because, at the time, HEW was the agency responsible for coordinating the implementation and enforcement of §504. *Con-*

solidated Rail Corporation v. Darrone, 465 U.S. 624, 634, (1984) [citing Exec. Order No. 11914, 3 CFR 117 (1976–1980 Comp.)]. The HEW regulations, which appear without change in the current regulations issued by the Department of Health and Human Services, define "physical or mental impairment" to mean:

"(A) any physiological disorder or condition, cosmetic disfigurement, or anatomical loss affecting one or more of the following body systems: neurological; musculoskeletal; special sense organs; respiratory, including speech organs; cardiovascular; reproductive, digestive, genito-urinary; hemic and lymphatic; skin; and endocrine; or

"(B) any mental or psychological disorder, such as mental retardation, organic brain syndrome, emotional or mental illness, and specific learning disabilities." 45 CFR § 84.3(j)(2)(i) (1997).

In issuing these regulations, HEW decided against including a list of disorders constituting physical or mental impairments, out of concern that any specific enumeration might not be comprehensive. . . . [material establishing that asymptomatic HIV infection is a disability is omitted]

In light of the immediacy with which the virus begins to damage the infected person's white blood cells and the severity of the disease, we hold it is an impairment from the moment of infection. As noted earlier, infection with HIV causes immediate abnormalities in a person's blood, and the infected person's white cell count continues to drop throughout the course of the disease, even when the attack is concentrated in the lymph nodes. In light of these facts, HIV infection must be regarded as a physiological disorder with a constant and detrimental effect on the infected person's hemic and lymphatic systems from the moment of infection. HIV infection satisfies the statutory and regulatory definition of a physical impairment during every stage of the disease.

2. The statute is not operative, and the definition not satisfied, unless the impairment affects a major life activity. Respondent's claim throughout this case has been that the HIV infection placed a substantial limitation on her ability to reproduce and to bear children. App. 14; 912 F. Supp., at 586; 107 F.3d, at 939. Given the pervasive, and invariably fatal, course of the disease, its effect on major life activities of many sorts might have been relevant to our inquiry. Respondent and a number of amici make arguments about HIV's profound impact on almost every phase of the infected person's life. See Brief for Respondent Sidney Abbott 24–27; Brief for American Medical Association as *Amicus Curiae* 20; Brief for Infectious Diseases Society of America et al. as *Amici Curiae* 7–11. In light of these submissions, it may seem legalistic to circumscribe our discussion to the activity of reproduction. We have little doubt that had different parties brought the suit they would have maintained that an HIV infection imposes substantial limitations on other major life activities.

From the outset, however, the case has been treated as one in which reproduction was the major life activity limited by the impairment. It is our practice to decide cases on the grounds raised and considered in the Court of Appeals and included in the question on which we granted certiorari. See, e.g., *Blessing v. Freestone*, 520 U.S. 329, 340, n. 3 (1997) (citing this Court's Rule 14.1(a)); *Capitol Square Review and Advisory Bd. v. Pinette*, 515 U.S. 753, 760 (1995). We ask, then, whether reproduction is a major life activity.

We have little difficulty concluding that it is. As the Court of Appeals held, "[t]he plain meaning of the word 'major' denotes comparative importance" and "suggest[s] that the touchstone for determining an activity's inclusion under the statutory rubric is its significance." 107 F.3d, at 939, 940. Reproduction falls well within the phrase "major life activity." Reproduction and the sexual dynamics surrounding it are central to the life process itself.

While petitioner concedes the importance of reproduction, he claims that Congress intended the ADA only to cover those aspects of a person's life which have a public, economic, or daily character. Brief for Petitioner 14, 28, 30, 31; see also id., at 36–37 [citing *Krauel v. Iowa Methodist Medical Center*, 95 F.3d 674, 677 (CA8 1996)]. The argument founders on the statutory language. Nothing in the definition suggests that activities without a public, economic, or daily dimension may somehow be regarded as so unimportant or insignificant as to fall outside the meaning of the word "major." The breadth of the term confounds the attempt to limit its construction in this manner.

As we have noted, the ADA must be construed to be consistent with regulations issued to implement the Rehabilitation Act. See 42 U.S.C. § 12201(a). Rather than enunciating a general principle for determining what is and is not a major life activity, the Rehabilitation Act regulations instead provide a representative list, defining term to include "functions such as caring for one's self, performing manual tasks, walking, seeing, hearing, speaking, breathing, learning, and working." 45 CFR § 84.3(j)(2)(ii) (1997); 28 CFR § 41.31(b)(2) (1997). As the use of the term "such as" confirms, the list is illustrative, not exhaustive.

These regulations are contrary to petitioner's attempt to limit the meaning of the term "major" to public activities. The inclusion of activities such as caring for one's self and performing manual tasks belies the suggestion that a task must have a public or economic character in order to be a major life activity for purposes of the ADA. On the contrary, the Rehabilitation Act regulations support the inclusion of reproduction as a major life activity, since reproduction could not be regarded as any less important than working and learning. Petitioner advances no credible basis for confining major life activities to those with a public, economic, or daily aspect. In the absence of any reason to reach a contrary conclusion, we agree with the Court of Appeals' determination that reproduction is a major life activity for the purposes of the ADA.

3. The final element of the disability definition in subsection (A) is whether respondent's physical impairment was a substantial limit on the major life activity she asserts. The Rehabilitation Act regulations provide no additional guidance. 45 CFR pt. 84, App. A, p. 334 (1997).

Our evaluation of the medical evidence leads us to conclude that respondent's infection substantially limited her ability to reproduce in two independent ways. First, a woman infected with HIV who tries to conceive a child imposes on the man a significant risk of becoming infected. The cumulative results of 13 studies collected in a 1994 textbook on AIDS indicates that 20% of male partners of women with HIV became HIV-positive themselves, with a majority of the studies finding a statistically significant risk of infection. Osmond & Padian, "Sexual Transmission of HIV," in *AIDS Knowledge Base* 1.9-8, and tbl. 2; see also Haverkos & Battjes, "Female-to-Male Transmission of HIV," 268 *JAMA* 1855, 1856, tbl. (1992) (cumulative results of 16 studies indicated 25% risk of female-to-male transmission). (Studies report a similar, if not more severe, risk of male-to-female transmission. See, e.g., Osmond & Padian, *AIDS Knowledge Base* 1.9-3, tbl. 1, 1.9-6 to 1.9-7.)

Second, an infected woman risks infecting her child during gestation and childbirth, i.e., perinatal transmission. Petitioner concedes that women infected with HIV face about a 25% risk of transmitting the virus to their children. 107 F.3d, at 942; 912 F. Supp., at 387, n. 6. Published reports available in 1994 confirm the accuracy of this statistic. Report of a Consensus Workshop, *Maternal Factors Involved in Mother-to-Child Transmission of HIV-1*, 5 J. *Acquired Immune Deficiency Syndromes* 1019, 1020 (1992) (collecting 13 studies placing risk between 14% and 40%, with most studies falling within the 25% to 30% range); Connor et al., "Reduction of Maternal-Infant Transmission of Human Immunodeficiency Virus Type 1 with Zidovudine Treatment," 331 *New Eng J Med* 1173, 1176 (1994) (placing risk at 25.5%); see also Strapans & Feinberg, *Medical Management of AIDS* 32 (studies report 13% to 45% risk of infection, with average of approximately 25%).

Petitioner points to evidence in the record suggesting that antiretroviral therapy can lower the risk of perinatal transmission to about 8%. App. 53; see also Connor, supra, at 1176 (8.3%); Sperling et al., "Maternal Viral Load, Zidovudine Treatment, and the Risk of Transmission of Human Immunodeficiency Virus Type 1 from Mother to Infant," 335 *New Eng J Med* 1621, 1622 (1996) (7.6%). The Solicitor General questions the relevance of the 8% figure, pointing to regulatory language requiring the substantiality of a limitation to be assessed without regard to available mitigating measures. Brief for United States as *Amicus Curiae* 18, n. 10 (citing 28 CFR pt. 36, App. B, p. 611

(1997); 29 CFR pt. 1630, App., p. 351 (1997)). We need not resolve this dispute in order to decide this case, however. It cannot be said as a matter of law that an 8% risk of transmitting a dread and fatal disease to one's child does not represent a substantial limitation on reproduction.

The Act addresses substantial limitations on major life activities, not utter inabilities. Conception and childbirth are not impossible for an HIV victim but, without doubt, are dangerous to the public health. This meets the definition of a substantial limitation. The decision to reproduce carries economic and legal consequences as well. There are added costs for antiretroviral therapy, supplemental insurance, and long-term health care for the child who must be examined and, tragic to think, treated for the infection. The laws of some States, moreover, forbid persons infected with HIV from having sex with others, regardless of consent. Iowa Code §§139.1, 139.31 (1997); Md. Health Code Ann. §18-601.1(a) (1994); Mont. Code Ann. §§50-18-101, 50-18-112 (1997); Utah Code Ann. §26-6-3.5(3) (Supp. 1997); id., §26-6-5 (1995); Wash. Rev. Code §9A.36.011(1)(b) (Supp. 1998); see also N. D. Cent. Code §12.1-20-17 (1997).

In the end, the disability definition does not turn on personal choice. When significant limitations result from the impairment, the definition is met even if the difficulties are not insurmountable. For the statistical and other reasons we have cited, of course, the limitations on reproduction may be insurmountable here. Testimony from the respondent that her HIV infection controlled her decision not to have a child is unchallenged. App. 14; 912 F. Supp., at 587; 107 F.3d, at 942. In the context of reviewing summary judgment, we must take it to be true. Fed. Rule Civ. Proc. 56(e). We agree with the District Court and the Court of Appeals that no triable issue of fact impedes a ruling on the question of statutory coverage. Respondent's HIV infection is a physical impairment which substantially limits a major life activity, as the ADA defines it. In view of our holding, we need not address the second question presented, i.e., whether HIV infection is a, *per se*, disability under the ADA.

B

Our holding is confirmed by a consistent course of agency interpretation before and after enactment of the ADA. Every agency to consider the issue under the Rehabilitation Act found statutory coverage for persons with asymptomatic HIV. . . .

[further discussion of various agencies' interpretation of disability and its application to asymptomatic HIV infection is omitted]

The regulatory authorities we cite are consistent with our holding that HIV infection, even in the so-called asymptomatic phase, is an impairment which substantially limits the major life activity of reproduction.

III

The petition for certiorari presented three other questions for review. The questions stated:

"3. When deciding under Title III of the ADA whether a private health care provider must perform invasive procedures on an infectious patient in his office, should courts defer to the health care provider's professional judgment, as long as it is reasonable in light of then-current medical knowledge?

"4. What is the proper standard of judicial review under Title III of the ADA of a private health care provider's judgment that the performance of certain invasive procedures in his office would pose a direct threat to the health or safety of others?

"5. Did petitioner, Randon Bragdon, D.M.D., raise a genuine issue of fact for trial as to whether he was warranted in his judgment that the performance of certain invasive procedures on a patient in his office would have posed a direct threat to the health or safety of others?" Pet. for Cert. i.

Of these, we granted certiorari only on question three. The question is phrased in an awkward way, for it conflates two separate inquiries. In asking whether it is appropriate to defer to petitioner's judgment, it assumes that petitioner's assessment of the objective facts was reasonable. The central premise of the question and the assumption on which it is based merit separate consideration. . . .

[material discussing whether Bragdon's health would have been endangered by treating Abbott is omitted]

We conclude the proper course is to give the Court of Appeals the opportunity to determine whether our analysis of some of the studies cited by the parties would change its conclusion that petitioner presented neither objective evidence nor a triable issue of fact on the question of risk. In remanding the case, we do not foreclose the possibility that the Court of Appeals may reach the same conclusion it did earlier. A remand will permit a full exploration of the issue through the adversary process.

The determination of the Court of Appeals that respondent's HIV infection was a disability under the ADA is affirmed. The judgment is vacated, and the case is remanded for further proceedings consistent with this opinion.

APPENDIX E

UNITED STATES V. KINCADE,
379 F.3D 813 (2004)

United States Court of Appeals for the Ninth Circuit
Argued and submitted March 23, 2004
Filed on August 18, 2004
Judge Diarmuid F. O'Scannlain delivered the opinion of the court. [Portions have been omitted, including most citations and footnotes.]

We must decide whether the Fourth Amendment permits compulsory DNA profiling of certain conditionally released federal offenders in the absence of individualized suspicion that they have committed additional crimes.

I

A

Pursuant to the DNA Analysis Backlog Elimination Act of 2000 ("DNA Act"), individuals who have been convicted of certain federal crimes and who are incarcerated, or on parole, probation, or supervised release must provide federal authorities with "a tissue, fluid, or other bodily sample . . . on which a[n] . . . analysis of th[at sample's] deoxyribonucleic acid (DNA) identification information" can be performed. Because the Federal Bureau of Investigation ("the Bureau") considers DNA information derived from blood samples to be more reliable than that obtained from other sources . . . Bureau guidelines require those in federal custody and subject to the DNA Act to submit to compulsory blood sampling. Failure "to cooperate in the collection of that sample [is] . . . a class A misdemeanor," punishable by up to one year's imprisonment and a fine of as much as $100,000. . . .

[B]lood samples are turned over to the Bureau for DNA analysis—the identification and recording of an individual's "genetic fingerprint." . . . The . . . [tested] loci [spots in the genome] are each found on so-called "junk

DNA"—that is, non-genic stretches of DNA not presently recognized as being responsible for trait coding—and "were purposely selected because they are not associated with any known physical or medical characteristics." ... Even so, DNA profiles generated by STR [short tandem repeat testing] are highly individuated: Due to the substantial number of alleles [variant forms] present at each of the 13 ... loci and widespread variances in their representation among human beings, the chance that two randomly selected individuals will share the same profile are infinitesimal—as are the chances that a person randomly selected from the population at large will present the same DNA profile as that drawn from crime-scene evidence.

Once STR has been used to produce an individual's DNA profile, the resulting record is loaded into the Bureau's Combined DNA Index System ("CODIS")—a massive centrally managed database linking DNA profiles culled from federal, state, and territorial DNA collection programs, as well as profiles drawn from crime-scene evidence, unidentified remains, and genetic samples voluntarily provided by relatives of missing persons. ...

CODIS can be used in two different ways. First, law enforcement can match one forensic crime scene sample to another forensic crime scene sample, thereby allowing officers to connect unsolved crimes through a common perpetrator. Second, and of perhaps greater significance, CODIS enables officials to match evidence obtained at the scene of a crime to a particular offender's profile. In this latter capacity, CODIS serves as a potent tool for monitoring the criminal activity of known offenders. ...

B

On July 20, 1993, driven by escalating personal and financial troubles, decorated Navy seaman Thomas Cameron Kincade robbed a bank using a firearm. He soon pleaded guilty to those charges and was sentenced to 97 months' imprisonment, followed by three years' supervised release. ... On March 25, 2002, Kincade's probation officer asked him to submit a blood sample pursuant to the DNA Act. He refused, eventually explaining that his objections were purely a matter of personal preference—in his words, "not a religious conviction." ... Kincade's probation officer informed the district court that Kincade had refused to submit the blood sample required by the DNA Act. He also recommended revocation of Kincade's supervised release, and reincarceration.

In briefing to the district court prior to a scheduled revocation hearing, Kincade challenged the constitutionality of the DNA Act on grounds that it violated the Ex Post Facto Clause, the Fourth Amendment, and separation of powers principles embodied in Article III and the Due Process Clause. On July 15, 2002, Kincade appeared at a revocation hearing before U.S. District Judge Dickran Tevrizian. ...

Judge Tevrizian rejected Kincade's constitutional challenges to the DNA Act . . . [and] sentenced [him] to four months' imprisonment and two years' supervised release. . . . [O]nce in custody, Kincade finally was forced to submit to DNA profiling. He persists in his challenge to the Act. [fn **13:** On appeal, Kincade raises only Fourth Amendment objections to the Act.]

II

[omitted]

A

[1] Pursuant to the Fourth Amendment, "[t]he right of the people to be secure in their persons, houses, papers, and effects, against unreasonable searches and seizures, shall not be violated, and no Warrants shall issue, but upon probable cause, supported by Oath or affirmation, and particularly describing the place to be searched, and the persons or things to be seized." U.S. Const. amend. IV. "The touchstone of our analysis under the Fourth Amendment is always 'the reasonableness in all the circumstances of the particular governmental invasion of a citizen's personal security.'"

Ordinarily, the reasonableness of a search depends on governmental compliance with the Warrant Clause, which requires authorities to demonstrate probable cause to a neutral magistrate and thereby convince him to provide formal authorization to proceed with a search by issuance of a particularized warrant. However, the general rule of the Warrant Clause is not unyielding. Under a variety of conditions, law enforcement may execute a search without first complying with its dictates. For instance, police may execute warrantless searches incident to a lawful arrest: It is reasonable for authorities to search an arrestee for weapons that might threaten their safety, or for evidence which might be destroyed. . . .

[3] A final category of suspicionless searches is referred to as "special needs," and in recent years, the Court has devoted increasing attention to the development of the accompanying analytical doctrine. . . .

[4] For the most part, these cases involve searches conducted for important non-law enforcement purposes in contexts where adherence to the warrant-and-probable cause requirement would be impracticable. . . .

1

Almost as soon as the "special needs" rationale was articulated, however, the Court applied special needs analysis in what seemed—at least on the surface—to be a clear law enforcement context. At issue in *Griffin* [*Griffin v. Wisconsin*, 1987] was a warrantless search of a probationer's home, instigated and carried out under the direction of law enforcement officials acting with what appeared to be pure law enforcement motives. . . .

On eventual appeal to the Supreme Court, the Justices explained:

A State's operation of a probation system, like its operation of a school, government office or prison, . . . presents 'special needs' beyond normal law enforcement that may justify departures from the usual warrant and probable-cause requirements. Probation, like incarceration, is a form of criminal sanction imposed by a court upon an offender after verdict, finding, or plea of guilty. . . . [I]t is always true of probationers (as we have said it to be true of parolees) that they do not enjoy the absolute liberty to which every citizen is entitled, but only conditional liberty properly dependent on observance of special probation restrictions. . . .

Carefully noting that these "special needs" . . . did not operate wholly to eliminate the Fourth Amendment rights of those subject to its strictures, the Court observed that the probation context nonetheless necessitated a relaxation of the usual warrant-and-probable cause requirement. . . . Thus, the Court concluded, the Constitution permits the execution of probation and parole searches based on no more than reasonable suspicion—even where the search at issue is triggered by law enforcement information and motivated by apparent law enforcement purposes.

2

[5] Notwithstanding *Griffin*'s apparent focus on the crucial law enforcement goals of probation and parole, however, the Court's more recent "special needs" cases have emphasized the absence of any law enforcement motive underlying the challenged search and seizure. . . .

4

While these recent cases may seem to be moving toward requiring that *any* search conducted primarily for law enforcement purposes must be accompanied by at least some quantum of individualized suspicion, the Court signaled the existence of possible limitations in *United States v. Knights* (534 U.S. 112, 2001). At issue there was a warrantless search of [the home of] a probationer long suspected of having committed crimes targeting Pacific Gas & Electric ("PG&E") facilities [done after observing suspicious behavior of a suspected accomplice leaving Knights's apartment]. . . . Rather than analyze the warrantless search of Knights's apartment within the special needs framework, the Court instead opted to "consider th[e] question [left open by *Griffin*] in assessing the constitutionality of the search of Knights's apartment."

To do so, it turned to the traditional totality of the circumstances test—balancing the invasion of Knights's interest in privacy against the State's interest in searching his home without a warrant supported by probable cause. Of central importance to our decision today, the Court explained that "Knights's status as a probationer subject to a search condition informs both sides of that balance." . . . The Court held [that] the government needs "no more than reasonable suspicion to conduct a search of [a] probationer's house.". . .

B

[7] . . . Confronted with challenges to the federal DNA Act and its state law analogues, our sister circuits and peers in the states have divided in their analytical approaches. . . . On one hand, the Second, Seventh, and Tenth Circuits, along with a variety of federal district courts and at least two state Supreme Courts, have upheld DNA collection statutes under a special needs analysis. . . .

[8] By contrast, the Fourth and Fifth Circuits, a Seventh Circuit Judge, numerous federal district courts, and a variety of state courts have approved compulsory DNA profiling under a traditional assessment of reasonableness gauged by the totality of the circumstances. . . . [fn. 25: To our knowledge, only two judges—besides, of course, the majority of the three-judge panel that first heard this case, *see United States v. Kincade*, 345 F.3d 1095, *vacated and reh'g en banc granted*, 354 F.3d 1000 (9th Cir. 2003)—have invalidated DNA collection statutes.]

[9] Finally, we observe that our own 1995 decision in *Rise v. Oregon*, 59 F.3d 1556 (9th Cir. 1995), upheld the constitutionality of a state DNA collection statute by applying a pure totality of the circumstances analysis. . . .

III

While not precluding the possibility that the federal DNA Act could satisfy a special needs analysis, we today reaffirm the continuing vitality of *Rise*—and hold that its reliance on a totality of the circumstances analysis to uphold compulsory DNA profiling of convicted offenders both comports with the Supreme Court's recent precedents and resolves this appeal in concert with the requirements of the Fourth Amendment.

A

. . . [11] . . . We begin our resolution of the issue by taking note of the well-established principle that parolees and other conditional releasees are not entitled to the full panoply of rights and protections possessed by the general public. . . .

[12] We believe that such a severe and fundamental disruption in the relationship between the offender and society, along with the government's concomitantly greater interest in closely monitoring and supervising conditional releasees, is in turn sufficient to sustain suspicionless searches of his person and property even in the absence of some non-law enforcement "special need"—at least where such searches meet the Fourth Amendment touchstone of reasonableness as gauged by the totality of the circumstances.

Let us be clear: Our holding in no way intimates that conditional releasees' diminished expectations of privacy serve to extinguish their ability to invoke the protections of the Fourth Amendment's guarantee against unreasonable searches and seizures. Where a given search or class of searches cannot satisfy the traditional totality of the circumstances test, a conditional releasee may lay claim to constitutional relief—just like any other citizen. . . .

We also wish to emphasize the limited nature of our holding. With its alarmist tone and obligatory reference to George Orwell's *1984*, Judge Reinhardt's dissent repeatedly asserts that our decision renders every person in America subject to DNA sampling for CODIS purposes, including "attendees of public high schools or universities, persons seeking to obtain drivers' licenses, applicants for federal employment, or persons requiring any form of federal identification, and those who desire to travel by airplane," "political opponents," "disfavored minorities," "all newborns," "passengers of vehicles," "arrestees,"—no, really, "the entire population." Nothing could be further from the truth—and we respectfully suggest that our dissenting colleague ought to recognize the obvious and significant distinction between the DNA profiling of law-abiding citizens who are passing through some transient status (*e.g.*, newborns, students, passengers in a car or on a plane) and lawfully adjudicated criminals whose proven conduct substantially heightens the government's interest in monitoring them and quite properly carries lasting consequences. . . .

B

[13] With this framework in mind, we can now appraise the reasonableness of the federal DNA Act's compulsory DNA profiling of qualified federal offenders. In evaluating the totality of the circumstances, we must balance the degree to which DNA profiling interferes with the privacy interests of qualified federal offenders against the significance of the public interests served by such profiling.

1

[14] As we have recognized, compulsory blood tests implicate the individual's interest in bodily integrity . . . . Nonetheless, it is firmly established that "the intrusion occasioned by a blood test is not significant" . . . .

[15] . . . [A]s we recognized in *Rise*, "[o]nce a person is convicted of one of the felonies included as predicate offenses under [the DNA Act], his identity has become a matter of state interest and he has lost any legitimate expectation of privacy in the identifying information derived from blood sampling."

Both Kincade and his supporting amici passionately protest that because the government does not destroy blood samples drawn for DNA profiling and because such samples therefore conceivably could be mined for more private information or otherwise misused in the future, any presently legitimate generation of DNA profiles is irretrievably tainted by the prospect of far more consequential future invasions of personal privacy. Judge Reinhardt's dissent likewise maintains that in light of the "nightmarish" possibilities CODIS portends, we must act immediately to halt the program—before the wolf enters the fold, rather than after.

[16] The concerns raised by amici and by Judge Reinhardt in his dissent are indeed weighty ones, and we do not dismiss them lightly. But beyond the fact that the DNA Act itself provides protections against such misuse, our job is limited to resolving the constitutionality of the program before us, as it is designed and as it has been implemented. In our system of government, courts base decisions not on dramatic Hollywood fantasies, but on concretely particularized facts developed in the cauldron of the adversary process and reduced to an assessable record. . . . As currently structured and implemented, . . . the DNA Act's compulsory profiling of qualified federal offenders can only be described as minimally invasive—both in terms of the bodily intrusion it occasions, and the information it lawfully produces.

2

[17] In contrast, the interests furthered by the federal DNA Act are undeniably compelling. By establishing a means of identification that can be used to link conditional releasees to crimes committed while they are at large, compulsory DNA profiling serves society's " 'overwhelming interest' in ensuring that a parolee complies with th[ ]e requirements [of his release] and is returned to prison if he fails to do so." . . . [T]he Supreme Court, too, has frequently stressed the pressing need to reduce recidivism among the offender population. . . . Finally, by contributing to the solution of past crimes, DNA profiling of qualified federal offenders helps bring closure to countless victims of crime who long have languished in the knowledge that perpetrators remain at large. Together, the weight of these interests is monumental. . . .

3

[18] In light of conditional releasees' substantially diminished expectations of privacy, the minimal intrusion occasioned by blood sampling, and the

overwhelming societal interests so clearly furthered by the collection of DNA information from convicted offenders, we must conclude that compulsory DNA profiling of qualified federal offenders is reasonable under the totality of the circumstances. Therefore, we today realign ourselves with every other state and federal appellate court to have considered these issues—squarely holding that the DNA Act satisfies the requirements of the Fourth Amendment.

IV

[19] Because compulsory DNA profiling conducted pursuant to the federal DNA Act would have occasioned no violation of Kincade's Fourth Amendment rights, the judgment and accompanying sentence of the district court are
AFFIRMED.

INDEX

Locators in **boldface** indicate main topics. Locators followed by *g* indicate glossary entries. Locators followed by *b* indicate biographical entries. Locators followed by *c* indicate chronology entries.

Index

Index

347

Index

Index

351

Index

353

Index

Index

Index

Index